Biomathematics

Volume 15

Managing Editor
S. A. Levin

D. L. DeAngelis W. M. Post C. C. Travis

Positive Feedback in Natural Systems

With 90 Figures

Springer-Verlag
Berlin Heidelberg New York Tokyo

Donald L. DeAngelis
Wilfred M. Post
Curtis C. Travis

Oak Ridge National Laboratory
Oak Ridge, Tennessee 37831, USA

Mathematics Subject Classification (1980): 92 A 17

ISBN-13: 978-3-642-82627-6   e-ISBN-13: 978-3-642-82625-2
DOI: 10.1007/978-3-642-82625-2

Library of Congress Cataloging in Publication Data.
DeAngelis, D. L. (Donald Lee), 1944– Positive feedback in natural systems. (Biomathematics; v. 15)
Bibliography: p. 1. Biological control systems. 2. System analysis. I. Post, W. M. II. Travis, C. C.
III. Title. IV. Series.
QH508.D43   1986   574.5   85-22830
ISBN 0-387-15942-8 (U.S.)

© Springer-Verlag Berlin Heidelberg 1986
Softcover reprint of the hardcover first edition 1986

2141/3140-543210

Preface

Cybernetics, a science concerned with understanding how systems are regulated, has reflected the preoccupations of the century in which it was born. Regulation is important in twentieth century society, where both machines and social organizations are complex. Cybernetics focused on and became primarily associated with the homeostasis or stability of system behavior and with the negative feedbacks that stabilize systems. It paid less attention to the processes opposite to negative feedback, the positive feedback processes that act to change systems.

We attempt to redress the balance here by illustrating the enormous importance of positive feedbacks in natural systems. In an article in the *American Scientist* in 1963, Maruyama called for increased attention to this topic, noting that processes of change could occur when a "deviation in any one component of the system caused deviations in other components that acted back on the first component to reinforce of amplify the initial deviation." The deviation amplification is the result of positive feedback among system components. Maruyama demonstrated by numerous examples that the neglect of such processes was unjustified and suggested that a new branch of cybernetics, "the second cybernetics," be devoted to their study.

It is now more than twenty years since the publication of Maruyama's article. Although "the second cybernetic" (or "positive feedback dynamics") has perhaps not achieved the full status of a new field, Maruyama's ideas are very much a part of current thinking. Scientists have become increasingly conscious of the positive feedback loops driving system changes. In physics the nature of phase transitions, those sudden, self-amplifying changes of matter from one state to another, is an active area of research. Climatologists are debating whether or not and in what direction positive feedback loops are altering our climate. Abandoning the purely negative feedback models of animal physiology and behavior, physiologists now favor models that also contain positive feedback. Social scientists are concerned with the accelerating, deviation-amplifying changes occurring in modern human organizations. Ecologists are becoming more interested in the positive feedback phenomena of mutalism and coevolution. Scientists in many fields are trying to understand how natural systems are organized through processes that involve both negative and positive feedback.

The current scientific literature contains abundant examples of the use of positive feedback as a component in explanatory principles. However, studies that deal at length with the subject of positive feedback are still relatively sparse. Books particularly worth mentioning for their inclusion of positive feedback ideas are *Positive Feedback*, edited by John Milsum (1968), and *Population Systems*, by Alan Berryman (1981). These books consciously hark back to Maruyama's article (1963). Among short articles, "The role of positive feedback" by Stanley-Jones (1970) independently made some of the same points Maruyama did.

The present book draws together many of these scattered examples to try to present a coherent picture of the role of positive feedback in nature. We are primarily concerned with its place in natural populations, ecological communities, and ecosystems, and these topics occupy most of the chapters of the book. However, early chapters in this book consider physical, chemical, and physiological systems, because an understanding of positive feedback on one level of complexity cannot be attempted without some discussion of its role in other levels of organization.

This book may have additional interest to some readers because of its emphasis on the mathematics appropriate to systems with positive feedback loops. Positive linear matrix theory, developed mostly by economists is recent decades, is appropriate to many ecological models involving positive feedback and is used rather extensively here. Several of our chapters are almost wholly discursive, with little mathematics, while in a few chapters we have analyzed ecological systems in considerable mathematical detail. We have tried to provide these latter chapters with enough verbal discussion to help those readers who are interested in familiarizing themselves with the importance of positive feedback in systems but who prefer to dispense with the mathematics.

We are indebted to numerous colleagues who have read and commented on all or parts of early drafts of the manuscript. We particularly thank Stan Auerbach, Larry Barnthouse, Steve Bartell, Hedley Bond, James Breck, Charles E. Clark, Sigurd Christensen, Joel Cohen, Mel Dyer, Steve Ellner, William Emanuel, Robert Gardner, Lou Gross, Tom Hallam, Katherine Keeler, Simon Levin, Dennis Newbold, Jerry Olson, Robert O'Neill, David Reichle, Frances Sharples, Hank Shugart, Robert Van Hook, Douglas Vaughan, Jack Waide, and Ronald Yoshiyama. Whatever clarity of exposition we have achieved is due in large part to the editorial help of Natalie Millemann and Ann Ragan. The typing and assembly of the manuscript was ably performed by members of the word processing staff at Oak Ridge National Laboratory, including Glenda Carter, Vicki Ewing, Karen Gibson, Linda Jennings, Linda Littleton, Donna Rhew, and Jennifer Seiber. This work was supported in part by the National Science Foundation's Ecosystem Studies Program under Interagency Agreement No. DEB 77-25781 with the U.S. Department of Energy under Contract No. DE-AC05-840R21400 with Martin Marietta Energy Systems, Inc.

Oak Ridge, February 1986
D. L. DeAngelis
W. M. Post
C. C. Travis

Table of Contents

1. **Introduction** 1

1.1 Homeostasis 5
1.2 Positive Feedback 7
1.3 Ecological Systems with Positive Feedback 8
1.4 Generalization 1: Increasing Complexity 10
1.5 Generalization 2: Accelerating Change 11
1.6 Generalization 3: Threshold Effects 11
1.7 Generalization 4: Fragility of Complex Systems 12
1.8 Summary and Conclusions 14

2. **The Mathematics of Positive Feedback** 15

2.1 Graphical Analysis of a Simple Dynamic Positive Feedback System 15
2.2 A System of Two Mutualists 16
2.3 A System of Two Competitors 20
2.4 Mathematical Analysis of Positive Feedback 21
2.5 Summary and Conclusions 25

3. **Physical Systems** 27

3.1 The Life History of a Star 27
3.2 Geophysical Systems 28
3.3 Autocatalysis in Chemical Systems 34
3.4 Summary and Conclusions 37

4. **Evolutionary Processes** 39

4.1 Early Evolution of Life 39
4.2 Evolution at the Species Level 43
4.3 Coevolution 47
4.4 Summary and Conclusions 49

5.   Organisms Physiology and Behaviour 51

5.1   Destructive Positive Feedback 52
5.2   Biochemical Processes in Cells and Organisms 53
5.3   Feeding and Drinking Behavior 55
5.4   Sleep 56
5.5   Movement and Motor-Sensory Relationships 56
5.6   Mind-Body Relationship 58
5.7   Summary and Conclusions 59

6.   Resource Utilization by Organisms 61

6.1   Energy Allocation Tactics 62
6.2   Territorial Defense Strategies 64
6.3   Chemical Defense Strategies 68
6.4   Growth Rate Strategy 70
6.5   Summary and Conclusions 72

7.   Social Behavior 75

7.1   Evolution of r- and K-strategies 75
7.2   Development of Social Strategies 77
7.3   Mating and Reproduction 80
7.4   Population Models Incorporating Sexual Reproduction 83
7.5   Small Group Dynamics 86
7.6   Castes In Insect Societies 87
7.7   Dominance Within Groups 89
7.8   Models of Group Formation and Size 90
7.9   The Schooling of Fish 93
7.10   Social Interactions and Game Theory 95
7.11   Summary and Conclusions 98

8.   Mutualistic and Competitive Systems 99

8.1   Dynamics of Mutualistic Communities 103
8.2   Limits to Mutual Benefaction 107
8.3   Multi-Species Mutualism 111
8.4   Models of the Evolution of Mutualism 114
8.5   Isolation and Obligate Mutualism 118
8.6   Limited Competition 121
8.7   Summary and Conclusions 125

9. Age-Structured Populations 127

9.1 Age Structure 127
9.2 Leslie Matrices 129
9.3 Compensatory Leslie Matrices 135
9.4 Interacting Populations 137
9.5 Coexistence of Two Interacting Populations 140
9.6 Other Compensatory Models 145
9.7 Life-History Strategies 148
9.8 Intrinsic Rate of Increase 149
9.9 Reproductive Strategies 150
9.10 Summary and Conclusions 154

10. Spatially Heterogeneous Systems: Islands and Patchy Regions . . . 157

10.1 Classical Theory of Island Biogeography 158
10.2 Island Clusters 162
10.3 Insular Reserves 163
10.4 Modeling the Patchy System 165
10.5 A Single Species in a Patchy Region 168
10.6 Time to Extinction on a Patch 170
10.7 Persistence of a Species in a Two-Patch Environment 171
10.8 Stability of a Single-Species, Two-Patch System 172
10.9 Persistence of a Species in an N-Patch Environment 172
10.10 Multi-Species, Multi-patch Systems with Competition and Mutalism 174
10.11 Persistence of a Species in a Two-Species, Two-Patch Environment 176
10.12 Persistence of a Species in an L-Species, N-Patch Environment . . 177
10.13 Stability of a Two-Species, Two-Patch Model 177
10.14 Stability of an L-Species, N-Patch Model 178
10.15 Relationship Between Reserve Design and Species Persistence . . 180
10.16 Summary and Conclusions 186

11. Spatially Heterogeneous Ecosystems: Pattern Formation 187

11.1 Spontaneous Emergence of Spatial Patterns 187
11.2 Diffusion Model 189
11.3 Pattern Formation Through Instability 192
11.4 Congregation of Colonial Organisms 194
11.5 Boundary Formation by Competition 196
11.6 Summary and Conclusions 199

12. Disease and Pest Outbreaks 201

12.1 Physiological Effects in the Host Species 201
12.2 Mutualistic Interactions of more than one Pathogenic Agent . . . 204

12.3 Models of a Directly Communicated Disease or Parasite 205
12.4 Effects of Spatial Heterogeneity
on Disease Outbreak Threshold Conditions 209
12.5 Design of Immunization Programs 213
12.6 Shape of the Contagion Rate Function 214
12.7 Comparison with other Spatially Heterogeneous Models 216
12.8 Host-Vector Models 217
12.9 Summary and Conclusions 219

13. The Ecosystem and Succession 221

13.1 The Ecosystem. 221
13.2 Succession as a Positive Feedback Process 221
13.3 A Clementsian Model 226
13.4 Markov Chain Models 228
13.5 A Model of a Fire-Dependent System 231
13.6 Positive Feedback Loops in Ecosystems 233
13.7 Nutrient Cycling 237
13.8 Selection on the Community or Ecosystem Level 242
13.9 Summary and Conclusions 244

Appendices 247

Appendix A: Positive Linear Systems 247
Appendix B: Stability of Positive Feedback Systems 248
Appendix C: Stability of Discrete-Time Systems 251
Appendix D: Positive Equilibria and Stability 252
Appendix E: Comparative Statics of Positive Feedback Systems 252
Appendix F: Similarity Transforms 254
Appendix G: Bounds on the Roots of a Positive Linear System 258
Appendix H: Relationship Between Positive Linear System Stability
Criteria and the Routh-Hurwitz Criteria 259

References 261

Subject Index 281

Author Index 285

1. Introduction

The concept of feedback must be classed with the seminal ideas of the century, and has become, as Judson (1980) expressed it, *"one of the chief themes of scientific understanding."* Actually, the idea was implicit in human technology long before the term "feedback" first appeared in print in 1920. Automatic feedback control devices were used in water clocks (Ctesibius of Alexandria, 3rd century B.C.) and later as regulators on furnaces (16th century) and governors on steam engines (Watt, 18th century). In the 19th century James Clerk Maxwell wrote the first paper describing the theory of feedback devices.

Before we try to describe how negative and positive feedback work in systems, let us state what we mean by a system. The delineation of a system is somewhat arbitrary. Ashby (1952) points out that there are an infinity of variables to choose from in any situation. But generally there is some logic for deciding what to include in the system and what to exclude because the components of a system tend to interact with each other in significant ways. These components of a system may be tangible objects such as planets in a solar system, gears in a machine, or individual organisms in a population, or they may be abstract properties like energy storages or amounts of information, for example.

In relation to feedback there are two types of systems, *"open loop systems"* and *"closed loop systems."* An example of an open loop system would be, under certain conditions, a plant in a flower pot to which water and plant nutrients are periodically added. The conditions are that the plant's nurturer or controller is not influenced in how much is added by the plant's condition at a given time. In this case the flow (of material) is all in one direction, and there is no return flow of information. If the controller varies the amount of water or nutrients according to whether the plant looks healthy or not, there is definitely an information flow from the plant that regulates the input. In this case the system is a closed loop system (Fig. 1.1).

All ecosystems on the earth are in some sense open loop systems, because they receive a flow of radiant energy from the sun that they cannot influence by feedback. However, there are countless ways in which ecosystems and subsystems of ecosystems can act as closed loop systems, regulating the flow of materials or the effect of various physical impacts on themselves.

The existence of a flow of information from the system *"output"* (in the example, the plant's appearance) to the *"input"* (the controller) that regulates the input to maintain a stable set point (a healthy plant), is the commonly accepted emblem of a cybernetic system. The systems we shall consider are more general than this in two ways. First, the feedback will not always be purely information flow, but may be a

1

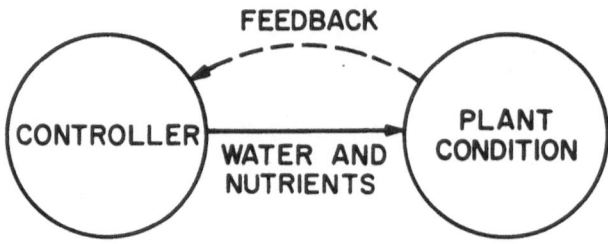

Fig. 1.1. An example of open loop and closed loop systems. A plant that is given water and nutrient over which it has no control is an open loop system. If information on the plant's condition influences the input of water and nutrients, then the system is a closed loop system

change in biomass, energy, or material that only incidentally conveys information affecting the system input. Secondly, the feedbacks in question will not necessarily maintain the system at a stable set point, but may cause the system to change from one state to another.

Let us formalize the description of open and closed loop systems mathematically with an example from population dynamics. First, consider a one-component system, a population with numerical size $X(t)$ at time t. The rate of change of population size through time will be described by the equation,

$$\frac{dX(t)}{dt} = B(t) - D(t),\qquad(1.1)$$

where $B(t)$ is the rate of input of population number and $D(t)$ is the rate of loss. For a one-component system to be an open loop system, the component $X(t)$ must exert no feedback influence on itself; that is, $B(t)$ and $D(t)$ must be independent of $X(t)$. this can be arranged if the input to the population, measured by $B(t)$, is controlled by purely external factors (e.g., immigration to the population in question from other populations) and if the loss rate, $D(t)$, is entirely controlled by an external agent that removes a certain number of members from the population at each time interval regardless of the size of $X(t)$. The rates $B(t)$ and $D(t)$ control the size of $X(t)$, but $X(t)$ exercises no reciprocal control on the size of $B(t)$ or $D(t)$. The control loop is thus open; there is no feedback (see Fig. 1.2a).

The above situation would be very unusual in real populations. Normally both $B(t)$ and $D(t)$ are strongly influenced by $X(t)$. For simplicity, assume that the influence of $X(t)$ on $B(t)$ and $D(t)$ is linear so that the equation takes the form

$$\frac{dX(t)}{dt} = (b - d)X(t),\qquad(1.2)$$

where b is the per capita reproductive rate and d is the per capita mortality rate (both assumed constant). The formulation represents the system as closed; that is, entirely regulated by feedback. Reproduction (assumed asexual here) augments the population and constitutes "*positive feedback*," whereas mortality diminishes the population and is, therefore, termed "*negative feedback*." The two feedbacks are represented graphically in Fig. 1.2b. Note that in our notation all feedbacks are represented by arrows with "+" and "−" signs representing positive and negative influences, respectively. The difference, $b - d$, determines the sign and magnitude of

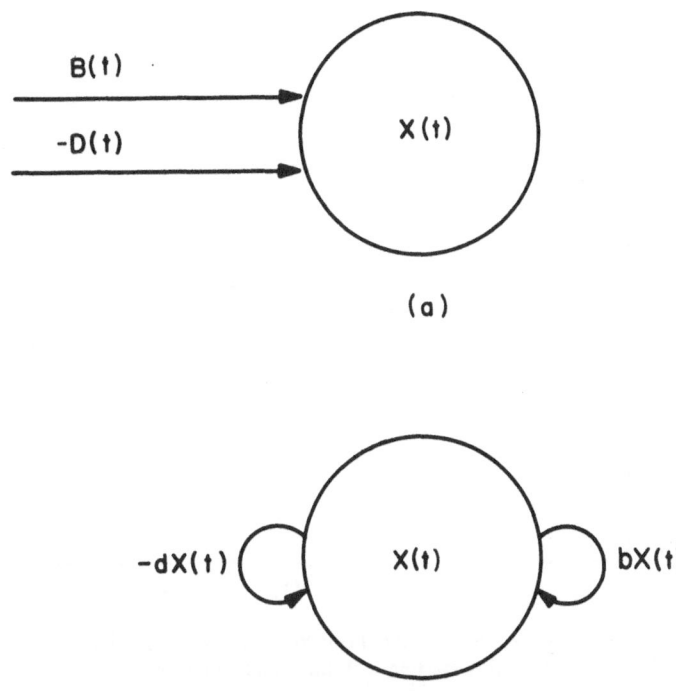

Fig. 1.2a, b. Graphical representation of two types of systems: **a** the system described by Eq. (1.1), with no feedback, and **b** the system described by Eq. (1.2), with both negative and positive feedbacks

the rate of change of the population size. If an initial nonzero population, $X(t)$, is present and if $b-d>0$, then the population increases because $(b-d)X(t)>0$; that is, the population has net positive feedback. In other words, the system is *"deviation amplifying."* If $b-d<0$, the population declines because the system has a net negative feedback. In other words, the system is *"deviation counteracting."* (If the population is initially zero, then $dX(t)/dt=0$ regardless of the value of $b-d$, so the population remains at zero.)

An equivalent terminology is to refer to the change in one time unit, $\exp(b-d)$, as the *"gain."* Net positive feedback corresponds to a gain greater than 1.0 while net negative feedback corresponds to a gain less than 1.0.

Seldom, if ever, are b and d constants in natural systems; they vary with time and with population size (or, more precisely, with population density). As a slight generalization of Eq. (1.2), let us assume that the per capita reproduction rate is $b_1-b_2X(t)$, where b_1 and b_2 are constants, and the per capita mortality rate is $d_1+d_2X(t)$, where d_1 and d_2 are constants. Then Eq. (1.2) becomes the Pearl-Verhulst equation,

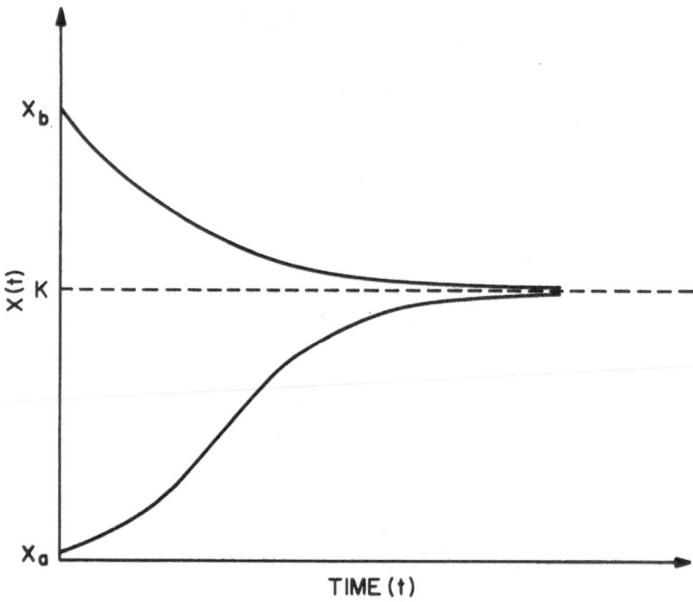

Fig. 1.3. Two solutions of the Pearl-Verhulst equation, one starting with an initial condition $X(0)=X_a$ and the other starting with the initial condition $X(0)=X_b$. Both solutions converge on K

$$\frac{dX(t)}{dt} = [(b_1 - d_1) - (b_2 + d_2)X(t)]X(t)$$

$$= r(1 - X(t)/K)X(t), \tag{1.3}$$

where $r = b_1 - d_1$ is the intrinsic rate of growth and $K = (b_1 - d_1)/(b_2 + d_2)$ is the maximum population density, or the carrying capacity.

The assumption made explicit by Eq. (1.3) is that as $X(t)$ increases, either reproduction diminishes, mortality increases, or a combination of the two occurs. The result is that the system exhibits net positive feedback when $X(t) < K$ and negative feedback when $X(t) > K$. Figure 1.3, depicting two solutions of the Pearl-Verhulst equation, illustrates how the population is driven from a very small value of $X(t)$ towards K by net positive feedback and from values of $X(t)$ larger than K towards K by net negative feedback. Because of this the population will tend towards a stable equilibrium, $\bar{X} = K$.

We can prove that the point $\bar{X}$ is stable by showing that the net feedback relative to the point $\bar{X}$ is negative for deviations of $X(t)$ either above it or below it. To do this, we introduce a new variable, $x(t)$, defined by $X(t) = \bar{X} + x(t)$. Substituting $X(t) = \bar{X} + x(t)$ into Eq. (1.3) and using the equality $\bar{X} = K$, as well as the fact that $d\bar{X}/dt = 0$, we obtain

$$\frac{dx(t)}{dt} = -rx(t). \tag{1.4}$$

If $x(t)$ happens to fluctuate to a negative value, Eq. (1.4) implies that $x(t)$ increases; if $x(t)$ fluctuates to a positive value, Eq. (1.4) implies that $x(t)$ will decrease.

Comparison of Eqs. (1.3) and (1.4) underscores an important "*relativistic*" effect in modeling feedback. In the reference frame of the original variable, $X(t)$ [Eq. (1.3)], the equilibrium point is at K and represents a balance between positive and negative feedbacks. However, in the reference frame of the substitute variable, $x(t)$ [Eq. (1.4)], the equilibrium point is at 0 and the feedback is purely negative. Often the choice of a particular frame of reference may obscure what is actually happening biologically. In this case the representation by Eq. (1.4) of the population as acted on simply by negative feedback is misleading. A stable, steady-state population represents a balance between positive feedback (births) and negative feedback (deaths). This is not an exercise in hairsplitting; the choice of a frame of reference that accentuates negative feedback mechanisms may cause one to ignore the positive feedbacks that may become important under changed circumstances.

A variation on the idea of feedback that should be mentioned here is "feedforward." While feedback is the effect, delayed in time and perhaps mediated by other systems components, that a change in a given component has on itself as a result of a prior change in that component, feedforward is an effect that a component has on itself, again perhaps mediated by other components, because of expected changes in the component. For example, investors may invest in a particular stock or commodity in anticipation that the price will rise. Such actions will tend to be self-fulfilling, since the price will go up if enough investors have the same idea. Feedforward phenomena are common in human social systems because humans have the ability to anticipate future occurrences and to behave accordingly. These phenomena are also frequently unstable because feedforward is often positive.

1.1 Homeostasis

Negative feedback regulation is often referred to as deviation-counteracting feedback, or homeostasis. The theory of homeostasis was first propounded explicitly for physiological systems by Cannon (1929), although the concept goes back at least as far as Descartes and has been implicit in the work of most major physiologists since Blagden (1775), especially that of the great 19th century physiologist Claude Bernard (see Langley, 1973).

Many ecological systems have generally been thought of as homeostatic systems: a stable predator-prey system, a balanced polymorphism in ecological genetics, or a climax state in ecological succession. Modern ecology has, in fact, been shaped in part by the debates over the nature of feedback regulation at the population level (Nicholson, 1957; Milne, 1957; Hairston et al., 1960; Wynne-Edwards, 1962; Murdoch, 1966; Ehrlich and Birch, 1967; McLaren, 1971; Tamarin, 1978) and the ecosystem level (Hutchinson, 1948; Margalef, 1963; Odum, 1969). Although there is still disagreement on the precise interactions responsible for the negative feedback and on how tightly it regulates various ecological systems, there is some concensus that negative feedback regulation is occurring at least to the degree that it normally

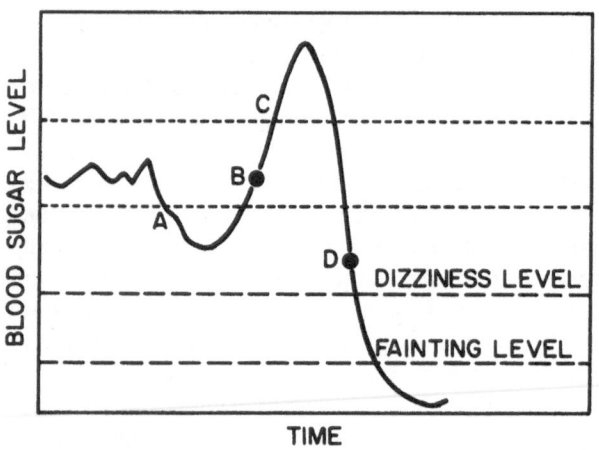

Fig. 1.4. Human blood sugar level in the case where a delay in the action of negative feedback causes overshooting and undershooting of the equilibrium. (Reproduced from Weinberg and Weinberg, 1979, with permission of John Wiley and Sons, Inc.)

keeps populations and communities from going completely out of control, although it may not always be strong enough to prevent sizable fluctuations.

Actually negative feedback regulation can fail badly if time delays exist that sufficiently slow down the action of negative feedback. Weinberg and Weinberg (1979) discuss an example of the regulation of the sugar level in human blood. Normally blood sugar fluctuates but stays within fairly narrow bounds because of a combination of food digestion, which provides material convertible to glucose, enzymes that convert stored fat into glucose, and insulin, which transports the sugar from the blood to cells. Let us suppose that by chance the blood sugar level falls below some lower threshold value. Enzymes to make more sugar from fat are produced, but because their production and diffusion through the body take time, a delay will occur. Meanwhile the person may be stimulated by the low blood sugar level to take corrective behavioral action such as eating candy or other foods with high sugar content. The net result of these two corrective actions could be an *"overshoot"* of blood sugar. This overshoot might then stimulate other regulatory mechanisms, such as insulin production, to eliminate some of the sugar. Again the delay before these can have an effect may lead to an overproduction of the regulators, which could lead to a later undershoot, possibly even to hypoglycemia (Fig. 1.4).

Regulatory delays such as the one described above are ubiquitous in homeostatic devices simply because regulation, whether it comes from the diffusion of molecules in an organism or from changes in the motion of the parts of a mechanical regulator, always takes time. A famous example in population dynamics is the oscillation of some predator-prey systems, for example, that of the Canadian lynx and snowshoe hare (Fig. 1.5). If the number of prey falls by chance to a low level in such a system, predators will die off rapidly, leaving too few to control the prey at an equilibrium level. The prey then undergo a population explosion, which is followed by an explosion of predators. The highly abundant predators then devastate the prey population, which had perhaps already approached in density the carrying capacity of its environment. The cycles can continue indefinitely. [The particular example of hares and lynx is somewhat dubious because many ecologists now believe that the

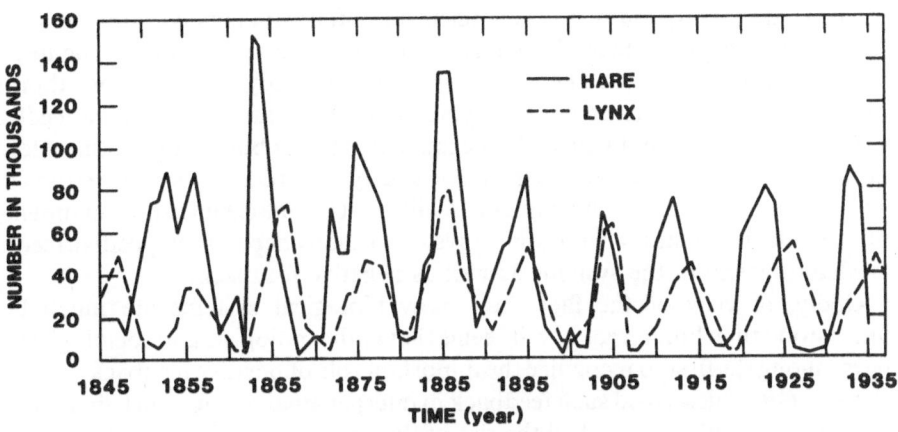

Fig. 1.5. Numbers of furs from trapped Canadian lynx and snowshoe hares over a long time period, indicating cycles in the two populations. (Reproduced from Odum, 1971, with permission of W. B. Saunders Company)

lynx may be a minor factor in driving the oscillations. The fundamental cause may be an analogous dynamical situation involving the hares and their plant forage. See Finerty (1981) for details on this and other possible explanations.]

The law of supply and demand operates somewhat similarly in economic systems. An undersupply of a certain product will result in increased demand. Because of the increased demand, the economic sector that makes the product will increase its production. However, once the demand is satisfied, it is usually difficult to slow down production immediately. This lag causes an oversupply of the product, which induces the manufacturers to curtail production drastically, leading eventually to another undersupply.

Many of the cyclical phenomena we see in nature and society are the result of this imperfect regulation. These examples of imperfect regulation demonstrate that negative feedback and homeostasis are not identical. Negative feedback is necessary for homeostasis, but may not always be sufficient.

1.2 Positive Feedback

Positive feedback occurs when the response of a system to an initial deviation of the system acts to reinforce the change in the direction of the deviation. Maruyama called attention to the fact that positive feedback had been relatively ignored: "*By focusing on the deviation-counteracting aspects of mutual causal relationships . . . the cyberneticians paid less attention to the systems in which the mutual causal effects are deviation-amplifying*" (Maruyama, 1963). He went on to say: "*In contrast to the progress in the study of equilibrating systems, the deviation-amplifying systems have not been given much investment of time and energy by the mathematical scientists on the one hand, and understanding and practical application on the part of geneticists, ecologists, politicians, and psychotherapists on the other hand.*"

The traditional emphasis by ecologists and other students of living systems on homeostasis reflects a bias of interest in the steady state and thus in the negative feedback mechanisms that keep systems close to steady state. A more thorough observation of ecosystems would accord more importance to positive feedbacks that are normally kept in check by negative feedbacks but that sometimes lead to instabilities, causing transitions from one steady state to another. These transitions will be a main theme of later chapters. It will also become clear that the maintenance of dynamic steady states in ecological systems usually depends on positive feedback mechanisms within the systems as well as negative feedback.

Oddly, in view of the fact that many biological systems are much better understood than human society, it seems that anthropologists and sociologists have been among the first to recognize the important role of positive feedback in systems. Bateson (1972) described such feedback in interpersonal relations in tribal societies, and Flannery (1972) described the rise of the centralized political state in terms of deviation amplification caused by positive feedback loops. Much earlier than this, Quesnay's "*tableau economique*" (18th century) embodied the notion of positive feedback, which Ricardo formalized mathematically in the early 19th century.

Today the time seems ripe for an intensified study of the mechanisms of positive feedback in ecological and other biological systems. Our later chapters will document conceptual and mathematical progress that is currently unfolding.

1.3 Ecological Systems with Positive Feedback

As was pointed out earlier, a simple population dynamics example of net positive feedback is exhibited by Eq. (1.2) in the case that $b - d > 0$. In this case there is an amplification of population size through time for any initial deviation from zero. Malthus' statement of the inherent tendency of populations to increase geometrically in the absence of limiting constraints is one of the early recognitions of positive feedback at work in populations. Although there has been a great deal of emphasis in the theoretical ecology literature on populations and communities near equilibrium or steady state and, hence, a concern with questions of homeostatic regulation of such systems, some population ecologists (e.g., Ehrlich and Birch, 1967) have pointed out that much of the living world is in a chronic state of nonequilibrium. Malthusian explosions or crashes of populations or of whole ecosystems are often the rule rather than the exception.

When a population or ecological community is perturbed slightly from equilibrium, the balance of negative and positive feedback mechanisms in the system tends to counteract the perturbation and restore the system to equilibrium (if this equilibrium is inherently stable). Neither positive nor negative feedbacks manifest themselves individually in a conspicuous way. However, when the population or community is driven far from equilibrium (by a temporary unusual environmental condition, for example), specific positive or negative feedbacks may become very obvious. The system could be driven into a regime where homeostatic forces no longer operate. Positive feedback loops may act to amplify the deviation, perhaps driving the system to extinction. On the other hand, other positive feedback loops may be activated that rescue the population from extinction and allow it to increase to the point where homeostatic forces restore it to equilibrium. Pest or epidemic outbreaks, invasions of communities by alien species, and the survival of rare species in environments of scattered habitat islands are all problems that involve positive feedback in important ways for systems far from equilibrium; these problems will be the topics of later chapters.

An ecological community is defined as a set of species living and interacting in the same area. A subset of two species populations of a community interacting competitively or mutualistically constitutes a positive feedback system (s. Figs. 1.6a and 1.6b for examples). In the mutualistic system (Fig. 1.6a), for example, an increase in species 1 causes an increase in species 2, which in turn causes an increase in species 1; therefore, the feedback is clearly positive. In the competitive system, an increase in species 1 causes a decrease in species 2, which, because species 2 competes with species 1, causes a further increase in species 1; hence, again the feedback is positive. In each case of positive feedback, the product of the signs around the feedback loop is positive. Mutualism and competition in populations are reviewed in detail in Chap. 8.

The systems mentioned so far involve mutual reinforcement of population numbers or biomasses. However, other ecological variables also exhibit mutual reinforcement. Let us consider the evolution of two species in a predator-prey

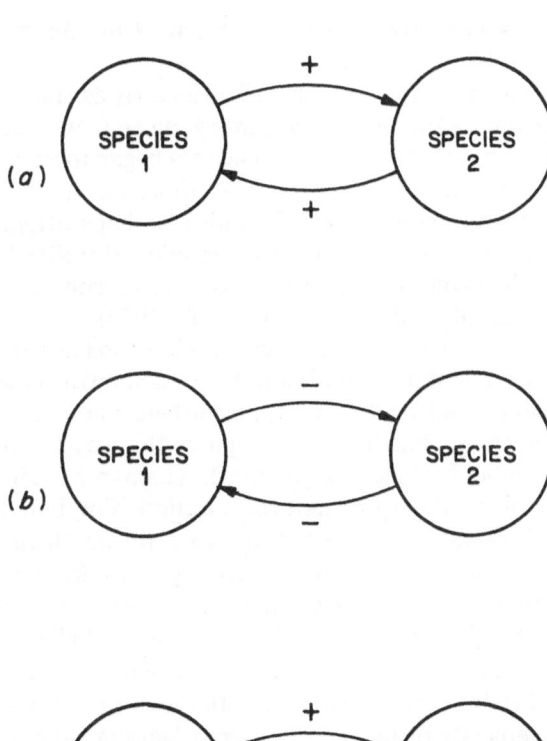

Fig. 1.6a–c. Diagrams of feedback systems: a a model of two-species mutualism; b a model of two-species competition; c a model of the coevolution of defensive and offensive ability in a predator-prey system

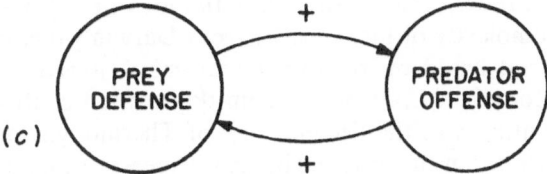

relationship. In its dynamic sense such a system is a negaive feedback system; however, in an evolutionary sense certain characteristics exhibit positive feedback (Fig. 1.6c). The prey evolves in such a way as to improve its defenses against the predator (or at least it has been hypothesized that it does), whereas the predator evolves to be more efficient at capturing prey. These two properties, defense and offense, can be looked at as variables that interact in a positive feedback over evolutionary time. An example of such coevolution is what Vermeij (1983) has termed an *"arms race"* between crushing crabs, fish, etc., and the molluscs on which they feed. The possible occurrence of such positive feedback coevolution will be discussed in Chap. 4.

Certain properties of whole ecosystems, such as structure and diversity, may have interacted over long time scales in deviation-amplifying ways such that profound changes from one state to another have taken place. We shall see later that information and complexity in the biosphere may be increasing through evolutionary time in a mutually reinforcing way. In order to prepare for this idea and other ecosystem properties, we shall examine in the next few sections four general properties associated with complex systems involving positive feedback loops.

1.4 Generalization 1: Increasing Complexity

Although fossils were described as early as Herodotus, an explanation of their true nature awaited the 18th century, when Cuvier and others recognized in them the remains of extinct species. The idea began to emerge that not only has the variety of life on earth changed through time, but also it has evolved in the direction of increasing complexity. This idea perhaps originated with the French biologists Robinet and Bonnet, but was significantly refined by Lamarck into a unified theory of the evolution of all life from initial simple forms to organisms of increasing complexity and perfection (Jacob, 1976).

Lamarck presumed the cause of the directional change to be the stepwise increase in the adaptation of organisms to their environment. According to Lamarck, each change that makes an organism better adapted to its environment brings about higher organization and prepares the way for further changes; hence, a sort of positive feedback is involved. (Lamarck's idea differed fundamentally from Darwin's theory of natural selection. For Lamarck, changes in organisms were purposeful and directed towards fitting them more perfectly to the existing environment. For Darwin, changes occurred spontaneously in organisms in a population, and those that by chance conferred greater fitness had a higher probability of surviving. However, as we shall see later, positive feedback lies at the bottom of the Darwinian view as well as the Lamarckian view.)

Modern science discounts the idea of any inherent purpose in the increasing complexity of life, because from Darwin's theory it is possible to explain such a trend based on random mutations followed by natural selection (s. especially Monod, 1972). One problem that caused confusion to theorists of the late 19th century was the Second Law of Thermodynamics, which predicts an inevitable increase in entropy, and hence a decrease in complexity, through time. This paradox

10

has been resolved in the 20th century with the recognition that the earth is an "open" thermodynamic system in which entropy can, in principle, decrease. In fact, the modern theory of nonequilibrium thermodynamics provides a conceptual framework for understanding the emergence of greater and greater complexity through time both in living and nonliving systems. We will discuss this in detail in Chap. 4.

1.5 Generalization 2: Accelerating Change

In his autobiography Henry Adams (1918) proposed that modern society was changing at an accelerating rate. This observation applied not only to such things as the world coal output, which, as Adams observed, had doubled every ten years between 1840 and 1900, but also to the acquisition of scientific knowledge and perhaps to the progress of society in general since about 1400 A.D.

Adams' law of acceleration has continued to apply since his time and has been the subject of both serious (e. g., Stent, 1978; Ellul, 1980) and popularized (e. g., Toffler, 1970) accounts. Discounting certain retrogressions, one can safely assert that nearly every index of human cultural complexity has increased at a growing rate through human history and, as far as can be fathomed, through human prehistory.

What has been true for human society has been no less true for the evolution of life in general. Procaryotic cells may have existed on earth for two billion years before the emergence of eucaryotic cells. Further major developments, such as the radiation of life forms on land, the amniotic egg, and homeothermy, occurred at ever shorter time intervals. Finally, the evolution of the human brain from the ordinary mammalian brain to one capable of abstract thinking took place over only a few million years.

What is the explanation for this apparently very basic property of evolving systems? Halle (1977), among others, points out that every increase in complexity in nature makes possible new changes that could not have occurred before. As a system becomes more complex, the number of possible novelties increases in proportion to this complexity. For example, a tropical rain forest, with its great diversity of species and, hence, great environmental diversity, offers many more niches for new species than does relatively monotonous environment of the boreal forest. Hence, a faster rate of evolutionary change can be expected in the rain forest. Acceleration is then an indication of a cumulative or positive feedback process taking place.

1.6 Generalization 3: Threshold Effects

Systems connected by positive feedback loops are often referred to as cooperative systems because each component produces effects that add positively to the other components either directly or indirectly. A characteristic of cooperative systems is the presence of "thresholds." When a threshold is crossed, presumably during slow changes in some of the system variables or parameter values, the mode of behavior of the system can suddenly change.

To take a familiar example, noted by Walker (1977) and originally due to MacLean (1959) and Hardy (1959), let us consider a cocktail party or some other indoor social gathering consisting of numerous small groups of people in conversation. When the density of groups is low, people can conduct conversation in normal voices. However, as the density of groups increases, the level of background noise may ultimately be reached, depending on the acoustic character-istics of the room, at which members of groups cannot hear each other when they speak normally. Gradually at first, everyone begins speaking more loudly; however, this further raises the general noise level. Once the threshold of normal conversational tone has been exceeded, the situation can then quickly become hopeless.

A more consequential example is the so-called Allee effect (Allee, 1938), which relates to the fact that many animal populations rely on strong social organization. If, for one reason or another, population density becomes too low, reproduction may be hampered, leading to further population reduction. A positive feedback cycle thus can result in population extinction. One can think of a population density threshold below which this extinction becomes inevitable. If somehow the population density could be restored to a level above the threshold, it could expand further until it reached the carrying capacity of the environment for that species.

Note that the thresholds in these two examples differ in a significant way. In the population example the system variable, population density, will decrease if initially below the threshold and increase if initially above the threshold. On the other hand, the system variable of the first example of the cocktail party, which we may choose as the loudness of individual voices, will increase when the threshold is exceeded but will stay at about the same normal conversational level below the noise threshold. This *"one-sided"* nature of the conversational threshold simply reflects the obvious fact that the ability to speak is not diminished by a paucity of listeners.

The two examples are similar in that what causes the thresholds to be crossed is changes in the system variables, population density and the volume of individual voices. Changes in externally controlled parameter values could also trigger a threshold effect, however. Suppose in the population example the population density, P, is slightly above the threshold value, P_T, or $P_T < P$, so that the population is increasing or at least stationary. If environmental factors, such as a change to a less favorable climate, decreased the reproduction rate in the population, the threshold value for population survival could be shifted upward to P_T', such that $P < P_T'$. The population would then go into decline.

1.7 Generalization 4: Fragility of Complex Systems

Closely related to the presence of thresholds in cooperative systems are phenomena that Weinberg and Weinberg (1979) have classified under the *"law of collapse"*. This law (which we prefer to call a generalization) states that any strong association among lifetimes (that is, the expected duration times of the individual components) leads to sudden depletions. For example, a nylon rope is made up of many individual fibers, any of which could break under the strain of a moderately heavy load. However, the woven cooperative aggregate can withstand enormous strains, the collective effect greatly increasing the expected lifetimes of each fiber. Eventually, however, some fibers will break. With each breaking of a fiber, the load shared by the remaining fibers increases, shortening their expected lifetimes. More fibers break, accelerating the process until the whole rope suddenly snaps in a cataclysm of fiber breakage.

Failure of structures is commonly catastrophic, particularly when the structure is designed to be resilient to stress. A ship's hull suddenly splits in two, or an airplane wing suddenly twists off. Part of the reason for catastrophic failures, as Gordon (1978) lucidly describes, lies in the very nature of structural design, which emphasizes resilience. Design for resilience stabilizes a structure under most conditions, but it can engender a vulnerability to positive feedback amplification of small cracks in the structure. When stressed by the environment, a resilient structure such as the steel hull of a ship can store enormous amounts of strain energy. This is fine because it allows the ship to undergo the extreme pounding of a storm at sea. However, let us suppose a crack develops somewhere in the hull. Any elongation of the crack requires an input of energy to break material bonds. To a good approximation, the energy required to elongate a crack with length, L, is directly proportional to L. This energy comes directly from the strain energy stored in the area of the material in the vicinity of the crack. The amount of energy that can be released is proportional not to the length of the crack, but to the square of its length.

An equation for the rate of increase in the length of the crack would have the form

$$\frac{dL}{dt} = \alpha L^2 - \beta L, \tag{1.5}$$

where α and β are coefficients depending on the resilience and bonding strength of the material. If the initial length of the crack, L_0, is greater than β/α (the Griffith critical length), then the positive feedback term, L^2, will dominate the negative feedback term and the crack will be self-amplifying.

Self-supporting columns can fail through elastic toppling. There is a ratio of height to diameter above which failure will occur. The mechanical theory of self-supporting columns has been applied to plant structure (McMahon, 1973; Givnish, 1982). One can calculate whether a small deflection from the vertical will increase or decrease the potential energy of an idealized plant ramet. Potential energy is defined as the sum of strain energy stored in the plant stem resulting from flexure, and the gravitational potential that results from the height of the distributed plant mass. If the potential energy of the ramet increases in response to a small deflection, it will tend to return to the vertical position of low potential energy. The ramet is stable in that case. If, on the other hand, the potential energy of the ramet decreases with increasing deflection, the vertical position is unstable, and small deflections will tend to grow.

This law of collapse also has relevance to advanced economic systems. According to input-output economic theory (Leontief, 1966), such economic systems are composed of a set of *"basic"* commodities, each of which is dependent on all the others. The loss or severe reduction of one commodity will generate a chain reaction, destroying or damaging all the others (Pasineti, 1977). A similar fragility afflicts advanced societies as a whole. Such societies have evolved a high density of

13

tight interactions among various sectors (Ellul, 1980). Disruption of a part of the society may propagate and could even trigger a total collapse, as has occasionally happened.

Does the law of collapse apply also to ecosystems? As we know, some communities are resilient to the loss of a few members; others "*completely disintegrate*" (Darlington, 1980). Darlington suggests that the delicate coadaptations that have evolved among species in certain ecosystems (e.g., tropical rain forests and coral reefs) may make the survival of most of these species outside of these networks of interactions impossible. Critical losses of a few component species could, therefore, lead to irreversible disintegration of the ecosystems.

We do not wish to overstate this point. The term "fragility" has been overused in describing ecosystems. Complex systems are not always fragile because there are ways in which they can compensate for the instability resulting from a high degree of interconnectedness. One way is by redundancy; i.e., if one critical pathway is severed another pathway, perhaps existing only latently, can take over for it. The human brain seems to employ a great deal of functional redundancy, and there is no reason to suppose that redundancy does not play a role in stabilizing ecosystems.

1.8 Summary and Conclusions

The concept of feedback is a key explanatory idea in our understanding of dynamic processes in the world. Although cybernetic research has focused largely on negative feedback, positive feedbacks are ubiquitous in nature, both in stable and unstable situations. In the latter case, deviation-amplifying positive feedbacks can drive a system to a new state.

At least four broad generalizations apply to the behavior of systems in which positive feedback or the potential of positive feedback occur. First, positive feedback may be essentially involved in processes within systems that bring about increasing complexity. Second, changes in the system tend to accelerate, so that rapid transitions from one state to the next can occur. Third, there are frequently thresholds such that small changes in a variable or parameter beyond a certain critical point can trigger large scale change in the system. Fourth, a network of potential positive feedbacks, perhaps not apparent at a given time because held in check, can make a system fragile to certain types of perturbations.

2. The Mathematics of Positive Feedback

Concomitant with the cybernetics revolution of the past half century has been the rapid development of mathematical modeling in the biological and ecological sciences, a methodology now firmly established. A model may be defined as an abstract description of the real world, a simple representation of more complex forms, processes, and functions of physical phenomena or ideas (Rubinstein, 1975). Models have aided in the understanding of ecological systems. When a model is put into explicit mathematical form, it can be analyzed to predict the effect on all parts of the system of external intervention on any single part.

Modeling is not a panacea, either for the understanding of natural systems or for predicting the effects of human impacts on ecosystems. Appropriate models are necessary before meaningful inferences can be drawn from them. Because each model can represent only one of many possible portraits of reality, it must be designed with a clear understanding of its purposes and limitations. The models we shall derive and analyze emphasize the feedback aspects of nature, often at the expense of other key properties of living systems.

The existence of underlying conceptual similarities among systems, as discussed earlier, suggests that similar mathematical representations are also appropriate for a functionally diverse set of systems. [Fascination with the fact that the same mathematics could be used to describe qualitatively diverse phenomena goes back to the Pythagoreans (Kline, 1980).] Consideration of simple two- and multicomponent systems will help separate out broad classes of models.

2.1 Graphical Analysis of a Simple Dynamic Positive Feedback System

In Chap. 1 some of the elements of the mathematical dynamics of positive feedback were discussed, at least with reference to a single population. However, feedback systems become much more interesting when two or more components are present; therefore, we need to introduce techniques appropriate to such systems. Minorsky (1962) presented one of the early comprehensive treatments of nonlinear mechanics. His description of a system of two nonlinearly coupled variables is lucid and intuitive. If $X(t)$ and $Y(t)$ are the two dependent variables as functions of time, the equations can be written in general as

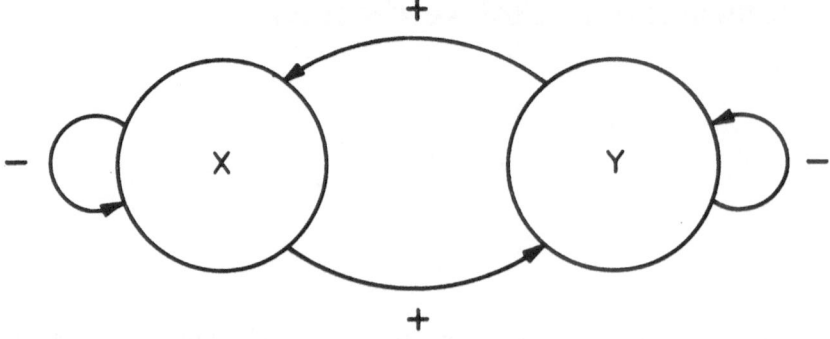

Fig. 2.1. Diagram of the system represented by Eqs. (2.1a) and (2.1b) assuming that X and Y interact through positive feedback while X and Y have internal negative feedback loops

$$\frac{dX(t)}{dt} = F_1(X, Y) \qquad (2.1a)$$

$$\frac{dY(t)}{dt} = F_2(X, Y), \qquad (2.1b)$$

where $F_1(X, Y)$ and $F_2(X, Y)$ are arbitrary functions of $X(t)$ and $Y(t)$.

The two variables, X and Y, are unspecified at present. They could, for example, represent prey and predator numbers in an ecological situation or supply and demand in an economic system. In these cases X and Y would be coupled by negative feedback. The variables could also represent the numbers of two competitors or two mutualists in an ecological setting or quantification of the respective defensive and offensive capabilities of a coevolving pair of prey and predator. In these cases X and Y would be coupled by positive feedback. Also, systems can be imagined in which the net feedback between the two variables would be positive for some values of X and Y and negative for other values. For example, two species may somehow benefit each other when their population densities are low but one may prey on the other when their densities are high.

Besides the feedback loop coupling the two variables, each variable may also have its own internal feedback, positive or negative, acting on it (see Fig. 2.1 for a schematic diagram). Specific cases of positive feedback will be discussed below.

2.2 A System of Two Mutualists

As an example that will be pertinent to the remainder of this book, let us suppose that X and Y represent the numbers of two obligate mutualist species. Let us further assume that each of the two species, in the absence of the other species, is governed by an equation similar to the Pearl-Verhulst equation discussed earlier. That is, for small values of X, dX/dt is positive while for large values it is negative. The same holds for the dependence of dY/dt on Y.

16

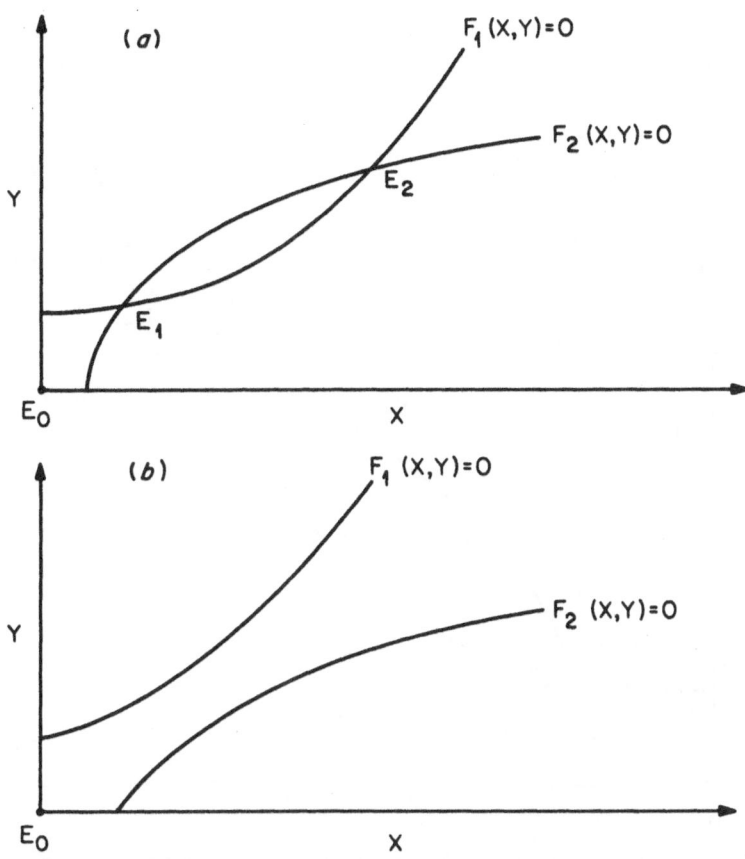

Fig. 2.2a, b. State-plane plot of Eqs. (2.1a) and (2.1b) when X and Y are mutualists. The two zero isoclines, $F_1(X, Y)=0$ and $F_2(X, Y)=0$, are plotted. In **a** the two zero isoclines intersect twice to form two equilibrium points; in **b** no equilibrium points exist for $X>0$ and $Y>0$

With these assumptions we can analyze the dynamics of the system graphically. First, let us relate these assumptions to the way in which $F_1(X, Y)$ and $F_2(X, Y)$ [s. Eqs. (2.1)] depend on X and Y. The fact that the species are obligate mutualists means that X cannot increase unless Y is larger than some finite level and that Y cannot grow unless X is larger than some finite value. In other words, $F_1(X, Y)$ can be positive only if $Y \geq Y_0$, and $F_2(X, Y)$ can be positive only if $X \geq X_0$, where Y_0 and X_0 are constant lower critical values. Because both species are self-regulating, $F_1(X, Y)$ and $F_2(X, Y)$ must change from positive to negative when X and Y, respectively, reach large enough values. In Fig. 2.2a and 2.2b we plot two possible configurations of the curves $F_1(X, Y)=0$ and $F_2(X, Y)=0$ in the X, Y plane. These lines, called the "*zero isoclines*," are very important because their intersections, where both $dX/dt=0$ and $dY/dt=0$, are the equilibrium points of the system. Note that in Fig. 2.2a there are two equilibrium points, E_1 and E_2, in the positive part of the plane, whereas in Fig. 2.2b there are none. Because the system in Fig. 2.2a is more interesting to us, we shall investigate it in detail and allow the reader to speculate on the system dynamics represented by Fig. 2.2b.

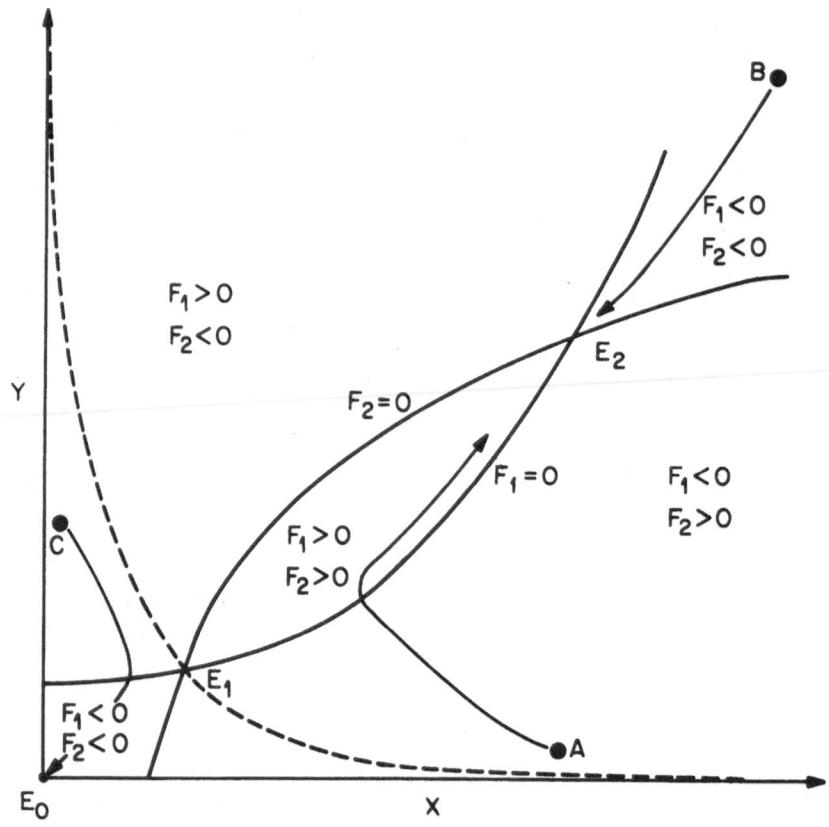

Fig. 2.3. State-plane plot of Eqs. (2.1a) and (2.1b) as in Fig. 2.2a. The signs of $F_1(X, Y)$ and $F_2(X, Y)$ are shown in the five regions formed by the zero isoclines. Three trajectories, A, B, and C, are sketched in, showing that E_1 is an unstable equilibrium and E_2 is a stable equilibrium. The dotted line is the separatrix

The two isoclines divide the X, Y plane of Fig. 2.2a into five regions, which we characterize in Fig. 2.3 according to the signs of $F_1(X, Y)$ and $F_2(X, Y)$. These signs, even in the absence of absolute magnitudes, give us some idea of the dynamics of the pair of mutualists. We have chosen to plot three trajectories that exemplify these dynamics. Let us start with a large value of X and a very small value of Y. At first X decreases rapidly, whereas Y stimultaneously increases (path A). When this path crosses the $F_1(X, Y)=0$ isocline, however, dX/dt changes sign and X starts to increase. The trajectory approaches the point E_2. If we start with large values of both X and Y, the self-regulation forces cause both to decrease (path B) until the trajectory approaches E_2. Finally we start with a moderate value of Y and a very small value of X. The trajectory (path C) ultimately approaches the origin, signifying extinction of both species.

The simple system pictured in Fig. 2.2a illustrates Generalizations 1, 2, 3 and 4 discussed in Chap. 1. Generalization 1 referred to the tendency of systems to increase in complexity through evolution, which this system illustrates trivially. The two variables X and Y are involved in mutualism, a state of greater complexity than

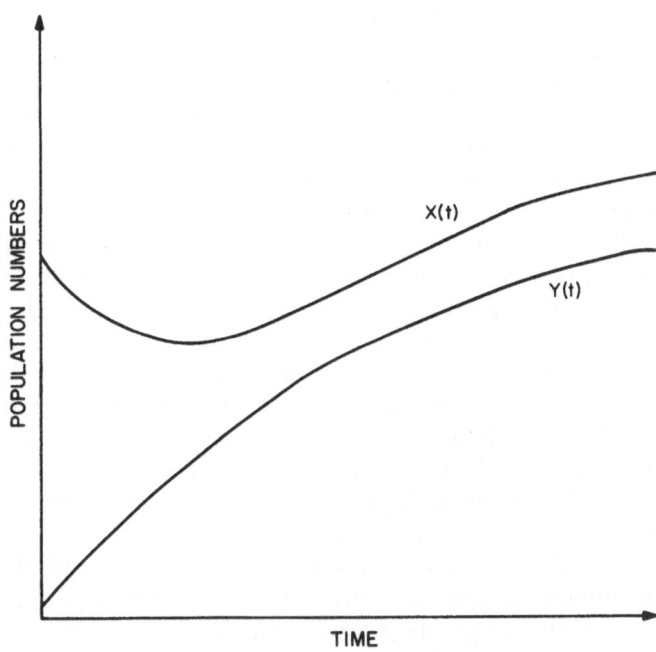

Fig. 2.4. The variables X and Y, shown as path A in Fig. 2.3, are plotted here versus time. Note that X is nonmonotonic

the two species would represent if not dependent on each other. This coupled system, arising through evolution, is capable of more complex behavior than either population by itself. For example, the trajectory of a single species, say X, described by the Pearl-Verhulst equation or something similar, would, in the absence of any time delays, exhibit only monotonically increasing or decreasing behavior through time. The variable X in the coupled system can behave more complexly, as shown in the plot of X and Y versus time in Fig. 2.4 (path A of Fig. 2.3).

Generalization 2 asserted that variables in systems with positive feedback characteristically exhibit acceleration. In path A, X is clearly accelerating because it changes from a negative to a positive rate of change. A plot of X and Y versus time (Fig. 2.4) makes this acceleration phase more obvious.

The threshold effect (Generalization 3) is very clearly demonstrated by a comparison of paths A and C. The values of X and Y that are the initial point on path A are above the threshold necessary for the trajectory to reach equilibrium point E_2. The values of X and Y that are the initial point on path C are below that threshold, so the system goes towards extinction. The threshold is denoted by a line (see dotted line in Fig. 2.3) called the separatrix. This example also incorporates the idea that complex systems can be fragile (Generalization 4). The presence of the separatrix arises from the fact that the two mutualists are obligatory, or dependent on each other's existence, as the components of a complex system are likely to be. Any perturbation that seriously depresses either X or Y could destroy the whole system.

19

2.3 A System of Two Competitors

The relationship between a pair of competitors, like that between two mutualists, is one of positive feedback (Fig. 2.5). This may seem paradoxical in that each species affects the other negatively. However, as pointed out in Chap. 1, a feedback loop is positive if the product of all the signs in the loop is positive. In the case of two competitors, this is simple: $(-)(-)=(+)$.

The positive feedback dynamics of two competitors will be obvious when the system is studied graphically in the same way that the system of two mutualists was analyzed above. Let us start with Eqs. (2.1a) and (2.1b) and assume again that both X and Y, in the absence of other species, are governed by a Pearl-Verhulst equation or something similar; that is, $F_1(0, Y)=0$, where $F_1(X, 0)>0$ for small X and $F_1(X, 0)<0$ for large X; and $F_2(X, 0)=0$, where $F_2(0, Y)>0$ for small Y and $F_2(0, Y)<0$ for large Y. Any increase in Y tends to decrease $F_1(X, Y)$, and any increase in X tends to decrease $F_2(X, Y)$. In Fig. 2.6 we plot a possible configuration of the zero isoclines, $F_1(X, Y)=0$ and $F_2(X, Y)=0$, that satisfies these conditions. Note that the zero isoclines divide the positive X, Y quadrant into four regions having different permutations of the signs of dX/dt and dY/dt. The isoclines cross at the equilibrium point E_1. Also note that from our assumptions above, E_0, E_2 and E_3 are equilibrium points because $dX/dt=0$ and $dY/dt=0$ at all these points.

Suppose we start the two populations at point A in Region I (see Fig. 2.6). At this point $dX/dt>0$ and $dY/dt>0$. The course of the trajectory takes the populations into Region II, where $dX/dt<0$ and $dY/dt>0$. In Region II, Y is growing, and as it grows, it causes X to decline, which increases the rate of increase of Y. The trajectory eventually approaches point E_3. By looking at other possible trajectories, one can show that they approach either point E_2 or E_3 depending on which side of the separatrix (dotted line) they start. The separatrix acts as a threshold on either side of which trajectories diverge from their initial conditions, because of positive feedback between X and Y, to entirely different terminal points.

Figure 2.6 is not the only configuration the zero isoclines can take, but we shall wait until Chap. 8 to consider other possible configurations.

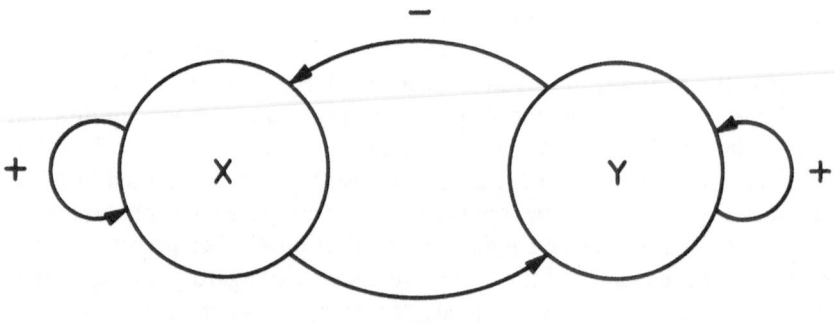

Fig. 2.5. Diagram of system of two competitors. The feedback loop between them is positive

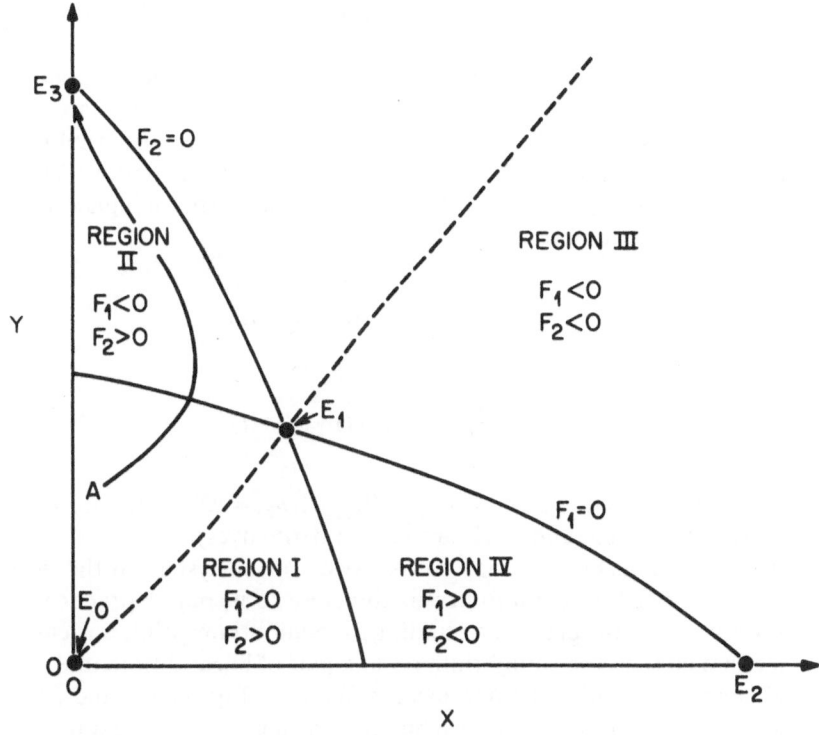

Fig. 2.6. State-plane plot of Eqs. (2.1a) and (2.1b) when X and Y are competitors. The two zero isoclines, $F_1(X, Y) = 0$ and $F_2(X, Y) = 0$, are plotted. The points E_0, E_1, E_2, and E_3 are equilibrium points. A typical trajectory, starting at point A, is shown

2.4 Mathematical Analysis of Positive Feedback

Although the graphical approach described in the preceding section is a useful, simple method for studying systems with two components, it is inadequate for dealing with systems with three or more components. In such cases mathematical analysis is necessary.

Even analytic methods are not totally satisfactory, however. For example, in general, Eqs. (2.1a) and (2.1b) will not be solvable analytically. At best one can usually obtain only approximate solutions valid in parts of the ranges of the variables. In particular, if the equilibrium points E_1 and E_2 are known, it will be possible to obtain analytic solutions to Eqs. (2.1a) and (2.1b) very close to these points, which will allow us to determine whether or not these points are stable.

From an examination of Fig. 2.3, we can see that all feasible trajectories move away from the vicinity of point E_1 and approach either point E_2 or the origin. We call E_2 "*stable*," or an "*attractor*," and E_1 "*unstable*," or a "*repellor*." Suppose, however, we know the equilibrium points E_1 and E_2 but do not have a figure like Fig. 2.3 at our disposal to tell us whether they are attractors or repellors. (This problem would certainly be the case in systems with more than two variables.)

Mathematical analysis can at least give us the answer to this question, provided we know the functional forms of F_1 and F_2, etc.

Let us consider one of the equilibrium points, say E_2, whose X and Y values are $(\bar{X}_2, \bar{Y}_2)$ [where $F_1(\bar{X}_2, \bar{Y}_2) = 0$ and $F_2(\bar{X}_2, \bar{Y}_2) = 0$]. In the immediate vicinity of the equilibrium point, Eqs. (2.1a) and (2.1b) can be expanded as a Taylor series in the variables $x(t)$ and $y(t)$, where $x(t)$, $y(t)$ are defined by the equations $X(t) = \bar{X}_2 + x(t)$ and $Y(t) = \bar{Y}_2 + y(t)$ [$x(t) \ll \bar{X}_2, y(t) \ll \bar{Y}_2$]. Dropping terms of higher than first order in $x(t)$ and $y(t)$, we obtain

$$\frac{dx(t)}{dt} = a_{11}x(t) + a_{12}y(t), \qquad (2.2a)$$

$$\frac{dy(t)}{dt} = a_{21}x(t) + a_{22}y(t), \qquad (2.2b)$$

where $a_{11} = \partial F_1/\partial X|_{X_2,Y_2}$; $a_{12} = \partial F_1/\partial Y|_{X_2,Y_2}$; $a_{21} = \partial F_2/\partial X|_{X_2,Y_2}$; and $a_{22} = \partial F_2/\partial Y|_{X_2,Y_2}$ (where $\partial/\partial X$ and $\partial/\partial Y$ are partial derivatives).

These equations characterize the behavior of the system in the vicinity of the equilibrium; strictly only infinitessimally close to it, but in practice often over a significant area of the plane surrounding the equilibrium point. Basically we are just saying that close to the equilibrium point a pair of linear differential equations is a good approximation of the nonlinear differential Eqs. (2.1a) and (2.1b).

Equations (2.2a) and (2.2b) are handily written in vector and matrix notation. Defining

$$\mathbf{z}(t) = \begin{vmatrix} x(t) \\ y(t) \end{vmatrix}$$

and

$$\mathbf{A} = \begin{vmatrix} a_{11} & a_{12} \\ a_{21} & a_{22} \end{vmatrix},$$

we can write

$$\frac{d\mathbf{z}(t)}{dt} = \mathbf{A}\mathbf{z}(t). \qquad (2.3)$$

Without even calculating the numerical values of a_{11}, a_{12}, a_{21}, and a_{22}, we can say something about their signs and relative magnitudes by looking at Fig. 2.3. Note that at E_2, $F_1(X, Y)$ decreases as X increases but increases as Y increases; also $F_2(X, Y)$ increases as X increases but decreases as Y increases. Hence, $a_{12}, a_{21} > 0$, and $a_{11}, a_{22} < 0$. The matrix of signs, $\mathbf{A}_s$, of $\mathbf{A}$ is

$$\mathbf{A}_s = \begin{vmatrix} - & + \\ + & - \end{vmatrix}.$$

The matrix $\mathbf{A}$ is a positive linear matrix because $a_{12}a_{21} = (+)(+) = (+)$; i.e., the feedback loop of length 2 has a net positive sign. From comparisons of the slopes of the two isoclines at E_2 in Fig. 2.3, one can also say that $|a_{22}| > |a_{21}|$ and $|a_{11}| > |a_{12}|$, where $||$ are absolute value signs.

22

Positive linear systems have some special properties, discussed in detail in Appendix A, that make determination of their stability rather simple, at least compared with standard methods of evaluating stability, such as eigenvalue calculation or application of Routh-Hurwitz criteria.

A positive linear system is defined in Appendix A as any linear system whose matrix has all nonnegative off-diagonals, that is, that can be represented as $A = B - cI$, where B is an essentially nonnegative matrix (all elements are non-negative), I is the unit diagonal matrix, and c is some real constant. Hence, the diagonal elements may be negative. (We are here talking about continuous systems, i.e., systems of differential equations. Discrete-time models, discussed in Chap. 9, are associated with matrices all of whose elements, including diagonal elements, are positive.)

A positive linear system matrix may be written with negative off-diagonal terms. For example, the sign matrix for a system of two competitors is

$$A_s = \begin{vmatrix} + & - \\ - & + \end{vmatrix}. \tag{2.4}$$

It is only necessary that a matrix S exist such that when it and its inverse, S^{-1}, are applied as a similarity transform, $A' = S^{-1}AS$ (see Appendix E), the off-diagonals of A' are all nonnegative. In the case of Eq. (2.4), the matrix

$$S = S^{-1} = \begin{vmatrix} 1 & 0 \\ 0 & -1 \end{vmatrix} \tag{2.5}$$

will transform A to a standard positive form.

By way of another example, note that the matrix

$$A = \begin{vmatrix} -1 & -1 & 1 \\ -1 & -1 & -1 \\ 1 & -1 & -1 \end{vmatrix} \tag{2.6}$$

is a positive linear system. The transform matrix

$$S = S^{-1} = \begin{vmatrix} 1 & 0 & 0 \\ 0 & -1 & 0 \\ 0 & 0 & 1 \end{vmatrix} \tag{2.7}$$

transforms A to

$$A' = S^{-1}AS = \begin{vmatrix} -1 & 1 & 1 \\ 1 & -1 & 1 \\ 1 & 1 & -1 \end{vmatrix}.$$

Perhaps the most convenient method for determining the stability of a positive linear system involves evaluation of the principal minors. If matrix A is an $n \times n$ matrix,

23

$$\mathbf{A} = \begin{vmatrix} a_{11} & a_{12} & a_{13} & \cdots & a_{1n} \\ a_{21} & a_{22} & a_{23} & \cdots & a_{2n} \\ \cdot & \cdot & \cdot & & \cdot \\ \cdot & \cdot & \cdot & & \cdot \\ \cdot & \cdot & \cdot & & \cdot \\ a_{n1} & a_{n2} & a_{n3} & & a_{nn} \end{vmatrix}, \qquad (2.8)$$

then the n principal minors are the determinants ("det")

$$D_{11} = \det \begin{vmatrix} a_{11} \end{vmatrix}$$

$$D_{22} = \det \begin{vmatrix} a_{11} & a_{12} \\ a_{21} & a_{22} \end{vmatrix}$$

$$\cdot$$
$$\cdot$$
$$\cdot$$

$$D_{nn} = \det \begin{vmatrix} \mathbf{A} \end{vmatrix}.$$

An equilibrium point is stable if and only if

$$(-1)^k D_{kk} > 0 \qquad (2.9)$$

for $k = 1, 2, \ldots, n$. The equilibrium is unstable if any one of these conditions is violated. Note that because any pair of rows and any pair of columns of a matrix can be simultaneously interchanged without affecting the determinant of the matrix, the above conditions are all that are required, no matter what the order of labeling on the components is.

The criteria on the D_{ii}'s are equivalent to the demand that negative feedback dominate at all levels of organization in the system (Berryman, 1981). An increase in the number of components or interactions in the system increases the number of conditions that must be satisfied for stability; hence, the likelihood of a randomly connected system being stable decreases as the size of the system increases.

For our example system [Eqs. (2.2a) and (2.2b)], given that $\mathbf{A}$ [Eq. (2.3)] is a positive linear matrix, the equilibrium point is stable if and only if

$$a_{11} < 0$$
$$a_{11} a_{22} > a_{21} a_{12}. \qquad (2.10)$$

From our discussion of the signs and relative magnitudes of the a_{ij}'s earlier, we can state that the point E_2 is stable.

The interest of ecologists in the subject of positive feedback grew from a recognition (Siljak, 1975; Goh, 1977; Travis and Post, 1979) that the mathematical theory of positive linear systems (see appendices) was ideally suited to the analysis of positive feedback phenomena. Most interesting models involving positive feedback are complex, involving more than two variables. The stability properties of these models, including many in the literature, have often proven difficult to solve analytically, so one must resort to numerical methods.

Maruyama (1963), then, was not entirely correct in saying that mathematicians have paid little attention to positive feedback systems. In fact, the advantage of this type of system from the point of view of analysis is that algebraic conditions determining stability that are much simpler than the well-known Routh-Hurwitz criteria can be applied even when the system is very large. Having algebraic relations at one's disposal makes it very easy to see what the effect of any system parameter on stability will be. This is very difficult to do from the complex Routh-Hurwitz criteria or from eigenvalue computation.

Our approach differs from the "*qualitative stability*" technique (May, 1973a; Levins, 1975). In this latter approach, conditions guaranteeing the stability amount to all feedback loops being negative. The conditions guaranteeing the stability of essentially nonnegative matrices, as described later, involve the negative (order unity) loops dominating the positive feedback loops. As will be shown later, using the stability criteria derived from matrix theory, one can deduce related ecological properties such as species persistence, susceptibility to invasion of communities, and so forth.

Because our book is concerned with positive feedback phenomena, we have stressed this aspect in selecting and developing models. Of course, all systems contain a variety of positive and negative feedback loops, but at a given scale one or the other may be more important. For example, if we consider modeling the feeding relationships of certain types of organisms, on the short time scale we see that feeding may stimulate more feeding (Toates, 1980) and hence result in a positive feedback on itself. Over slightly longer times, feeding leads to satiation, producing negative feedback on further feeding. On a somewhat longer time scale, increased feeding causes an improvement in the size or physiological condition of the feeder that allows the organism to feed at a greater rate or on more desirable prey, producing a positive feedback effect in improving the organism further. Finally, on time scales of the order of the lifetime of the organism, genotypic limits on size exert a negative effect on growth and hence limit feeding. This sort of "*nesting*" of positive and negative feedbacks has been discussed with respect to relationships between bees and flowers by Heinrich (1984) and for population growth phenomena by Berryman (1981).

2.5 Summary and Conclusions

Natural systems can be analyzed by building mathematical models, which are highly abstract representations of certain features of real systems deemed significant. Graphical methods often suffice in the analysis of simple two-compartment dynamic positive feedback models, such as those for two mutualistic species or two competitive species. For larger systems matrix methods are necessary to determine whether or not systems are stable at particular equilibria. The theory of positive linear matrices can be very helpful in this regard, providing criteria that are easily used.

3. Physical Systems

The behavior of complex biological and ecological systems seems at first sight rather remote from the dynamics of inanimate matter. Yet common threads run from the simplest of physical interactions to the most intricate processes in an ecosystem. One such thread, positive feedback, is prevalent in all manner of every day and exotic physical and chemical phenomena. We will follow this strand through a diversity of inanimate systems in this chapter and will see that it leads us to the biological and ecological systems that are the subjects of later chapters.

3.1 The Life History of a Star

The progenitors of stars are clouds of gas and dust that begin to contract through mutual gravitational attraction of the particles. Normally the gravitational attraction is not enough to overcome the increase in gas pressure resulting from this contraction. Just as gas in a tightly sealed cylinder will resist the pressure of the piston past a certain density, so the pressure of the intersteller gas can resist its own gravitational forces up to a certain point and maintain an equilibrium density. However, if a large enough amount of gas and dust come together at sufficient density, the gravitational force can dominate the subsequent dynamics. The precise trigger of the gravitational collapse is not well understood, though compression by some sort of shock wave, for example a supernova shock or a collision between two gas clouds, has been suggested (Gehrz et al., 1984). About 10,000 solar masses of matter are necessary for collapse to take place (Kippenhahn, 1983). Once the collapse threshold is passed, every increment of shrinkage increases the mutual gravitational forces more than it does the opposing gas pressure. In feedback terms, the positive feedback of the gravitational pressure of the gas surpasses the negative feedback of thermodynamic gas pressure. This spontaneous condensation process is the first stage in the formation of a cluster of stars.

A number of smaller condensing clouds form within a large cloud, each an incipient star. The next stage in stellar development is the triggering of nuclear fusion in the ionized stellar interior. Light nuclei fuse together into heavier nuclei, releasing energy in the process. Before this can happen, the temperature of the gas must be raised to at least 10 million degrees Kelvin through conversion of gravitational energy to thermal energy. This build-up of thermal energy is promoted by the increasing optical opacity of the condensing gas cloud.

There are many paths by which the fusion can proceed, but the stars that formed early in the history of the universe, where there was only hydrogen as a raw material, had to rely on proton-proton fusion. Through a series of steps, four protons fuse to form a helium nucleus. During the reaction a certain amount of matter is converted into energy, which, as additional thermal energy, raises the internal temperature still higher, stimulating further fusion. The increased thermal energy also eventually balances the gravitational forces that were collapsing the star. The positive feedback between the fusion energy release and this thermal energy of the stellar material can sustain some stars for billions of years in steady state, the excess thermal energy being radiated into space.

The fusion reaction does not necessarily stop with the creation of helium nuclei. Given sufficient velocities, helium nuclei can fuse together to form carbon nuclei, which, at higher temperatures, can fuse together to form still heavier nuclei. Not only does the diversity of species of nuclei increase, but so does the diversity of fusion pathways. For example, carbon and nitrogen nuclei can catalyze further fusion of hydrogen to helium. Precisely what evolutionary course a star follows depends on its initial mass and other factors. The creation of heavier nuclei cannot proceed indefinitely, however, because creation of elements heavier than iron requires a net input of energy. In a very massive star a sphere of gaseous iron may form at the core and collapse on itself, releasing enormous amounts of energy. The star then explodes as a supernova. The extremely high temperatures produced may be enough to form, by fusion, the nuclei of many chemical elements above iron. This and other possible scenarios are discussed by Kippenhahn (1983).

The autocatalysis of atoms in stellar interiors has been called "*nuclear evolution*" (Calvin, 1956). The term evolution here is perhaps not accidental, but reflects an intuitive feeling one has of an analogy, if only a tenuous one, between the proliferation of nuclear species in a star and that of biological species in the earth's biosphere. The star is an open thermodynamic system, in which heavy nuclei are continuously created through fusion. The more kinds of heavier nuclei exist, the more pathways there will be for the creation of still newer and heavier nuclei. This is a positive feedback process in which complexity (the variety of nuclei) engenders greater complexity (see, for example McClendon, 1980). We shall argue later that a similar positive feedback process underlies the biological evolution and successional processes that lead to diversification of biological species and to greater ecosystem complexity.

3.2 Geophysical Systems

Many of the geophysical occurences on and within the earth result from the interplay of positive and negative feedback loops. Important examples are the convective processes within the earth's atmosphere. How some of these work may be explained in terms of experiments carried out around the turn of the century by the French physicist Bénard.

In these experiments a thin layer of fluid (spermaceti oil) was heated uniformly from below while its free upper surface was kept relatively cool by contact with free air at a constant ambient temperature (Fig. 3.1). As long as the temperature gradient was sufficiently small, nothing dramatic happened. However, if the applied heat was increased enough to raise this gradient above some critical value depending

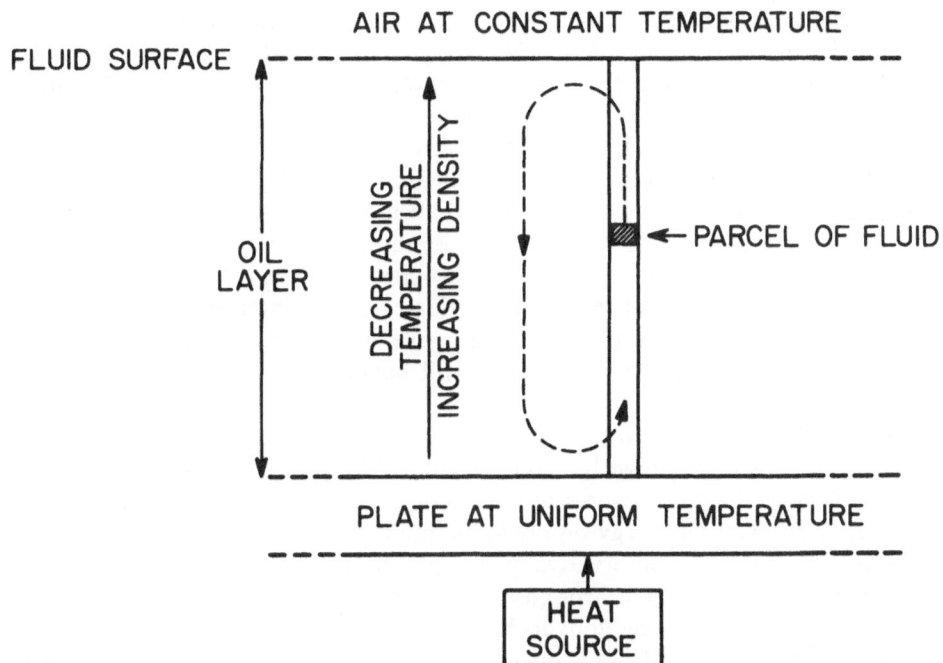

Fig. 3.1. Cross section of oil layer in Bénard's experiment. A fluctuation in the depth of the parcel of fluid upward or downward can be deviation amplifying, and result in closed convective circulation loops. The vertical dimension is magnified relative to the horizontal dimension here

on the type of fluid, its depth, and other conditions, the uniform surface would suddenly change to a tesselated pattern, usually a mosaic of hexagons. The visible polygons, called Bénard cells, were signs of the emergence of complex but spatially regular convective cells.

A detailed description of how Bénard cells develop is useful not only because this instability is a special case of the more general phenomenon of convective instability, but also because it is analogous to many other processes in nature. In Bénard's experiment one can conceive of the fluid being made up of vertical columns, as is pictured in Fig. 3.1. A high temperature gradient has been applied across the fluid but initially no convective current is flowing. Consider a given parcel of fluid at some depth in a given column of fluid. Velarde and Normand (1980) succinctly describe the onset of instability: "*Suppose now that through some random perturbation the parcel of fluid is given a slight upward motion. What effect does the displacement have on the balance of forces? The parcel is now surrounded by cooler and denser fluid. As a result it has positive bouyancy, so that it tends to rise. The net upward force is proportional to the density difference and to the volume of the parcel. Thus an initial upward displacement of the warm fluid is amplified by the density gradient, and the amplification gives rise to forces that cause further upward movement. A similar analysis could be made for a slight downward displacement of a parcel of cool, dense fluid near the top of the layer. On moving downward the parcel would enter an environment of lower average density, and so the parcel would become heavier than its surroundings. It would therefore tend to sink, amplifying the initial perturbation. Natural convection is the result of these combined upward and downward flows, and it tends to overturn the entire layer of fluid."*

Thus, what is occurring at the onset of convective motion is a positive feedback relationship. An incremental convective fluctuation slightly increases the positive bouyancy of a fluid parcel, which in turn enhances the convection. One can say that the initial convective motion catalyzes more motion, or that the motion is self-amplifying. Ultimately the convective movement will reach some limiting flux, depending on the amount of energy dissipation caused by internal viscous drag and the energy loss due to radiation processes relative to the energy being supplied to the fluid column. These negative feedbacks will balance the positive feedback and create a new steady state.

Because energy dissipation is a key characteristic of open systems such as the Bénard experiment, they have been referred to as *"dissipative structures"* (Prigogine et al., 1972a, b). Such structures are distinguished from conservative structures, which need no external energy supply to maintain them. For example, the rotating earth-moon system is a conservative system. The total energy of this system, kinetic plus potential, is conserved through time (discounting tidal friction).

Convective patterns analogous to Bénard cells can occur in many geophysical systems (Whitehead, 1971). For example, the influx of solar energy to the land surface heats this surface, thus creating a negative temperature gradient from lower to higher altitudes. If this gradient is steep enough, convective cells may develop. These are sometimes detectable by the presence of well-ordered cumulus patterns called *"streets,"* lying parallel to the direction of the prevailing wind (e. g., Haken, 1984). The clouds are formed as columns of upward-moving air cool so that their moisture condenses (Fig. 3.2).

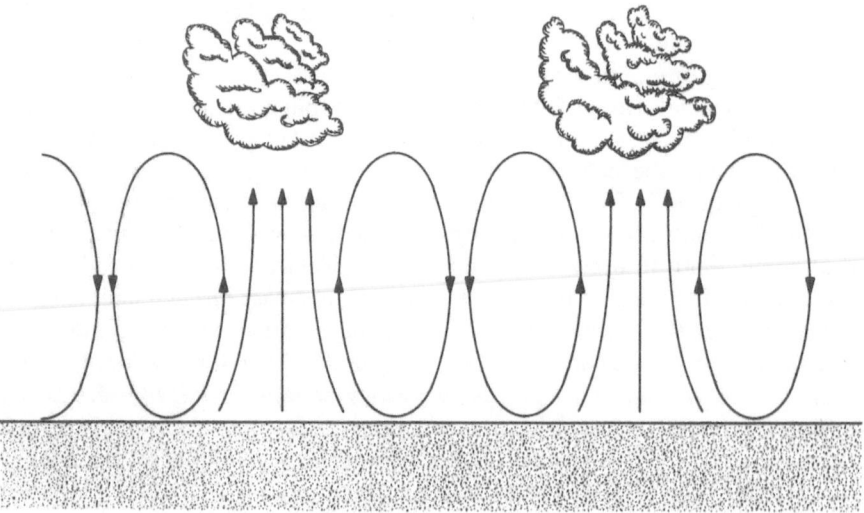

Fig. 3.2. Cloud streets formed by convective cells in the atmosphere

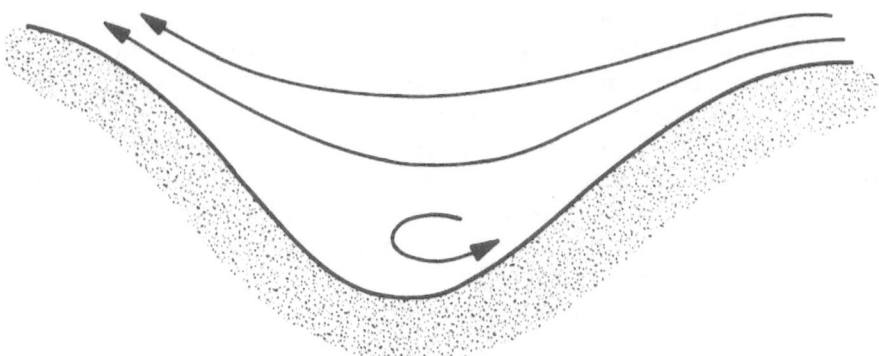

Fig. 3.3. Formation of waves in water subjected to horizontal wind along to the surface. This is an example of a convective instability

The formation of waves in water due to wind is another case of an instability driven by positive feedback. Given the chance occurence of some unevenness in the surface of the water, air blowing over the wave is forced to flow up over the high part of the irregularity and down into the incipient trough behind it. At the peak of the developing wave, the air velocity is greater; hence, by Bernoulli's law the air pressure is less than in the trough (Fig. 3.3). The air in the trough circulates in the reverse direction and transfers energy to the water in a manner that increases the wave amplitude. As the wave becomes larger, its structure increases the pressure differential over the water created by the wind, and the wave continues to grow until gravitational forces prevent further growth (Walker, 1977).

The Bernoulli effect creates the water waves because the increasing amplitude of the wave feeds back positively on the force tending to increase the amplitude of the wave. The Bernoulli effect can act in a similar manner when increasing velocity feeds back positively on the force causing a displacement. This can occur, for example, when air is blown forcefully across a suspended semicircular rod. Any slight movement of the rod perpendicular to the direction of the wind is amplified, leading to self-excited oscillations (s. Bishop, 1979). The reed of a clarinet being played is a case of self-excited oscillations stimulated by positive feedback drawing energy supplied by human lungs. Abraham and Shaw (1983) describe the dynamics of this phenomenon in detail.

Other similar instabilities include river meander, vortex and eddy formation in fluid streams when the fluid velocity exceeds critical values, instabilities in shear layers that form between jets and the surrounding medium of fluid (Lugt, 1983), the "*flute instability*" in a column of plasma contained by a magnetic field, the formation of sand dunes, and many other patterns in the dynamics of solids, fluids, and plasmas.

A geophysical phenomenon that has been the subject of much interest is the origin of the earth's magnetic field. The most convincing theories to date consider this field the result of a convective instability, though of a more complex sort than we have discussed so far because the interaction of electric currents with fluid motions is involved. The physicist Larmor first suggested that if the rotating earth was assumed to have a core of conducting fluid (molten iron and nickel), an initially very weak magnetic field along the axis caused by some fluctuation would induce electric currents that tend to amplify the magnetic field (Jacobs, 1963). Thus, the earth would resemble a self-generating dynamo.

The details of the theory are very complex, as might be expected for interactions between conducting fluids and magnetic fields in a rotating sphere. Early formulations of the theory were made by Elsasser (1939, 1946a, b). Bullard (1949) showed that even a simple model of the phenomenon includes two types of fluid motion and four stages of magnetic field.

Despite the great mathematical difficulty of the dynamo theory of the earth's magnetic field, some highly simplified models can give insights into its workings. Rikitake (1958), building on the work of Bullard (1955), presented one such model. Rikitake illustrated the process of inductive coupling by considering two rotating discs of conductive fluid. These are intended to represent only a part of the chain of magnetic couplings in the earth's core. Rikitake's example is interesting to us because the coupling between the disks via induced magnetic fields is analogous to that for a system of mutualistic species (s. Chaps. 2 and 8 of this book) and belongs to the class of positive feedback models. Below we sketch Rikitake's general formulation, although we do not follow his analysis in all its detail.

Let us consider a rotating electrically conducting disk. In the presence of some small initial magnetic field in the direction of the axis, a potential difference (voltage) is set up between the axis and the periphery of the disk. This potential difference produces an electric current if a conducting circuit connects the axle and the periphery by wire and brushes. This is just an example of Faraday's law, which states that a current will be generated in a conductor moving across a magnetic field.

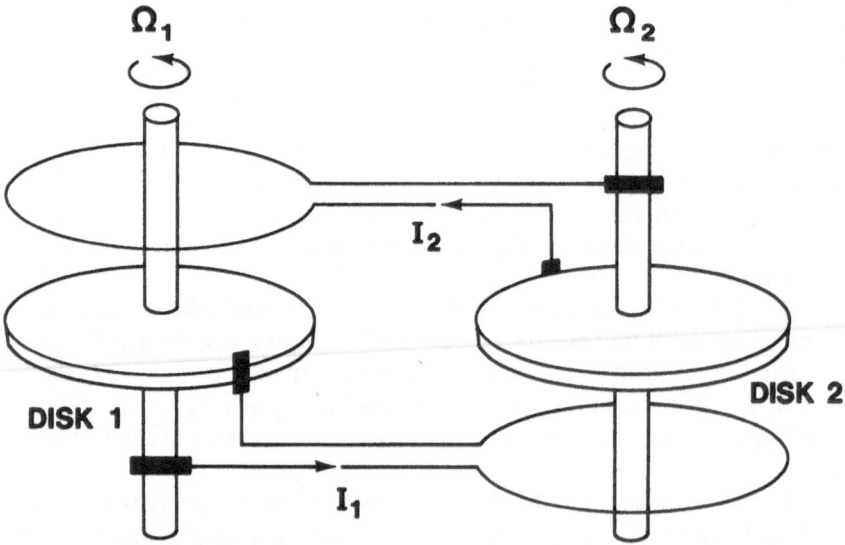

Fig. 3.4. Two rotating, electrically conducting discs coupled by electric currents, I_1 and I_2, which are generated by axial magnetic fields. The currents and the magnetic fields of the two disks form a positive feedback system. (Reproduced from Rikitake, 1958, with permission of Cambridge University Press)

Now consider two disks representing two sections of the earth's core (Fig. 3.4). These disks are assumed to be spinning freely with no external energy inputs or frictional forces to slow them down. The electric current induced from the axis to the periphery of each disk is led by a wire looped about the axis of the other disk as shown in Fig. 3.4, simulating the hypothesized coupling between sections of the earth's core. The current through each loop will induce, by Ampere's law, additional magnetic fields as the two disks feed back positively on each other. If the feedback effect is strong enough, the two axial fields grow simultaneously in a mutualistic fashion.

It will be helpful to see how this system is represented mathematically. The equations describing the induced currents, I_1 and I_2, generated in the two loops are

$$L_1 \frac{dI_1}{dt} = -R_1 I_1 + \Omega_1 M I_2, \tag{3.1a}$$

$$L_2 \frac{dI_2}{dt} = -R_2 I_2 + \Omega_2 N I_1, \tag{3.1b}$$

where L and R denote, respectively, the self-inductance and resistance of the loops. The angular velocity of the disks is represented by Ω while M and N are, respectively, the mutual inductances between loop 2 and the periphery of disk 1 and between loop 1 and the periphery of disk 2.

A complete description of the system of disks would require equations not only for the changes in the induced currents, I_1 and I_2, but also for the changes in the angular velocities, Ω_1 and Ω_2, which must slow down because their energy is transferred to the growing magnetic field. However, if the disks are sufficiently massive that their rotational energy is initially large compared to the energy transferred to the magnetic fields, the negative feedback from the changing rotational velocities can be ignored. We need consider only Eqs. (3.1a) and (3.1b) to determine whether or not the magnetic field is self-amplifying.

The stability matrix for the linear system (3.1) is

$$\mathbf{A} = \begin{vmatrix} -R_1 & \Omega_1 M \\ \Omega_2 N & -R_2 \end{vmatrix}. \tag{3.2}$$

This is a positive feedback system as defined in Chap. 2, where we demonstrated that simple criteria exist for the determination of the stability of linear systems of the form Eq. (3.1). Instability of the matrix $\mathbf{A}$ is equivalent to the condition that the initial small magnetic field (and equivalently the electric current) will grow in time. This criterion is simply (s. Chap. 2)

$$(-1)^2 \det(\mathbf{A}) < 0, \tag{3.3}$$

(where "*det*" means determinant) or

$$\Omega_1 \Omega_2 > R_1 R_2 / MN. \tag{3.4}$$

This two-disk system is only a simple representation of the n-disk system that would be needed to accurately model the self-generating dynamo of the earth. However, the n-disk system will also be a positive feedback system, and the criterion for magnetic field generation can be found by the methods used to study communities of mutualistic species presented in Chap. 8.

Phase changes in matter are another class of phenomena that involve positive feedback. The most familiar phase change in everyday life is the liquid-solid transition, which can occur abruptly. The freezing of water at 0 °C begins with a few nuclei around which crystals start to grow. As the surface area of the crystals increases, the number of places available for additional molecules to attach increases, causing a rapid acceleration of the process. Crystal growth in an abstract sense has some features similar to growth in a living organism. The accretion of new molecules "*creates*" structure from the random motion of the ions or molecules in solution. A flow of energy also occurs; the ions or molecules, when they attach themselves to the crystal, release increments of energy (heat of fusion).

Liquid droplets in a supersaturated vapor grow in an analogous manner once they have passed a critical threshold of size. Molecules in a droplet are attracted to each other by weak electromagnetic (van der Waals) forces. This is somewhat analogous to the process described earlier of how stellar and planetary bodies are formed by dust clouds drawn together by gravitational attraction. When a droplet becomes enlarged enough, it will break in two, that is, reproduce, but not in the sense that a living organism reproduces.

The process of dissolution of a crystal occurs when the crystal is subjected to increasing temperature. As the temperature is increased, defects occur in the structure. The number of defects increases until many are close at each other. "*At this point disorder can rise catastrophically because of cooperative effects of the interactions*" (Careri, 1984).

Theorists have attempted to develop a "universal" theory for all types of phase transitions (Brush, 1983). The zone of transition between two phases is characterized as being in a delicate balance between interaction forces that tend to impose order and the disordering effects of thermal agitation. A slight temperature change can be pictured as triggering a positive feedback induced landslide from one phase to the other.

3.3 Autocatalysis in Chemical Systems

As we have seen above, physical structures such as convective flows can arise from self-amplifying processes. These structures are evanescent and disappear when the external energy they feed on is removed. Other structures are more durable and can form building blocks on the path to higher levels of organization.

Nuclear evolution provided the earth with a variety of stable elements, many of which were able to interact chemically. The stage was set for "*chemical evolution*." Chemical evolution may be defined as the progressive building up of more complex chemical structures. This can occur when there is an external source of free energy, which can support self-amplifying processes. The particular sort of self-amplification inherent in chemical systems is called autocatalysis, in which a product of a

34

reaction acts as a catalyst for the reaction. Of course, for autocatalysis to drive a reaction, it must overcome the reverse reaction or negative feedback that results from the buildup of reaction products (Le Chatelier's principle). Calvin (1956) gives several examples of chemical autocatalysis.

Let us suppose a chemical reaction takes place in a chemostat into which energy in one form or another (for example, by means of an influx of high-energy compounds) is continually pumped. The chemostat can act as a dissipative structure. As a concrete example, we take a system discussed by Calvin (1956). Assume that a constant supply of hydrogen and cupric ions is provided to the reactor. In the presence of slight amounts of cuprous ions, an exothermic reaction can take place between the hydrogen and cupric ions to form more cuprous ions. Initially, with no cuprous ions present, the concentrations of the hydrogen and cupric ions will build up through time. Eventually a random fluctuation at the atomic level will cause a cupric ion to change spontaneously to a cuprous ion. This cuprous ion will catalyze further reactions until a new steady state is reached in which energy dissipation proceeds at a higher rate than it did before the transition.

Berry et al. (1980) give the example of a chain reaction of methane and fluorine described by the equations

$$CH_4 + F_2 \rightarrow CH_3 + HF + F \qquad \text{(initiation)}$$

$$CH_3 + F_2 \rightarrow CH_3F \qquad \text{(propagation)}$$

$$CH_4 + F \rightarrow CH_3 + HF$$

$$CH_3 + F + M \rightarrow CH_3F + M. \qquad \text{(termination)}$$

In this reaction the initiation step produces the free radicals, CH_3 and F, which are highly reactive. These participate in the propagation steps, which together produce as many new free radicals as they consume. These propagation steps are positive feedback reactions because they perputuate themselves. A termination phase (negative feedback) removes the free radicals from the reaction. (M denotes either a molecule in the gas phase or on the surface of the reaction vessel.)

Another chain reaction given by Berry et al. (1980) is that between gaseous O_2 and H_2. The propagation steps in this reaction are

$$OH + H_2 \rightarrow H_2O + H$$

$$O + H_2 \rightarrow OH + H$$

$$H + O_2 \rightarrow OH + O.$$

For every free radical OH consumed in these steps, two are produced, causing the reaction to accelerate.

Berry et al. (1980) refer to reactions of the $CH_4 - F_2$ type as stationary chain reactions because the balance of very reactive species going into and coming out of the propagation steps constrains the reaction to proceed at a constant rate. Reactions of the $O_2 - H_2$ type, however, are called nonstationary or branching-type reactions. This latter type of reaction is responsible for explosions. Explosions may, in fact, be engendered by an exothermic reaction if the heat produced is trapped so as to raise the temperature during the reaction. The increasing temperature acts as part of a positive feedback loop that drives the reaction faster, which in turn raises the temperature further.

From the bioligical viewpoint the most interesting reactions are those that involve enzymes. Many such reactions contain positive feedback loops. An example given by Berry et al. (1980) is that of an enzyme E that can undergo two stages of ionization:

$$E \rightleftharpoons E' + H^+$$

$$E' \rightleftharpoons E'' + H^+ .$$

In the particular enzyme, papain, considered by Berry et al., E' and E'' are inactive but E is catalytically active, catalyzing the hydrolysis of an ester (S):

$$S + E \rightarrow E + P^- + H^+ .$$

If E is initially quite small compared to E' and E'', the reaction at first proceeds slowly. However, the hydrolysis produces H^+ ions, shifting the enzyme equilibrium toward E, which further accelerates the hydrolysis. If S is continually added to the system, all of the enzyme in the system will eventually be converted to E.

Molecules that could catalyze their own synthesis would soon outnumber molecules that could not. Furthermore, if by some spontaneous fluctuation a new molecule appeared that could catalyze its own production more effectively than other catalysts, this new molecule would "outcompete" the others. For instance, suppose chemical A is transformed with equal probabilities into chemicals B and C. If chemical C mutates into a similar chemical C', which tends to catalyze its own production, C' will eventually dominate B. Calvin (1956) gives the example of porphyrin, which is an iron ion surrounded by a tetrapyrrole ring or a similar structure. Because this molecule facilitates reactions leading to its own synthesis, iron ions are likely to convert to porphyrin under the right circumstances. Porphyrin, when combined with a specific protein, becomes hemoprotein and has a much higher catalytic ability. Hence, chance fluctuations that lead to hemoprotein would have triggered positive feedback that increased this chemical species.

Another case is the evolution of the catalyst of the decarboxylation reaction (Calvin, 1956). The increasingly complex molecules methylamine, glycine, phenyl-glycine, etc., are all catalysts of this reaction, and they increase in catalyzing efficiency in the same order. This catalytic ability also happens to help in the formation of the catalytic molecules themselves, a case of autocatalysis. Therefore, if a steady-state reaction begins with only methylamine as a catalyst, eventually there will be a successional trend toward the more complex catalysts and hence usually toward a more efficient dissipative steady state.

The autocatalytic chemical cycles described above come closer to what we think of as "*life*" than do the physical convective systems described earlier in the chapter. Some speculations on how chemical autocatalysis may have developed into life are examined in the next chapter.

3.4 Summary and Conclusions

It is in the nature of many physical systems that certain processes, once initiated, tend to perpetuate themselves through positive feedback mechanisms. Such mechanisms occur at all physical levels. Nuclear fusion is involved in the positive feedback reaction that sustains stars and creates the diversity of chemical elements. Solar energy provides the energy that can initiate positive feedbacks that drive convective processes in the earth's atmosphere. Energy from the earth's rotation excites the axial magnetic field through a number of mutually reinforcing feedbacks. The class of chemical reactions termed autocatalytic also sustain themselves through positive feedback. All of these processes are members of the general class of dissipative structures. They require energy to maintain the positive feedback reactions, while dissipation acts as a negative feedback tending to balance the positive feedback.

4. Evolutionary Processes

4.1 Early Evolution of Life

The evolution toward life on earth involved processes occurring in its primeval waters or mud. The examples in the preceding chapter show that evolution toward more and more complex organic molecules under these primal conditions was not only possible but favored. Prigogine et al. (1972a, b) and Nicolis and Prigogine (1977) have speculated on the relevance of dissipative structures to prebiotic evolution. Short-wavelength solar radiation provided a flow of free energy into the original "soup" of inorganic compounds constituting the earth's primeval seas. This resulted in the continual synthesis of simple carbon-based compounds. These simple compounds would have occasionally joined to form stable molecules of greater complexity.

As the number and density of complex organic molecules increased, competition would have occurred between more complex compounds for the simpler monomer building blocks. When a new autocatalytic process appeared that was more efficient at utilizing the monomers, it could replace some of those already present. In turn, each new state, more complex than its predecessor, could have created conditions to make the system more vulnerable to instabilities from new fluctuations. Each new autocatalytic process can be referred to as a "fluctuation." Then the creation of greater and greater molecular complexity can be pictured as a loop (Fig. 4.1, adapted from Prigogine et al., 1972a, b). This is a positive feedback loop, because the more complex the system becomes, the greater becomes the probability of new fluctuations occurring to which the system is unstable, producing greater complexity. The sequence in Fig. 4.1 was referred to by Prigogine and his colleagues as the fluctuation-dissipation cycle. The dissipation of energy (along with entropy production) initially increases during each new fluctuation, but then settles down to a lower level as the system adjusts itself to its new constraints. The continued working of this fluctuation-dissipation cycle drives the system further and further from thermal equilibrium (see also Peacocke, 1983).

The theoretical possibility of systems producing more complicated systems was first demonstrated by von Neumann using the theory of automata developed by Turing (see Poundstone, 1985). Von Neumann showed that an automaton below a certain level of complexity, because of inherent unreliabilities of its parts, can itself produce only less complicated automata. However, if an automaton exceeds a certain threshold complexity, so that its parts collectively have enough reliability for it to meet the criterion for a Turing machine, then theoretically it can create automata as complicated as or more complicated than itself. Von Neumann

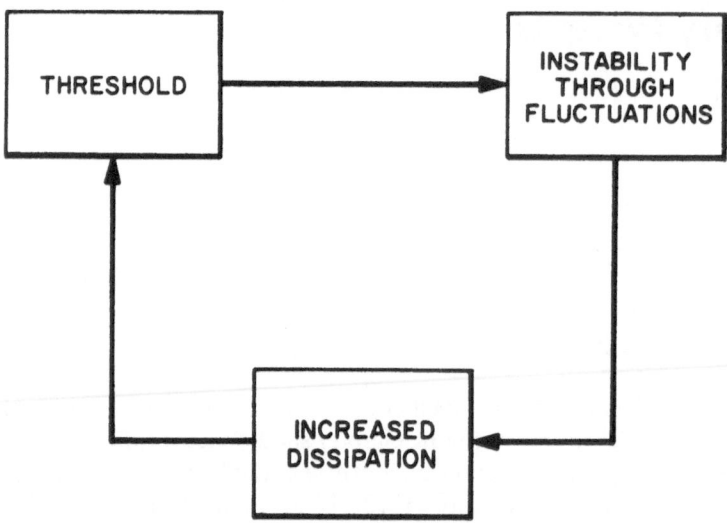

Fig. 4.1. A positive feedback process of evolution in an open physico-chemical system as envisioned by Prigogine et al. (1972a, b)

suggested that this principle should hold for chemical autocatalysis and that there was therefore no logical infeasibility in higher levels of organization, including life, evolving from lower levels.

A long series of instability-induced transitions was necessary before this self-organization reached the level we call life. It has often been asserted that living systems differ from nonliving systems primarily in being subject to the additional conditions of natural selection. However, the theoretical work of Prigogine and others in irreversible thermodynamics extends the action of natural selection down to purely physico-chemical systems.

Some sort of simple gene, probably a polynucleotide, would have been the first level of self-replicating organization identifiable as life. A fundamental question is how these individual genes could have assembled themselves into the groups of genes that carry the genetic information of living organisms. Mutations would have occurred from time to time in the genes, leading to a "community" of genes competing with each other for resource material and perhaps even specializing and dividing up niche space much as biological species do. In a relatively mature community, as in highly evolved ecological associations, mutualisms among some of these primitive genes must also have developed. Certain genes (Type A) replicated better in the presence of other genes (Type B) because of the way Type B altered the environment of Type A. Mutualism, the subject of a later chapter, can be a powerful positive feedback phenomenon, and pairs or groups of mutualists, each member specializing in particular functions, would have been efficient exploiters of their environment and hence strong competitors. Mutations that promoted spatial proximity of mutualists, perhaps through attractive forces of some sort, would have been favored.

Eventually the cooperative gene groupings would have become large enough to form their own internal environment. A threshold was crossed beyond which mutant genes that could not have existed on their own were successful and survived in the sheltered environment of the gene groupings. Mutual coadaptations through mutation would take place in the groups until the genes could survive only in the environment of the group (also called "supergene" or phenotype). This phenotype would, of course, be much more efficient than any particular member gene in its competitive and replicative abilities. Thus, the member genes could have traded their ability to replicate themselves independently for greater collective efficiency.

These primitive gene groupings or phenotypes may have resembled the catalytic hypercycles proposed by Eigen and his colleagues (Eigen and Schuster, 1979; Schuster and Sigmund, 1980; Eigen et al., 1981). Hypercycles are conceived of as positive feedback systems, each molecular species ("quasi-species") composing a given hypercycle playing a part in catalyzing the production of the next quasi-species. These hypercycles would have used as building blocks the organic compounds continually produced by physical processes in primitive waters or clay.

Eigen and his co-workers have shown that the rate of autocatalytic growth of these hypercycles would be proportional not just to the current populations of the quasi-species, but to the populations raised to a power greater than unity. They have also described the way in which hypercycles compete with each other. If two hypercycles that share no quasi-species are competing, the first to establish itself will inevitably outcompete the other. However, if the two hypercycles have some quasi-species in common, then the outcome of the competition will depend on the rate coefficients of various catalytic reactions of the hypercycles. If hypercycle 2 shares some quasi-species with hypercycle 1 and if hypercycle 2 is more efficient than hypercycle 1, then the former will "kill" the latter, or rather starve it out, inheriting the shared molecular species. This is how prebiotic evolution, a sort of Darwinian natural selection at the inanimate level, may have taken place (Haken, 1984). Each new, more efficient hypercycle would represent a fluctuation to which the dissipative system is unstable.

In primitive hypercycles, simple proteins and nucleic acids would have been logical partners. Cairns-Smith (1971) suggests that one of the many genes in such a phenotype may have been ribonucleic acid (RNA). This gene probably performed some trivial function at first, but its structure gave it a property of fundamental importance for the long term; it could aid in the synthesis of protein from amino acids. Protein is so useful as a building material that phenotypes with great amounts of RNA would have been more successful, on the average, than those with little RNA. Only a few more steps would likely have been necessary for deoxyribonucleic acid (DNA) to be synthesized by RNA. The central process of life as we know it could then have been inaugurated. DNA became the master controller of the phenotype, replicating itself and serving as the template for the production of messenger RNA, which produced the proteins for use in forming a binding structure about the gene group and providing necessary control enzymes. The positive feedback cycle of polynucleotides and polypeptides thus became established (s. Fig. 4.2).

The subsequent evolution of life as we know it was based on the establishment of polynucleotide-polypeptide hypercycles taking place in a nutrient-rich medium

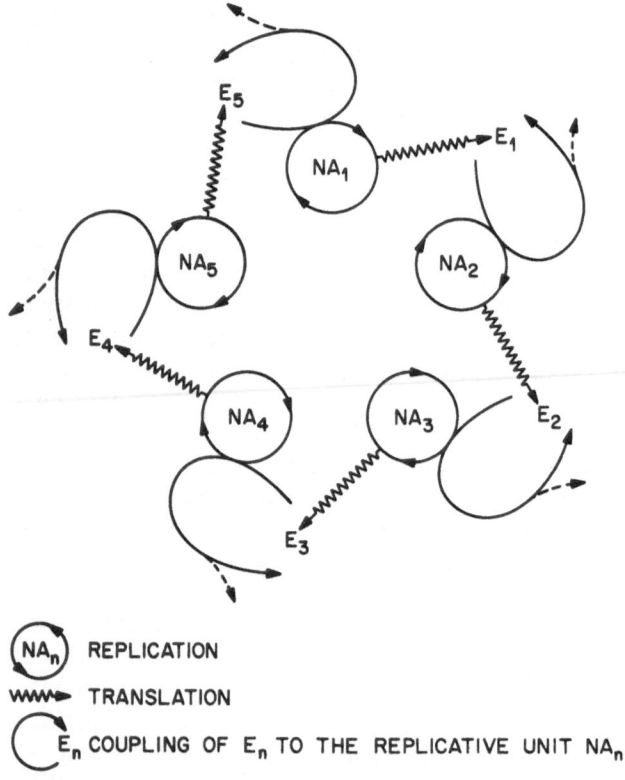

NA$_n$ REPLICATION

wwww➤ TRANSLATION

E_n COUPLING OF E_n TO THE REPLICATIVE UNIT NA$_n$

Fig. 4.2. A hypercycle consisting of nucleic acids and proteins. Each nucleic acid unit, NA_n, replicates itself and produces enzymes, E_n, that are involved in other parts of the cycle. (Reproduced from Cohn et al., 1980, with permission of Plenum Press)

surrounded by cell walls. Polypeptides and other molecules produced under the direction of DNA were used to form cell walls, creating an internal environment within which the conditions favorable for life processes could be maintained. The earliest cells were procaryotic cells and were relatively simple in structure compared with the eucaryotic cells that evolved later. Though surrounded by a cell wall, these organisms are open thermodynamic systems capable of maintaining a controlled flux of energy and matter. They are able to grow in size and reproduce at certain time intervals, after they have reached genetically predetermined size. This reproduction is the biological equivalent of chemical autocatalysis.

Evolution did not stop with the creation of simple procaryotic cells. The pattern of formation of mutualistic communities and their competition with other such communities continued at higher levels. Higher organisms consist not of the simple procaryotic cells, the possible origin of which was discussed in the preceding section, but of eucaryotic cells. Unlike procaryotic cells, whose genetic material is dispersed throughout the cell, the eucaryotic cells centralize most of theirs in a nucleus. In addition, eucaryotic cells contain organelles; for example, chloroplasts, where photosynthesis takes place, in plants, and mitochondria, where food is oxidized, in both plants and animals.

The next level of development of life was the emergence of multicellular organisms. In these organisms, the cells are now themselves specialized to perform particular functions. Some cells look after nutrition, while others specialize in locomotion, and so forth. No cell can survive and reproduce on its own, but each is dependent on all other cells. The pattern again is one of cooperative or mutualistic groups (cells, in this case) which were superior in efficiency to individuals in exploiting certain environmental resources. As had happened earlier to the mutualists composing the lower levels of organization, each member of the cell group now became more vulnerable in the sense that its fate depended entirely on the fate of the multicellular organisms of which it was part.

One can argue that each new level of complexity alters conditions in a way that makes the next level possible. Hence, one may picture a positive feedback loop somewhat similar to Fig. 4.1 but with the terms "fluctuation" and "dissipation" made more specific. The former term now means mutations to a higher organizational stage, and the latter term now means the more efficient use of both old and new environments. Whether or not the broad evolutionary trend toward greater complexity of organisms has reached an end can not be determined, but it is clear that evolutionary changes within species and groups of biological species can continue. Possible mechanisms involving positive feedback are discussed next.

4.2 Evolution at the Species Level

How strong a factor is positive feedback in the evolution of particular phylogenetic lines and in the process of speciation? Progressive evolution along phylogenetic lines (in contrast to branching into new species) has been termed "anagenesis" by Rensch (1959). Among the anagenetic trends in organisms are increasing complexity, increased plasticity of structure and function, improvement permitting further improvement (positive feedback of a sort), and increased independence from the environment.

The elaboration of complexity through evolutionary time has, of course, provided scope for a variety of trends in adaptation for particular species. Perhaps this progressive increase in complexity over evolutionary time is related to the fact that the maximum size of existing animals on earth has shown a tendency to increase; e.g., the largest mammal is larger than the largest dinosaur (Bonner, 1974). Bonner suggests that each step in increasing complexity enabled organisms to solve certain problems, among them the maintenance of larger size, better than previously evolved taxonomic groups. For example, although the development of vascular plants may have had causes quite unrelated to size, the increase in complexity allowed these plants ultimately to become much taller than non-vascular plants.

It has sometimes been stated that positive feedback is of relatively little importance in evolutionary trends within species compared, say, to its role in changes within human societies. The gist of the argument is as follows. The sophistication of the human brain makes possible the direct and rapid link between "need and novelty" (Alexander, 1979). Humans are capable of foresight and of developing tools, weapons, and methods that can quickly improve the relative fitness of individual human groups with respect to other such groups.

Unlike human cultural evolution, the argument continues, biological evolution results from the accumulation of genetic changes, each of which is the result of a random mutation. There is no foresight on the part of the genes as to whether the effect of a mutation will be beneficial or not to the individual or species. Only by natural selection in the environment are the detrimental mutations winnowed out. There is feedback, then, from the environment to the genome, though it is "indirect, complex, and usually slow" (Simpson, 1964). This feedback has historically been assumed to be negative feedback that winnows out mutations that are disadvantageous in a particular environment. Unlike human culture, which, as Rapoport (1974) and Wilson (1975), among others, have pointed out, can deviate considerably from the norm without hampering the survival value of a population, changes within natural populations will always be limited by the degree of their adaptive value in the current environment. If a species is nearly optimally suited to its environment, feedback acts almost exclusively in a negative fashion, selecting against deviations from the norm.

As counterexamples to this argument, we shall cite three mechanisms of genetic change that clearly incorporate positive feedback;

1) sexual selection,
2) coupling of behavioral change with genetic change,
3) and some types of frequency-dependent natural selection.

Darwin (1859) and later Fisher (1930) proposed that the sexual dimorphisms observed in many animal species could be the result of sexual selection. Male competition for females is especially prevalent. Barash (1979) describes the implications of this sort of selection: "*Once females of a species show a preference for a certain male trait — whatever its original utility — the trait can develop further reproductive utility of its own.*" The mechanism has presumably led to the development of bizzare and otherwise seemingly disadvantageous features as the elaborate plumage of the peacock and the huge claws of male fiddler crabs. For example, the peahen that mates with a peacock well-endowed with plumage increases her inclusive fitness because her male offspring are also likely to be well-endowed and to attract mates.

A species in which strong sexual selection is occurring could well be caught in a "*genetic positive feedback loop*" (Gould, 1982) that could be strongly "*dysgenic*" (Ghiselin, 1974). One might wonder whether a run-away selection for some paraphenalia of sexual competition, such as antlers on deer, could so hinder the males as to threaten the species with extinction. Critics of this idea reply that once such features as antlers became enough of a disadvantage, negative selection pressures from predators would have balanced the positive feedback of sexual selection and would have stabilized antlers at some size for which the species would still be viable. Other less dramatic hypotheses have been advanced for the evolutionary trends and ultimate extinction of the Irish Elk (see Gould, 1977).

Perhaps, however, it is not impossible for sexual selection to have at least created conditions that led to the extinction of the presumably polygynous species. All that would have been necessary is for males with larger antlers to have had more successful total life cycles (probability of survival to maturity multiplied by the expected number of offspring they would sire, given that they reached maturity) than those with smaller antlers. If the proportion of offspring sired by the larger antlered bucks (due to their greater ability to compete for females) more than compensated for their lower survival probability, then the trend towards larger antlers would have continued even though the species was developing characteristics that would make it vulnerable to high mortality rates if a sudden shift in environmental conditions occurred. Genetic models of O'Donald (1980) and Lande (1981) have corroborated the suggestions that sexual selection could *"become self-reinforcing and accelerate until checked by strong selection against the more extreme forms of the male traits"* (Mayo, 1983).

A more insidious threat to species survival could result from sexual competition at the gene or gamete level. If, for example, a mutant male Y-chromosome that caused males to have only sons appeared in a population, the proportion of males could amplify each generation, assuming they were equally successful as normal males at mating, until at some time all remaining females would mate with males carrying the mutant male chromosome, and the next generation would be all males (Hamilton, 1967, cited by Dawkins, 1982). This effect has been demonstrated in laboratory populations of Drosophila (Lyttle, 1977, cited by Dawkins, 1982). Competition between sperm to fertilize the egg could also lower the fitness of a species. This could occur if the sperm best able to reach the egg also carried genes that caused some debility to the offspring that showed up only later in life. Fisher (1941) had already used a similar type of model for the takeover of a particular genotype, regardless of whether it is better or not (see Mayo, 1983). If there are three genotypes, *gg*, *Gg*, and *GG*, and the *G* ensures self-fertilization, then the three genotypes are *"open-pollinated,"* *"half-open half-self,"* and *"all-self."* The presence of *G* will lead to its own inevitable increase.

A species is seldom a passive inhabitant in its environment, but interacts with its surroundings through its behavior. This interaction can result in a second type of positive feedback mechanism that can propel evolutionary change. A single species colonizing a new environment or migrating along a spatial gradient may display dramatic changes, since it must evolve rapidly to be successful. Take the case of an invading species that finds itself in an environment in which it has little or no competition. Adaptation to a new habitat may trigger a series of changes in body size, demographic strategy, mating system, and group size that may reinforce the adaptation (Eisenberg, 1981), or may result in the rapid elaboration of certain preadaptations (Mayr, 1960). In either case, positive feedback can promote rapid evolution.

A classic example of this idea is the case of the finches of the Galapagos Islands studied by Darwin. The behavior changes of the finches in sampling new foods quickly reinforced any genetic changes (because of the small size of the populations), which in turn promoted further dependence on new foods. Adaptive radiation of the original founding finches resulted. Mayr (1960) presented another example of evolutionary change in feeding. He suggested the idea of a population of fish acquiring the habit of eating small snails. If this behavioral trait increased fitness, then evolution in the population toward stronger and flatter teeth would be expected, leading to more feeding on snails. Bush (1975) posed a hypothetical model to explain a switch that actually occurred of a fruit fly from apple to cherry hosts. The model assumed that two mutant genes, one for feeding on apple, H_1, and one for adaptation to survival on apple, S_1, would have to be replaced by the equivalent genes for cherry, H_2 and S_2, for the change in the fruit fly population to come about. The occurrence of either of the mutant genes in the population favored the increase in the other mutant gene, which fed back positively on the increase in the first.

As another example, Grant (1963) described the radiation of the plants of the *Polemonium* genus believed to have orginated in the tropics or subtropics. In northern California *P. carneum* inhabits the moist coastal region, *P. caeruleum* the coniferous forest, *P. californicum* the upper coniferous forest and subalpine zone, *P. pulanerrium* the alpine zone, and *P. eximum* the high mountain peaks. This and similar patterns of adaptive radiation are examples of positive feedback in which, because of genetic variation within the original population, an incremental spread along the environmental gradient is possible. Development of new species at the margins of the range can occur through reproductive isolation. Natural selection then acts on the new species to reinforce its environmental adaptations. Eventually, some member of this new species will have the genetic capability of spreading farther along the gradient into new regions (Fig. 4.3). This phenomenon has been analyzed for *"fringe"* populations of pest insects such as the corn borer (Chiang, 1961). There is, then, the possibility for creativity in evolution, by organisms discovering and exploiting new opportunities that are made available to them by their invasions of new habitats and by whatever preadaptations they may have developed during their evolutionary history (Corning, 1983).

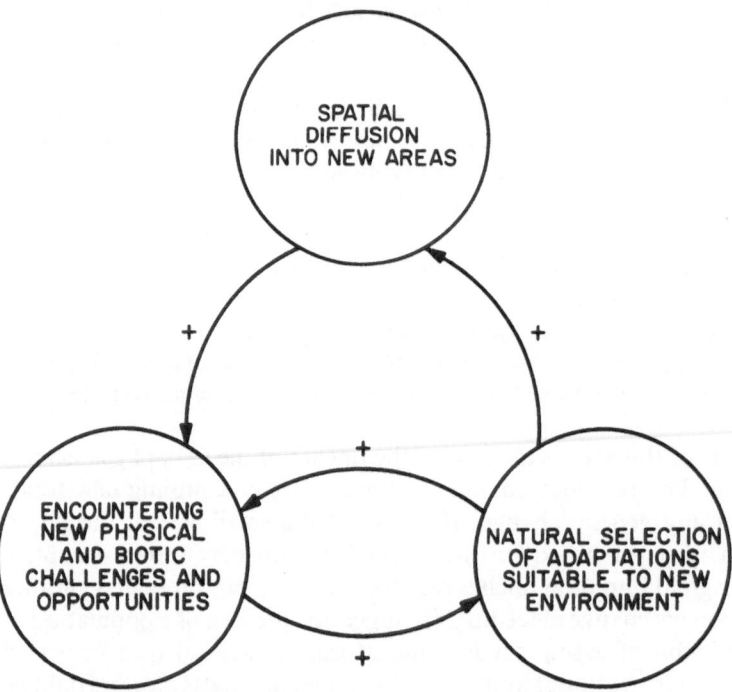

Fig. 4.3. Systems diagram of a phylogenetic line evolving as it migrates along an environmental gradient

A third type of positive feedback mechanism that may at least stimulate local genetic shifts encompasses a range of frequency-dependent genetic selection phenomena. An example given by Dawkins (1983) illustrates the mechanism. Certain East African cicadas seek to escape predation by clustering together on plant stems so that they closely resemble inflorescences of lupin. At this point Dawkins imagines, for the sake of the discussion, that the lupin occurs in two colors, pink and blue, and that the inflorescences are never mixed. Further, he assumes that they are uniformly spread over the region and that the cicada mimics occur as morphs of these two colors. Dawkins argues that any initially mixed distribution of the two cicada color morphs in a given locale will change to complete dominance by one color. This happens through positive feedback as follows. A mimic inflorescence that consists of mixed cicada morphs may be easily spotted as a fake by predators and attacked. If there are initially slightly more pink than blue cicadas, the blue ones will be slightly more likely to occur in mixed inflorescences and hence slightly more vulnerable than pink ones. As more blue cicadas are removed, the blue ones become rarer, and thus still more likely than pink ones to be in mixed inflorescences. Selection against blue morphs accelerates.

It should be added that frequency-dependent selection need not always be a positive feedback process as above. In cases of polymorphic populations where predators form search images of the more common morph, the predation feedback effect on the population will be negative.

4.3 Coevolution

Most organisms are influenced as strongly by their biotic environments as by their physical environments. The difference is that organisms are much more likely to have strong reciprocal actions on aspects of their biotic environment than on their physical environment. The possibility of mutual effects is, therefore, present to a greater degree in the first category. When these mutual effects lead to mutual genetic changes, one may call this coevolution. A concise definition of coevolution was given by Vermeij (1983) as "*the reciprocal adaption involving the heritable traits of two or more species.*" Evidence of coevolution has been zealously sought in recent years and some idea of its generality is beginning to emerge (see Futuyma and Slatkin, 1983; Dawkins and Krebs, 1979; Levin, 1983; Nitecki, 1983).

Coevolution, as we normally think of it, is a positive feedback process. A classic example is what Slobodkin (1974) called the "*evolutionary arms race*" between predators and prey. Natural selection favors both the predators that are slightly better than other predators at capturing prey and the prey that are slightly better than other prey at avoiding predators. This, in principle, should result in a sequence of offensive and defensive adaptations and counter-adaptations. As long as neither side gains a decisive advantage over the other, the positive feedback will lead to the gradual improvement of the two strategies.

Coevolution need not involve an antagonistic relationship. Mutualisms (discussed in Chap. 8) also arise through coevolutionary interactions. Many features of mutualism that at first sight seem puzzling can be plausibly explained by tracing positive feedback causal loops. For example, it has been noted that most hummingbirds that breed in North America gather nectar from red flowers there. This does not seem to reflect an innate color preference of the hummingbirds, because they visit flowers of other shades in their winter homes in the tropics. Grant and Grant (1968) suggested that, if, by genetic accident, a plurality of the hummingbird's plants in their summer homes in North America were red, this could have caused the nectar feeders to associate red flowers with rich nectar sources, and to be more easily attracted to them. This preference would have put selective pressure favoring red flower color in the plants that required hummingbird pollination. Over evolutionary time the tendency for most of the plants to be red would have amplified.

Not all coevolutionary interaction will necessarily involve positive feedback leading to unidirectional change. It is possible to imagine scenarios where an improvement in prey defense simply leads to decreased feeding by the predator on that particular prey. The decreased predation might cause an evolutionary relaxation of the prey's defenses (which are usually acquired only at some cost to other aspects of fitness). Increased predation might then ensue. Cyclical processes of this type would indicate negative, not positive, feedback. The changes might not be dramatic enough to be noticed in fossil records.

What is the actual evidence of coevolution? Futuyma (1983) remarked that it may be *"extraordinarily difficult"* to demonstrate reciprocal evolutionary interaction between plants and herbivorous insects. Some insects seem to respond to changes in particular plants, but plants show little evidence of response to particular insects. What coevolution there is, then, is diffuse, with the plants evolving generalized defenses. Bakker (1983), in analyzing the fossil record of ungulates and their distant-pursuit predators, has also found the reciprocal coupling to be very loose. It appears that over long periods of evolutionary time the predators lagged behind the prey in improvement of speed. Vermeij (1983) discusses the possibility of predator-prey arms races among marine organisms such as drilling gastropods and their bivalve prey. It is not known for certain in this case how much the drilling apparatus of the gastropod has improved in response to increases in shell thickness. Vermeij suggests that the effect of the prey on the predators may be weak. On the other hand, coevolution between bacteria and their viruses has been observed in the laboratory (Levin and Lenski, 1983). When the phage is highly virulent, a directional positive selection can occur causing increased virulence in the phage and increased resistance in the bacterium.

A strong case can be made for the occurrence of positive feedback in the evolution of mimicry. For example, Batesian mimicry occurs frequently in ecosystems. Batesian mimicry is the resemblance of a harmless, nontoxic organism to one that is toxic or otherwise dangerous. Darlington (1980) conjectured that a potential Batesian mimic initially probably bore only a faint resemblance to its model. This slight resemblance may have been enough, however, to give differential protection from predators, and hence higher survival rates, to those members of the mimic species that were slightly better mimics. *"As the resemblance became closer, selective advantage would increase and* (the mimic species') *rate of evolution would increase. Its evolution would then be both self-propagating and self-accelerating. But as the resemblance became very close, both the possibility and the advantage of further improvement would decrease and so would the rate of evolution* (Darlington, 1980).

This evolution of Batesian mimicry is not truly coevolutionary because only the mimic is changing through time. Müllerian mimicry, in which different distasteful species evolve towards the same color pattern, is a better candidate for coevolution (Gilbert, 1983). If both of the distasteful prey, A and B, are relatively abundant, the predators will associate a bad taste with both. Hence, as prey A approaches the pattern of B, the predator that has tasted A will also avoid B. This will select for individuals of prey B that more closely resemble A.

The mechanisms by which pollination and seed dispersal is carried out by insects and vertebrates for plants are often cited as products of long coevolutionary histories. Janzen (1983) believes that such coevolution has been rampant, but generally diffuse, and between similar suites of animals and plants.

The above examples, then, are a brief survey of possible types of coevolution now under study. While few unequivocal examples of tight coevolution between two species are at hand, the number of probable cases of diffuse coevolution is enormous. Whether the processes of coevolution have been specific or diffuse is not important, of course, to our argument that positive feedback in some form has brought about the evolutionary change.

4.4 Summary and Conclusions

This chapter has surveyed some of the positive feedback phenomena in evolution. The survey ranged from the evolution of prebiotic chemical complexes to speciation and phylogenetic change within complex living organisms. Our view of early evolution pivots around ideas of Neumann, Prigogine, and Eigen. In particular, in an open thermodynamic system, where a continuing supply of usable energy is available, complex organization can beget further complexity. Some of the complex structures so produced are more likely to survive than others. Therefore, a positive feedback cycles can ensue, yielding more and more complex structural forms.

The phylogenetic changes of biological species, on the other hand, seem able to follow a diversity of paths, depending on the particular situation. The changes may or may not be adaptive. The consequent genetic variation within a population, the differential success of the variants, and the possibility of behavioral modifications on the part of the organisms, together form a basis for genetic changes propelled by positive feedback. The types of behavior involved may be sexual selection, movement into new environments, or attempted escape from predators. All of these can feed back into selective success and then to gene frequencies, which further influence behavior.

Coevolution involves two or more species reciprocally interacting in a way that can drive both in a certain direction due to positive feedback. Possible examples including predator-prey interactions and mimicry were discussed.

5. Organism Physiology and Behavior

The life cycle of an individual living organism involves initial growth in size and complexity, along with maintenance of its fundamental integrity until its reproductive capability is completed. In this section we shall discuss some of the feedback loops involved in such life cycles, particularly as they relate to the physiology and behavior of animals.

Positive feedback is a basic element in the morphogenesis of an organism, not only with respect to growth in size but also in tissue differentiation. Martinez (1972), for example, following up the classic work of Turing (1952), showed how a system of cells can change from one distinct state to another through a series of instabilities and cell divisions, leading to a final, differentiated pattern. This can be described as a positive feedback loop similar to Fig. 4.1 (Prigogine and Lefever, 1975; Peacocke, 1983).

In mature (steady-state) organisms, however, it has been widely believed that organism physiology and behavior could be largely explained in terms of negative feedback regulation. If body temperatures fell too low, the animal burned more glucose (if it was warm-blooded) or moved from the shade into the sunshine (if it was cold-blooded); if its energy level fell too low, the animal searched for more food, etc. For each requirement, the animal was assumed to have some preferred level that it tried to stay close to.

In a highly respected textbook by Schmidt-Nielsen (1975), negative feedback is frequently referred to, but positive feedback is discussed only through a contrived example, which is worth mentioning since it underscores the prevalent view that positive feedback is always associated with things going wrong: "*Assume that a husband and wife have separate electric blankets, each thermostatically controlled through negative feedback. Say that the husband prefers a rather cool blanket and the wife a higher temperature. Let us now assume that the thermostats inadvertently get interchanged. The husband will set 'his' thermostat at his preferred low temperature, and his wife who now finds her blanket cooler than she likes, will turn up 'her' thermostat. The husband soon finds his blanket much too warm, and turns the thermostat further down, whereupon the wife turns 'hers' up even more. This is positive feedback, in which a deviation leads to an every-increasing augmentation of the deviation.*" Schmidt-Nielsen's example suggests that positive feedback is somewhat artificial and not at all common in real systems, except perhaps in destructive circumstances. It should be added, however, that Schmidt-Nielsen does briefly mention another, more useful, role for positive feedback in the courtship rituals of many animals.

5.1 Destructive Positive Feedback

The notion that positive feedback is a destructive process in an organism is often quite correct. The organs of the body are a tightly cooperative system. Damage to one of them usually has adverse effects on the others. For example, in higher animals a massive loss of blood weakens the heart by decreasing its blood supply. This slows down the rate of pumping, and hence the circulation of blood, which weakens the heart still further. Once this vicious cycle has begun, death soon follows (Toates, 1980). Explosive heat death is an analogous runaway situation where an initial loss of water from the circulatory system causes a thickening of the blood, reducing the body's cooling rate and accelerating the heat build-up (Murchie, 1978).

The circulatory system seems, in fact, suspectible to many positive feedback disorders. Witzmann (1981) reports on a blood pressure disorder in persons with cardiac or kidney defects or liver cirrhosis. These kidney or liver defects can cause edema, or the loss of fluid from the blood vessels to the surrounding tissues. Receptors in the blood sense the loss of water, but do not "*perceive*" the cause. They stimulate secretion of the hormone aldosterone from the adrenal gland, causing water to be retained in the kidneys. This places a greater burden on the heart, increasing the edema. Up to 30 liters of additional fluid may be retained in the body due to this vicious cycle.

A positive feedback loop has been explicitly implicated in a particular type of cancer, melanoma B16. In studies on mice, Bajzer et al. (1984) showed that the melanoma tumor produced and secreted into the bloodstream a substance or substances, called SICRI's, that are accompanied by a decrease in blood glucose and an increase in a growth hormone that stimulates increased tumor growth.

The multicellular organism that escapes death from the above or any number of other "*unnatural causes*" must perish all the same because of inevitable aging of its cells. One of the many theories of aging relates it to the accumulation of errors in protein synthesis (Orgel, 1973, cited by Calow, 1978). Orgel postulated that positive feedback eventually accelerates the build-up of malformed molecules after a threshold is passed when cellular mechanisms can no longer repair or remove the damage as fast as it is produced.

Sometimes a potentially destructive positive feedback loop is not wholly self-contained within the organism, but also involves interaction with the surrounding environment. Consider the case of a fish, which constantly works against positive feedback acting to change its depth. If a fish is initially at neutral bouyancy at a given depth in the water, a slight movement either downward or upward will tend to upset the neutral bouyancy by causing the swim bladder of the fish to be compressed or to expand, respectively. Hence, a downward perturbation away from x will give the fish negative bouyancy since the fish is now denser, while an upward perturbation will give the fish a positive bouyancy since the fish is now less dense than initially. The perturbation in either case will tend to grow since the deviation in density reinforces the initial deviation in position of the fish. The fish must prevent this cycle from developing by secreting additional gas into the swim bladder to counter the downward perturbation or by removing gas from the swim bladder to counter the upward perturbation. The fish will be stable if its negative feedback regulation is stronger than the positive feedback acting on it.

The above examples show that positive feedback can be dangerous to an organism when the feedback is not controlled by negative feedback and thus leads to a runaway situations. However, because it allows rapid responses by an organisms to contingencies, limited positive feedback is very useful. Stanley-Jones (1970) likens the safe use of positive feedback by an organism to the way in which a ramjet engine employs such feedback. A ramjet works by taking in air through the front of the jet, compressing it in a chamber, and mixing it with injected fuel. The mixture is ignited and the products are released through the rear nozzle, producing forward thrust. The faster the input of air through the front end is, the greater is the possible compression in the chamber and, therefore, the greater the thrust from the exiting gases. But with increasing thrust, the speed of the jet increases, thereby increasing the velocity of the air entering through the front. This positive feedback loop would cause the jet to fly faster and faster, out of control, if negative feedback, in the form of limiting the fuel injected into the chamber, were not at the pilot's disposal. With the control, the ramjet engine is usable as a mode of flight. It is reasonable, then, to look for examples of suitably controlled positive feedback in normal processes of organisms. Many such examples exist, a few of which are described below.

5.2 Biochemical Processes in Cells and Organisms

Both negative and positive feedback controls exist in cell metabolic systems. Cells produce the products they need through enzyme chains of the type shown in Fig. 5.1a. In this figure, a metabolite such as glucose is used to produce an enzyme E_1 from M_1, which produces enzyme E_2, and so forth, until the desired product P_1 is formed. In this pathway the product P_1 acts to repress the production of E_1, limiting, through the chain of reactions, its own further production. In other enzyme chains, however, it may be advantageous for the cell to produce a surge of

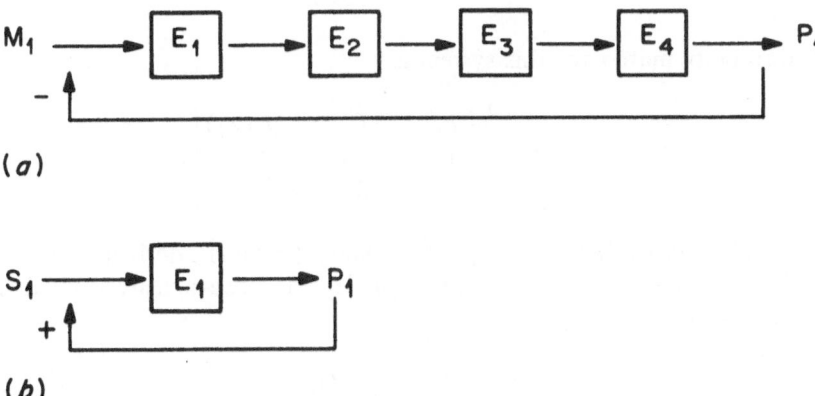

(a)

(b)

Fig. 5.1a, b. Examples of the cell metabolic feedback loops: a A negative feedback loop, where M_1 is the entering metabolite, $E_1 - E_4$ the metabolic chain, and P_1 the end product, which inhibits further production of E_1. b A positive feedback loop, where S_1 is an entering substrate (ATP), E_1 is phosphofructokinase and P_1 is one of the end products (ADP), which stimulates further production of E_1

some product. Because positive feedback is capable of producing such bursts, it makes sense in this case for the product to catalyze further production of E_1. A particular example of an enzyme chain is glycolysis, which is the conversion of the 6-carbon glucose to two 3-carbon fragments (pyruvate or lactate) that occurs during energy metabolism. Phosphofructokinase, an allosteric enzyme, plays the role of a regulator in this pathway. Its rate of reaction is enhanced by one of the products of its action, ADP, and is inhibited by one of its substrates, ATP. The positive feedback loop associated with this process is shown in Fig. 5.1b, where E_1 = phosphofructokinase, S_1 = ATP, and P_1 = ADP (Peacocke, 1983).

Rapp (1980) analyzed mathematically the cellular biochemical system consisting of messenger RNA (mRNA), Z_1, an enzyme coded by the mRNA, Z_2, and a product catalyzed by the enzyme, Z_3. The input to this system is the pool of nucleotides necessary for synthesis of mRNA. The positive feedback in this system results from the induction by Z_3 of further production of Z_1 (by combining with a repressor). Here we omit the details of the derivation and merely present the equations in their dimensionless form;

$$\frac{dZ_1}{dt} = f(Z_3) - k_1 Z_1$$

$$\frac{dZ_2}{dt} = Z_1 - k_2 Z_2 \tag{5.1}$$

$$\frac{dZ_3}{dt} = Z_2 - k_3 Z_3,$$

where

$$f(Z_3) = \frac{1 + Z_3^{\varrho}}{K + Z_3^{\varrho}}.$$

The stability matrix for this system is

$$\mathbf{A} = \begin{vmatrix} -k_1 & 0 & f'(\bar{Z}_3) \\ 1 & -k_2 & 0 \\ 0 & 1 & -k_3 \end{vmatrix},$$

where $\bar{Z}_3$ is the equilibrium value of Z_3, and $f'(\bar{Z}_3)$ is the derivative of $f(Z_3)$ at $\bar{Z}_3$.

The necessary and sufficient conditions for stability are easily found from (2.9) of Chap. 2 to be

$$k_1 k_2 k_3 > f(\bar{Z}_3)$$

$$k_1 > 0 \tag{5.2}$$

$$k_1 k_2 > 0.$$

There are positive feedback loops occurring both in the production of the female hormone estrogen and the male hormone testosterone (Fraser, 1979). Among the many hormonal feedbacks that have been analyzed is the positive feedback between the estrogen estradiol-17β and luteinizing hormone (LH) in the female. A surge in LH appears to be necessary for ovulation. When estradiol-17 is present in the bloodstream at sufficiently high concentrations, an LH surge can be triggered, the surge being part of a feedback loop causing a concomitant increase in estradiol-17β. A positive feedback loop also occurs in the sequence of reactions governing the clotting of blood, enabling a rapid response when needed (Esnouf and MacFarlane, 1968).

5.3 Feeding and Drinking Behavior

Feeding and drinking, while they serve overall homeostatic functions of keeping fluids, energy, and nutrients at proper levels in the body, contain aspects that are properly called positive feedback. For example, Toates (1980), in citing the work of Wiepkema (1968) on mice, points out that the first effect of a drink of water is to incite the animal's drive to drink more. Only later will the negative feedback of satiation cause drinking to terminate. Presumably the same sort of positive feedback acts will respect to eating, causing, for example, large carnivores to gorge themselves at meals.

Recently, Geiselman and Novin (1982) found that when glucose is absorbed in the duodenum of a mouse at a rapid rate, it induces hunger. Such positive feedback, by causing the animal to drink or eat beyond its immediate needs, would have a functional value in situations where water and food supplies were available only sporadically.

This same sort of positive feedback guides the animal in deciding which of its several requirements to satisfy at a given time. An animal may have equal needs for both food and water at a given moment, which could cause internal conflicts over which to seek. However, any initial move towards either "*cue*," the supply of food or of water, seems to amplify as the cue strength increases, overriding the other need until the first is satisfied (Toates, 1980; see also Harth, 1982). Ludlow (1982) developed a model of this effect.

Feeding often involves learning in higher animals, since they are not born with precise innate knowledge of what is edible and what is not. Learning is often viewed as a homeostatic process: "Learning by trial and error can be considered an example of negative feedback, as has been pointed out" (Bonner, 1980). But learning also has strong positive feedback components. Suppose an animal enters a new environment in which its familiar types of food are scarce but in which some nourishing though unfamiliar type is abundant. The animal may at first cautiously sample a small portion of the new type of food. If the animal is physiologically rewarded, it may take some more and finally switch entirely to the new type. This is a positive feedback type of learning because the initial slight deviation in its feeding habits amplifies through time.

A similar thing happens when animals (rats) are allowed the choice between two different foods, one more nourishing than the other. As the rats through sampling of the foods gradually sense that one of the foods has higher nutritional value (some delay time is necessary for the physiological response to be felt), they switch entirely to this food (Booth and Simson, 1971). Negative feedback, of course, acts to keep the animal's diet from straying to harmful foods. A small taste of toxic food, if the animal is lucky, will only make it sick, and this will discourage further ingestion of that food or food with similar taste.

5.4 Sleep

The phenomenon of sleep is one of the puzzles of the physiology of higher animals. The homeostatic view is that sleep is either brought on by fatigue of neurons, which must be periodically rested, or by the buildup of undesirable substances which can only be cleared from the body when it is resting. This explanation cannot be the entire story. Some animals such as dolphins sleep little, if at all, while others, including large carnivores, sleep a great deal. To explain these inconsistencies, other theories have stressed the functional importance of sleep. Sleep may be useful in conserving energy in some cases, or it may protect potential prey from predators by immobilizing them in hiding places during times of maximum danger (Toates, 1980). Since dolphins have few hiding places, sleep would not have this benefit for them.

Whatever the precise functional and homeostatic functions that sleep fulfills, there is speculation that positive feedback is involved in the transitions between sleep and wakefulness. According to some theories, the reticular system of the brain is primarily concerned with arousing the cerebral cortex to receive and interpret incoming sensory signals. Parsegian (1972) states: "*As activation of the reticular system arouses the cerebral cortex, the latter further increases the activity of the reticular system. It would seem that the interrelationship between the reticular system and the cortex represents positive feedback up to the state of full wakefulness or down to the state of deep sleep; that is, the states of full wakefulness and deep sleep represent terminal states, and transition from one state to the other is usually accomplished by progressive stages in which positive feedback interrelationships play a role. The feedback is assumed to be positive, because negative feedback would hold the system in its existing state.*" This loop between the reticular system and the cerebral cortex is not the only feedback involved in the sleeping/waking transition, according to Parsegian. He also suggests a positive loop by which the reticular system sends signals down the spinal cord to enhance muscle tone and is itself further excited by return impulses from muscle proprioceptors. A final positive feedback loop is hypothesized to join the reticular system with the sympathetic nerves, involving a release of adrenalin during waking.

5.5 Movement and Motor-Sensory Relationships

Feedback control processes regulating sensor-motor interactions also contain certain positive feedback aspects. Weis-Fogh (1949, cited by Stanley-Jones, 1970) noted that flying locusts regulate their flight by a system that, cybernetically at least, strongly resembles the ramjet engine mentioned earlier. The locust has patches of sensory hairs on its head. Wind impinging these hairs stimulates flight. The greater the spend of flight, the greater the input of wind speed and, thus, the more strenuously the locust flies. The speed of flight will automatically increase until it is opposed by the negative feedback of fuel limitation. This positive feedback mechanism is useful in that it causes the locust to fly at its maximum possible speed.

A similar case is the thermoregulation of the sphinx moth, studied by Heinrich (1971). Moths increase their thoracic temperature in order to fly. When this temperature is low, they fly slowly. The activity of the flight muscles warms the moths' thorax, allowing them to fly faster. Faster flight and higher thoracic temperature form a positive feedback loop, which causes both to increase until heat loss, governed be Newton's law of cooling, is great enough to stabilize the temperature.

Another motor-sensory interrelationship where positive feedback has been identified is in the walking locomotion of organisms. Pearson (1976) summarizes models of walking that may describe a range of animals from cockroaches to cats. In particular, transitions from stance, where the leg is down, to swing, where the leg is up, are guided by sensory information from stress receptors in the feet of the cat or the cuticle of the cockroach. Impulses from the stress receptors during the stance phase feed back positively on the neurons that control extension of the leg, which in turn increases stress. This feedback loop helps compensate for possible increases in load on a leg due to walking on irregular terrain, since an increased load, and thus additional stress, on a leg, will be countered by an increased rate of extension.

A particularly difficult problem for animal physiologists has been that of explaining the directed movement of very simple organisms, such as bacteria, along gradients of attractive or repulsive substances or fields. Explaining this movement has been difficult because one must explain first how the gradient is detected. Koshland (1977, see also Staddon, 1983) proposed a model based on a positive feedback loop to explain this kinesis. It is observed that there are two types of organism movement; random tumbling and straight-line swimming. Koshland assumed that the concentration, X, of a regulator chemical in the organism determines which response the organism exhibits. Suppose X suppresses tumbling above a certain threshold and that X is formed at a rate V_f from a precursor W, and decomposes at a rate V_d. Then, for an attractant, A, X is described by

$$\frac{dX}{dt} = V_f(A)W - V_d X, \qquad (5.3)$$

where $V_f(A)$ increases with the level of the attractant. As long as the organism is swimming in the direction of increasing levels of A, the concentration of A will feed back positively on X to keep X at levels that cause the organism to continue swimming in the same direction. However, if the concentration of A levels off or starts to decline, then X will decrease to below the threshold value because of the negative feedback term, $V_d X$, and the organism will cease to progress.

The nerve cells connecting motor and sensory systems transmit messages by a method that depends on positive feedback. When the neuron is not transmitting a message, a difference in electrical potential between the inside and the outside of the cell membrane, the "*resting potential*," is maintained. This electrical gradient occurs because sodium is actively pumped across the membrane so that potassium ions outnumber the sodium ions inside the cell and vice versa outside. When a stimulus reaches a spot on the nerve-cell dendrite, a rapid increase in permeability of the membrane to sodium ions occurs. This increase in permeability seems to involve a positive feedback loop in which an increase in Na^+ permeability leads to increased Na^+ inflow, leading to still greater permeability to Na^+ (Aidley, 1978). The result is that an electric disturbance is transmitted along the nerve axon. When it reaches the synaptic end of the cell, this impulse may be transmitted by chemical messengers to the dentritic membrane of the next cell where the process may repeat. Recently, other neuronal positive feedback loops, involving phosphoproteins, have been described (Nestler et al., 1984). These feedback loops appear to play an important role in neuronal function.

5.6 Mind-body Relationship

One of the deepest mysteries of nature, both for biologists and philosophers, is the mind- body relation. How does purposeful behavior come about? Does one run away because one is afraid, or is one afraid because one runs away? As with many of the other "*modern*" problems in biology, it was stated very clearly by Darwin: "*When the heart is affected it reacts on the brain; and the state of the brain reacts through the pneumo-gastric nerve on the heart; so that under any excitement there will be much mutual action and reaction between these*" (Darwin, 1871).

The mind-body question poses difficult problems to scientific understanding. Nonetheless, it is believed that feedback, both negative and positive, is pivotal in this relation, as was hypothesized Rosenblueth et al. (1943). For example, the emotions of fear, anger, and anxiety are associated with the production of adrenalin by the adrenal gland when stimulated by the sympathetic nervous system. This adrenalin circulates in the bloodstream and contacts the posterior hypothalamus, which is sensitive to low concentrations of adrenalin (Porter, 1952, cited by Stanley-Jones, 1970). When stimulated by the adrenalin in the bloodstream, the posterior hypothalamus sends signals through the sympathetic nervous system that trigger further production of adrenalin. A feedback cycle of intensifying rage or fear can result if this cycle is not counteracted by the parasympathetic nervous system, a system based wholly on negative feedback.

A set of studies, interesting for both its theoretical and medical implications, was performed by Sperry and his colleagues on sufferers of grand mal epileptic seizures. When the scientists cut the main bundles of neural fibers (the corpus callosum) connecting the right and left hemispheres of the neocortex, the frequency and intensity of the seizures diminished. Sperry interpreted this procedure as possibly having interrupted a positive feedback of epileptic electrical energy between the two hemispheres (Sagan, 1977). There are numerous other cases from the literature of neurophysiology and psychology that hint at connections between mind and body that scientists are only beginning to understand.

5.7 Summary and Conclusions

The idea of an organism as a purely homeostatic device regulating itself only through negative feedback is incorrect. Positive feedback is a component of many essential processes, from the cellular level to the complex interactions of organs in higher organisms. Models of feeding and drinking, sleep, motor activity, and the excitation of emotions, all involving positive feedback, have been reviewed here. Other processes, thermoregulation for example, also can involve self-reinforcing mechanisms (Lovelock, 1979). While runaway positive feedback is destructive to an organism, properly controlled positive feedback can allow the organism to respond quickly to some of the contingencies that confront it.

All in all, it is fruitful to consider Land's (1973) assessment of positive feedback as a natural and often beneficial (though also often pathalogical) process in all living organisms. He claims: "*The basic causal machinery of a living organism is not homeostatic. It is, in fact, quite the reverse, it is heterostatic. In the absence of positive and nutritive feedback, it can slow or lose the essential character of life, growth.*"

6. Resource Utilization by Organisms

"The annual produce of the land and labour of any nation can be increased in its value by no other means, but by increasing either the number of its productive labourers, or the productive powers of those labourers who had before been employed. The number of its productive labourers, it is evident, can never be much increased, but in consequence of an increase in capital, or of the funds destined for maintaining them. The productive powers of the same number of labourers cannot be increased, but in consequence either of some addition and improvement to those machines and instruments which facilitate and abridge labour; or of a more proper division and distribution of employment. In either case an additional capital is almost always required. It is by means of an additional capital only, that the undertaker of any work can either provide his workmen with better machinery, or make a more proper distribution of employment among them." The quote is from *The Wealth of Nations* (Adam Smith, 1776), and it introduces the idea of the relationship between reinvestment in capital and increased production.

In the *"economy of nature"* this principle of positive feedback has been used all along by organisms. It is hardly necessary to point out how often the metaphor of the human economy has used by biologists in describing the growth of an organism (or how often, too, the reversal of the metaphor has been employed by political theorists like Burke, for whom human society was like an organism). The nineteenth century physiologist, Claude Bernard, likened an organism to a society or a factory. More recently, Michael Ghiselin (1974) applied the laissez faire economic model to evolutionary theory. Arthur Galston (1981) noted the paradox that the word *"plant"* referred to both a *"green organism growing quietly in the sun and a noisy factory consuming fuel and discharging smoke as it turns out its products."* This is not an accident, Galston goes on the say, for these two plants are similar in a profound way.

A factory uses energy, usually from fossil fuels, to turn raw materials into manufactured products that can be sold. If the owner can sell these finished goods for more than the cost of the energy, raw materials, labor and factory upkeep, he or she will have some surplus to invest in expansion or capital improvements. Goods can then be produced at a faster rate. An organic plant uses solar energy, through photosynthesis, to produce sugar and, indirectly, proteins and other materials from the water, carbon dioxide, and nutrients at its disposal. Some of these products are invested in further growth of the plant, including creation of more photosynthetic material, so that the plant can produce sugar and other products at a faster rate. Hence, both plants, when expanding, exhibit positive feedback. (This argument applies to all biota, animals as well as plants. An animal operates in the same way, except that it derives its energy and nutrients from plants and other animals.)

The differences as well as the parallels between the two "*plants*" are instructive. Ultimately, the organic plant diverts its sugar and protein product to seeds for the purpose of perpetuating its genes. The capitalist owner of the factory diverts surplus production (profit), beyond that which is reinvested, for his or her own benefit (as with the organic plant, however, this profit often translates into enhanced reproduction). Another dissimilarity is that, while the human factory owner controls the operation of the factory through conscious planning, the organic plant is genetically programmed to behave in the way it does.

The term "*strategy*" is commonly used to imply a conscious plan of action, which can certainly apply to the factory owner, but not to primitive life forms. But here we shall make free use of the term to mean any behavior pattern to attain a specified goal, whether reasoned or instinctive. Thus the term can be applied to describe the behavior of any life form.

A strategy may be short-term, say the time scale of a brief encounter between two individuals; it may be intermediate, say over a day or a season; or it may be long-term, say a duration comparable to the life span of an organism. The ultimate purpose of any strategy, whether short-term or long-term, should somehow involve the increase in "*genetic fitness*", or relative representation of the individual's genotype in the population. In this chapter we shall be concerned with a category of questions sometimes referred to as ergonomics and involving the effective exploitation and use of resources by an organism, primarily over the short term. In Chap. 7, social strategies will be discussed.

6.1 Energy Allocation Tactics

Every organism requires physical space, energy, water, and a variety of nutrients to carry on its life processes. Although all of these resources are vital, energy is usually perceived to play a special, fundamental role. One of the essential characteristics of a living organism is its ability to extract free energy (energy from which useful work can be obtained) from its environment. It uses this energy to build and maintain its own organic structure from the raw materials at its disposal. Energy can be stored chemically as sugars and lipids and utilized when necessary through the medium of adenosine triphosphate (ATP).

Energy is a currency that can be expended to obtain, within limits, any of the organism's other needs, including more energy. In meeting its needs, an organism always expends time as well as energy. Hence time is commonly thought of as a resource itself, to be used in the most efficient manner possible.

An organism can only be successful if it stores and expends energy and uses time efficiently so that it will have time and energy to continue meeting its needs, plus some extra time and energy to devote to reproduction. A predator that captures a large prey item has procured for itself the energy it needs to improve its physiological condition. It therefore improves its ability to hunt more effectively for its next prey, to protect its resources from competitors and to defend its own life against other predators. If we consider the stored excess energy of the predator and

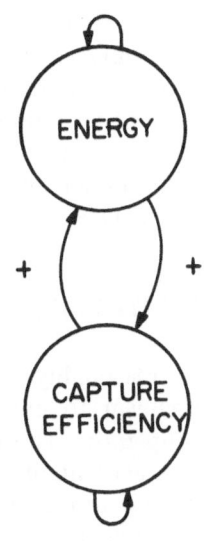

Fig. 6.1. A simple positive feedback model, showing that past success in capturing prey, by providing a predator with surplus energy, increases present prey-capturing ability, which will provide more surplus energy, and so forth

its *"prey capture efficiency"* as variables, then they should form a positive feedback loop (Fig. 6.1). Of course, there are genetically prescribed limitations both on the ability to store energy and on the predator's capture efficiency, and these limitations act as negative feedback, shown as the loops of unit length acting on the variables in Fig. 6.1. When these negative feedbacks balance the positive feedback loop, the system is in steady state. Stability of the steady state depends on details of the predator's physiology and the accessibility of its prey.

As a specific example, Heinrich (1979) describes a bumblebee colony (which can in some sense be regarded as a *"superorganism"*) in terms suggestive of a positive feedback system. If foraging worker bees extract more energy from flowers than they expend in foraging, they can *"invest"* the surplus energy both in supplying energy for further exploratory foraging trips and in producing new workers. The new workers add to the number of foragers and increase the inflow of energy. This positive feedback cycle continues until the colony's growth is limited by various internal and external factors, or the colony diverts its energy to the production of reproductives.

Another way in which animals can increase their returns depends on their feeding on more than one type of resource. How they allocate their time and energy to each particular resource depends on this positive return. Again, Heinrich (1979) draws a cogent analogy with respect to bumblebee colonies: *"The bees play a game analogous to the stock market. They do not know beforehand which is the most upcoming commodity (flower), and their best strategy is to invest primarily in the flower that appears to be the most remunerative, while simultaneously investing some energy in several minor species. When the rewards of a minor species go up, investments can be shifted accordingly."* The bees concentrate on whichever resource gives the greatest positive return. A similar example of changing resource allocation, in this case over an evolutionary time scale, was the shift in the human diet from incidental to intense use of agriculture. Rindos (1984) hypothesized that, during this shift, humans broadened their diets to feed on all available resources in direct proportion to their perceived abundances. The increased feeding on certain resources, those that were preadapted for easy domestication, led to their proportional increase near human settlements (e.g., through the scattering of seeds). This acted back positively on the human utilization of these domesticates.

6.2 Territorial Defense Strategies

No reasonably accessible resource in nature is uncontested for long. For an organism to exploit a resource, it must often defend the resource against competitors. This requires both time and energy, although the diversion of energy may take a number of forms, ranging from mechanical threat displays, fighting and chasing, to the production and release of toxicants.

Mathematical models can be useful in helping to explore the viability of various strategies an organism may have to select from in allocating time and energy. In particular, models can take into account the complex feedback relationships involved in such allocation. For example, is an animal better off searching for food full time, and consequently using energy in the process, or is it better off sleeping most of the time and foraging only occasionally? Both choices may be viable, but to predict which is more advantageous in a particular case, one must analyze the circular causal chain of how allocation of energy and time to one activity affects energy and time available for another activity and how this activity in turn influences the amount of energy and time that can be expended in more of the first activity.

As an example of a mathematical model based on these ideas, consider a simple model of an animal exploiting a single prey resource. We shall limit the example to include energy only. The system can be reduced to three variables: the animal's stored energy, E, territorial defense energy (in whatever form, say, production of toxicants or territorial behavior), D, and available prey biomass within the territory, P (Fig. 6.2). The equations relevant to this system might take the form

$$\frac{dE}{dt} = f_1(E, D, P), \tag{6.1}$$

$$\frac{dD}{dt} = f_2(E, D), \tag{6.2}$$

$$\frac{dP}{dt} = f_3(E, D, P,), \tag{6.3}$$

where it is reasonable to expect, for small values of E and D, the following inequalities to hold;

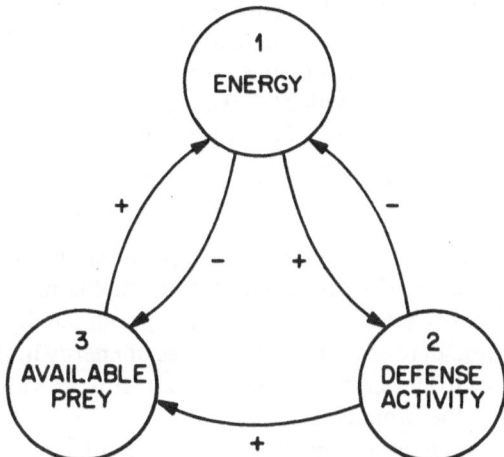

Fig. 6.2. Systems diagram for an animal. Allocation of energy to territorial defense may increase available prey, forming a positive feedback loop

$$\frac{\partial f_1}{\partial E}<0, \quad \frac{\partial f_1}{\partial D}<0, \quad \frac{\partial f_1}{\partial P}>0, \quad \frac{\partial f_2}{\partial E}>0$$

$$\frac{\partial f_2}{\partial D}<0, \quad \frac{\partial f_3}{\partial E}<0, \quad \frac{\partial f_3}{\partial D}>0, \quad \frac{\partial f_3}{\partial P}\lessgtr 0.$$

These inequalities imply that both stored energy, E, and defense activity, D, have some associated intrinsic loss, that the prey decrease with the size of the predator, represented by E, but increase with defense, D, and that the prey may experience either positive or negative feedback from itself in the absence of predation. Let us assume that the point $(0,0,P_0)$ is an equilibrium point, where P_0 is the prey equilibrium level in the absence of feeding and defense by the resident. We call a particular energy allocation strategy "*viable*" if this equilibrium point is unstable for an incremental insertion of E and D into the system. Letting $a_{ij}=|\partial f_i/\partial X_j|$, where X_1, X_2, and X_3 represent E, D, and P, respectively and $\|$ are absolute value signs, we then have for the stability matrix near the equilibrium $(0,0,P_0)$,

$$\mathbf{A}=\begin{vmatrix} -a_{11} & -a_{12} & a_{13} \\ a_{21} & -a_{22} & 0 \\ -a_{31} & a_{32} & \pm a_{33} \end{vmatrix}, \tag{6.4}$$

where the a_{ij}'s are all positive.

Note that since we have left unspecified for the present whether or not the available prey have a tendency of increase on their own near equilibrium, a_{33} may have either sign in front of it.

To begin with, let us make some assumptions that will simplify the problem. We can reasonably postulate that the rate of energy input to territorial defense, D, say some toxin or aggresive behavior, depends only on the amount of stored energy, E.

A less reasonable assumption, but one that is convenient for the time being, is that the prey biomass is relatively unaffected by feeding of the territorial resident. From these two assumptions, we have $a_{12}=a_{31}=0$. The matrix $\mathbf{A}$ is now essentially nonnegative, which allows us to analyze its stability characteristics quite easily. Using the principal minor stability criterion (Chap. 2), we find that E and D will increase from zero if $\mathbf{A}$ is unstable, or if

$$(-1)^3 \det(\mathbf{A}) = -a_{11}a_{22}(\pm a_{33}) - a_{13}a_{21}a_{32} < 0. \tag{6.5}$$

It is obvious that if $(\pm a_{33})$ is positive, the system is always unstable, and E, and D, increase. If $(\pm a_{33})$ is negative, this is not necessarily the case, and the positive feedback loop $1 \to 2 \to 3 \to 1$, represented by the term $a_{13}a_{21}a_{32}$, must be strong enough to counteract the losses of energy in the system. Note that this method does not tell us anything about which strategies are optimal, only about which are viable for the organism.

Now let us consider the more realistic case where a_{31} is nonzero. The matrix $\mathbf{A}$ now has negative off-diagonal terms. However, it can be converted by the similarity transform (see Appendix F), $\mathbf{A}' = \mathbf{S}^{-1}\mathbf{A}\mathbf{S}$, where

$$\mathbf{S} = \begin{vmatrix} 1 & 0 & 0 \\ 0 & 1 & 0 \\ \alpha & 0 & 1 \end{vmatrix} \quad \mathbf{S}^{-1} = \begin{vmatrix} 1 & 0 & 0 \\ 0 & 1 & 0 \\ -\alpha & 0 & 1 \end{vmatrix}, \tag{6.6}$$

and where α is an arbitrary paramter, to the matrix,

$$\mathbf{A}' = \begin{vmatrix} -a_{11}+\alpha a_{13} & 0 & a_{13} \\ a_{21} & -a_{22} & 0 \\ \alpha a_{11}-a_{31}+\alpha a_{33}-\alpha^2 a_{13} & a_{32} & \alpha a_{13} \pm a_{33} \end{vmatrix}. \tag{6.7}$$

If there are values of α for which the term $a'_{31} = \alpha a_{11} - a_{31} + \alpha a_{33} - \alpha^2 a_{13} > 0$, then the new matrix, $\mathbf{A}'$, is essentially positive. It can be seen that a'_{31} is most likely to be positive when a_{11} is large, a_{33} is large and positive, and a_{31} and a_{13} are relatively small. Suppose there are values of α for which $a'_{31} > 0$. Then from the principal minor criterion for the stability of essentially positive matrices, it follows that the system is unstable if $(-1)^3 \det(\mathbf{A}') < 0$.

The quantitative methods used here become especially useful when the positive feedback occurs in a more indirect manner. Consider a slightly more complex model in which invading competitors are explicitly included in the model of the animal (Fig. 6.3). Additional energy to the resident permits more defense activity, which decreases the invader density that would tend to reduce available prey. If we can assume, as earlier, that the resident does not have a significant effect on prey density, and that energy input to defense, D, depends only on available energy, E, the stability matrix near the equilibrium point, $(0,0,I_0,P_0)$, where I_0 is the equilibrium invader density, is

$$\mathbf{A} = \begin{vmatrix} -a_{11} & 0 & 0 & a_{14} \\ a_{21} & -a_{22} & 0 & 0 \\ 0 & -a_{32} & -a_{33} & 0 \\ 0 & 0 & -a_{43} & -a_{44} \end{vmatrix} \tag{6.8}$$

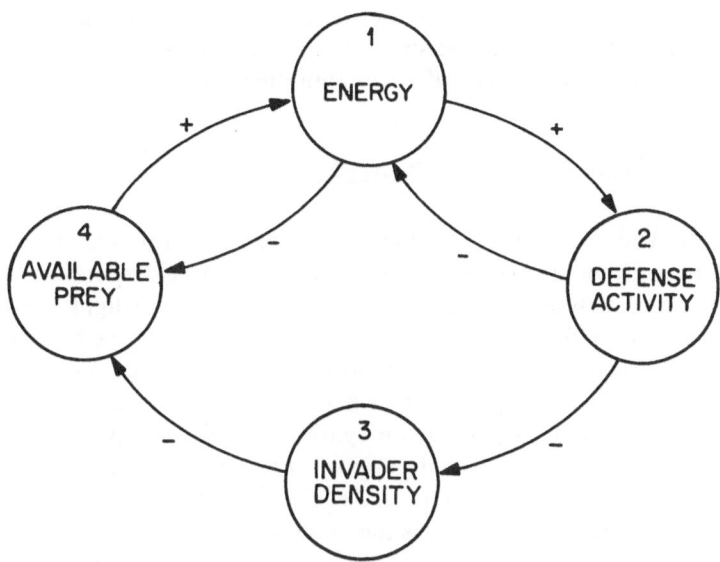

Fig. 6.3. Systems diagram for a territorial animal in which the invader density is now explicitly included

Matrix **A** can be transformed to the matrix

$$\mathbf{A}' = \begin{vmatrix} -a_{11} & 0 & 0 & a_{14} \\ a_{21} & -a_{22} & 0 & 0 \\ 0 & a_{32} & -a_{33} & 0 \\ 0 & 0 & a_{43} & -a_{44} \end{vmatrix} \qquad (6.9)$$

by use of the similarity transform

$$\mathbf{S} = \mathbf{S}^{-1} = \begin{vmatrix} 1 & 0 & 0 & 0 \\ 0 & 1 & 0 & 0 \\ 0 & 0 & -1 & 0 \\ 0 & 0 & 0 & 1 \end{vmatrix}. \qquad (6.10)$$

Because **A**′ is essentially positive, the criterion for viability is easily obtained and is

$$a_{21} a_{32} a_{43} a_{14} > a_{11} a_{22} a_{33} a_{44}, \qquad (6.11)$$

again implying that if E and D are to increase, then the one positive feedback loop must exceed in strength the product of the four negative feedback loops of unit length.

These examples illustrate that application of the mathematical methods of positive linear systems can become complicated when negative feedback loops of length greater than unity are involved. This forces one to make simplifying assumptions if simple analytic criteria are desired. Still, this is a useful procedure for weighing the relative efficacy of one more positive feedback loops in overcoming the negative self-feedbacks of each component of a system on itself.

6.3 Chemical Defense Strategies

The above strategies involved competition for resources through defense. Competitive strategies can be very complex. Competition between plants, for example, can include aboveground competition for light as well as belowground competition for water and nutrients (Grime, 1979). The belowground competition may also be waged chemically, since competitive strategy can involve production of toxins to be used against other organisms in the vicinity. Consider a vegetatively spreading plant and assume it synthesizes biomass, X_b, at a rate proportional to existing biomass, which is apportioned in fixed ratios of photosynthetic and structural material. Assume further that some fraction, u_x, of this biomass goes into toxin production, X_t. If r_x is the maximum rate of biomass production and g_x is a coefficient of self-regulation, then we obtain, following Vincent and Pulliam (1980),

$$\frac{dX_b}{dt} = r_x(1 - u_x)X_b - g_x X_b^2 \tag{6.12}$$

for the growth rate of the plant, and

$$\frac{dX_t}{dt} = r_x u_x X_b - d_x X_t \tag{6.13}$$

as the equation for toxin concentration, where d_x is the loss of toxin through degradation and leaching or other forms of transport out of the system.

Now suppose another plant of biomass Y_b is competing with the first plant by using its own toxins. If we let c_{xy} represent the rate of depression of the biomass, X_b, per unit of toxin, Y_t, then equation (6.12) is modified to read

$$\frac{dX_b}{dt} = r_x(1 - u_x)X_b - g_x X_b^2 - c_{xy} X_b Y_t. \tag{6.14}$$

There is a similar equation for Y_b;

$$\frac{dY_b}{dt} = r_y(1 - u_y)Y_b - g_y Y_b^2 - c_{yx} Y_b X_t. \tag{6.15}$$

In addition to Eqs. (6.13), (6.14), and (6.15), there is also an equation for the amount of toxin from species Y, Y_t.

$$\frac{dY_t}{dt} = r_y u_y Y_b - d_y Y_t. \tag{6.16}$$

The matrix describing the linearized two-plant competition near equilibrium is

$$
\mathbf{A} = \begin{vmatrix}
-g_x \bar{X}_b & 0 & 0 & -c_{xy} \bar{X}_b \\
r_x u_x & -d_x & 0 & 0 \\
0 & -c_{yx} \bar{Y}_b & -g_y \bar{Y}_b & 0 \\
0 & 0 & r_y u_y & -d_y
\end{vmatrix}. \tag{6.17}
$$

The above matrix $\mathbf{A}$ has the form of a positive feedback matrix since the only feedback loop longer than one, $(-c_{xy}\bar{X}_b)(r_x u_x)(-c_y \bar{Y}_b)(r_y u_y)$, is positive. Using the conditions on the principal minors, one can determine under what conditions the two plants will coexist. If this matrix is unstable, one of the two model species will be driven to extinction, while the other increases in biomass.

Alternatively, one can look at the problem from the point of view of the invasibility by each species individually, given that the other species is initially at the equilibrium values it would take in the absence of the first species. If $\bar{Y}_t$ is the equilibrium value of the toxin of species Y when species X is near zero, then the relevant criterion for the ability of plant X to invade is determined from the submatrix

$$
\mathbf{A}_{11} = \begin{vmatrix}
r_x(1 - u_x) - c_{xy} \bar{Y}_t & 0 \\
r_x u_x & -d_x
\end{vmatrix}. \tag{6.18}
$$

Species X can invade if $(-1)^2 \det(\mathbf{A}_{11}) < 0$, or

$$
r_x(1 - u_x) > c_{xy} \bar{Y}_t. \tag{6.19}
$$

This inequality simply means that the growth rate of X (including the loss due to toxin production) must exceed the competitive effect on X by Y.

Although our example explicitly considers chemical defense, one could, with minor adjustments, also let u represent some other form of competition, so the model has quite general implications. Grime (1979) presents verbal arguments that give credibility to the mathematical assumptions of this model. He points out that in productive, undisturbed habitats, "*there is a positive feedback between the ability of the stronger competitors to capture resources and the tendency to subject weaker competitors to 'plant-induced' or 'plant-intensified' stresses, principally by shading and by the depletion of water or mineral nutrients in the rhizosphere. This phenomenon may be described as competitive dominance . . .*" (p. 22). Further, "*if the growth of the larger species remains unchecked, a process of exclusion occurs and this may eventually result in the vegetation approaching a state of monoculture. It seems reasonable to conclude that the dramatic reduction in species densities in meadows observed over the last 30 years in Europe is to a large extent the result of an increasing intensity of competitive dominance brought about by stimulating the yield of the more robust and productive species and genotypes through the application of high rates of mineral fertilizer*" (p. 125).

6.4 Growth Rate Strategy

It is not uncommon for populations of certain fish species in small lakes or ponds to exist in either of two extreme states; a small population consisting primarily of large fish, or a large population consisting of small, stunted fish. While there are many possible explanations for this dichotomy, one interesting hypothesis is that in some cases a small subset of fish in an age class cohort of young fish are able to reach a size that allows them to exploit an abundant supply of large prey, including sometimes their own siblings. This advantage is most often due to their being spawned earlier than other fish, but may also relate to faster growth rates. The result is that fish in this small subset can continue to grow so fast that they have available to them an increasing food supply despite the gradual exhaustion of smaller prey. Fish that are able to achieve this sort of self-amplifying growth are sometimes called "*jumpers*". These fish will continue to grow until physiological or environmental factors limit their growth, forming a fairly small population of large fish. Others that do not reach the critical size range to have abundant prey at their disposal, may be eaten or perhaps starve. This phenomenon is often called "*growth depensation*," reflecting the idea that the fish that have grown the largest in size by a certain time will have the fastest rate of growth in the future. Observations of this effect in fish populations have been made by Keast (1965) and Shelton et al. (1979), among others. In other cases, perhaps by chance, there is no small group of fish of sufficient size to "*jump;*" what results is a large population of stunted individuals. It is obviously genetically advantageous for a fish to attain large size as quickly as possible when young, so as to increase its chances of being a jumper (of course, this advantage of large size, often gained by being spawned early, must be bought at the expense of the risks inherent in being spawned too early into unfavorable conditions).

A very simple mathematical model can be used to illustrate this conceptual model graphically. Let W denote the average weight of a small group of young, but relatively large fish, potential jumpers, and let F denote the food supply available to these fish. This average weight will increase as some function of current weight and assimilated food supply, $\alpha f(F, W)$, where α is an assimilation coefficient, minus a respiration term, $r(W)$;

$$\frac{dW}{dt} = \alpha f(F, W) - r(W). \tag{6.20}$$

The available food supply is then diminished at the rate $f(F, W)$, but is at the same time replenished at a rate, $g(F, W)$, that depends on the amount, F, already present and on the size of the fish, W. Hence we can write

$$\frac{dF}{dt} = g(F, W) - f(F, W). \tag{6.21}$$

To explore the dynamics of Eqs. (6.20) and (6.21) we must make some assumptions concerning the forms of $f(F, W)$, $g(F, W)$, and $r(W)$. One can imagine a hypothetical situation where the isoclines $dF/dt = 0$ and $dW/dt = 0$ have roughly the shapes shown in Fig. 6.4. The shape of the $dF/dt = 0$ isocline can be interpreted as follows. The food available to small fish (small W), is assumed to consist of zooplankton species that replenish themselves rapidly, so it is difficult for dF/dt to become negative. For somewhat larger values of W, the desired prey are assumed fewer in number and $dF/dt < 0$ even at low values of F. But if W increases far enough (i.e., into the area of the peak of the $dF/dt = 0$ isocline), a more sizeable prey budget is available, perhaps including small fish. Ultimately, W can become so large that appropriate food is not plentiful for fish of that size, so the $dF/dt = 0$ isocline approaches to $F=0$ axis. The shape of the $dW/dt = 0$ isocline signifies that the growth of the fish becomes less efficient as the weight increases until W becomes large enough that no further weight increase is possible.

In Fig. 6.4 three nontrivial equilibrium points, E_1, E_2, and E_3, occur. Points E_1 and E_3 are stable, while E_2 is unstable. Note that the relationship between F and W shown in the graph is analogous to the two-species mutalistic system (Chap. 2). Positive feedback will propel the fish to E_3 or to E_1, depending on the starting value of W.

In the analysis of this simple model there is no need for sophisticated mathematical techniques. However, a more detailed model of the jump phenomenon may involve more complex feedback loops, and determination of occurrence of jump phenomena would then require the more elaborate methods.

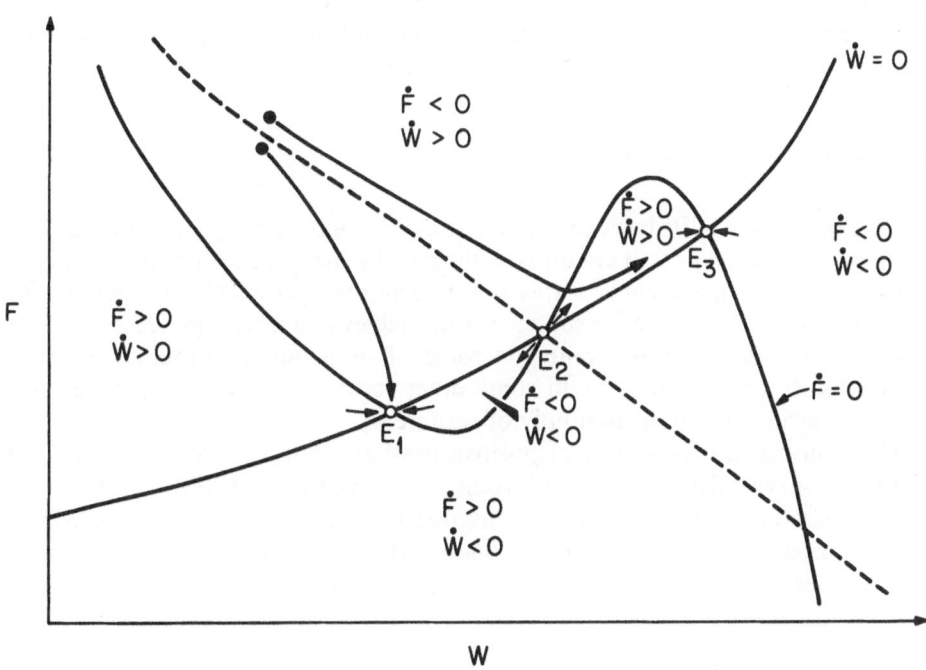

Fig. 6.4. Zero isoclines of Eqs. (6.20) and (6.21) ($dW/dt = 0$, $dF/dt = 0$). There are two possible stable equilibria for consumer sizes, W, one at E_1 and one at E_3. A strategy of rapid initial growth of consumers may allow them to reach the more favorable point, E_3

In plants, as in animals, there is a premium on rapid growth. Because the ultimate purpose of an individual organism is to reproduce, growth is subordinated, over the long term, to that purpose. Over certain time scales within its life span, however, the strategy of a plant may be viewed as that of maximizing growth, subject, of course, to constraints imposed by the environment. Maximizing growth over any extended period is not a straightforward problem. At any given moment the plant has the choice of allocating biomass among its structural components; the photosynthetic system, the support and conductive system, and the root system, as well as among the subsystems of these systems. The rate of growth may be assumed roughly proportional to the amount of photosynthetic matter already present in the plant. This is a "*compound interest law of plant growth*" (Blackman, 1919). Long term maximization of growth depends on the way in which biomass in "*invested*" at given moments. A simple representation of plant growth (from Mooney, 1972) is shown in Fig. 6.5. For growth to occur, the positive feedback of the large feedback loop must outweigh the negative feedback involved in the synthesis and maintenance of each plant component.

Mooney (1972) has noted that different types of plants may employ markedly different biomass allocation strategies. For example, in Arctic, desert, and grassland communities 50% of the plant biomass may be in the roots compared to 25% in temperate forests. In an evergreen forest 8% of the biomass is in the photosynthetic system, while only 1% is in a deciduous forest.

To study the outcome of competition between plant species employing different growth strategies, one could develop a more detailed version of the model used for plant competition in an earlier section, this time explicitly modeling the apportioning of biomass onto various structural components.

6.5 Summary and Conclusions

In the exploitation of its basic resources, an organism or social group or organisms resembles a business firm. It attempts to allocate the energy and nutrient resources it wins from the environment in a way that maximizes a product-reproduction. But maximization of reproductive success is only achievable after a period of growth, during which the organism reinvests some of its acquired resources either in increasing its own biomass or in some other form of storage (e.g., honey in a beehive), or in defense of its supply of resources.

This reinvestment is a form of positive feedback. On the average, only those genotypes survive that produce the strongest positive feedback loops. The types of possible strategies are manifold. As discussed here, they can include territorial defense, production of toxins, and various growth rate strategies. Of course, the list of strategies described here only scratches the surface.

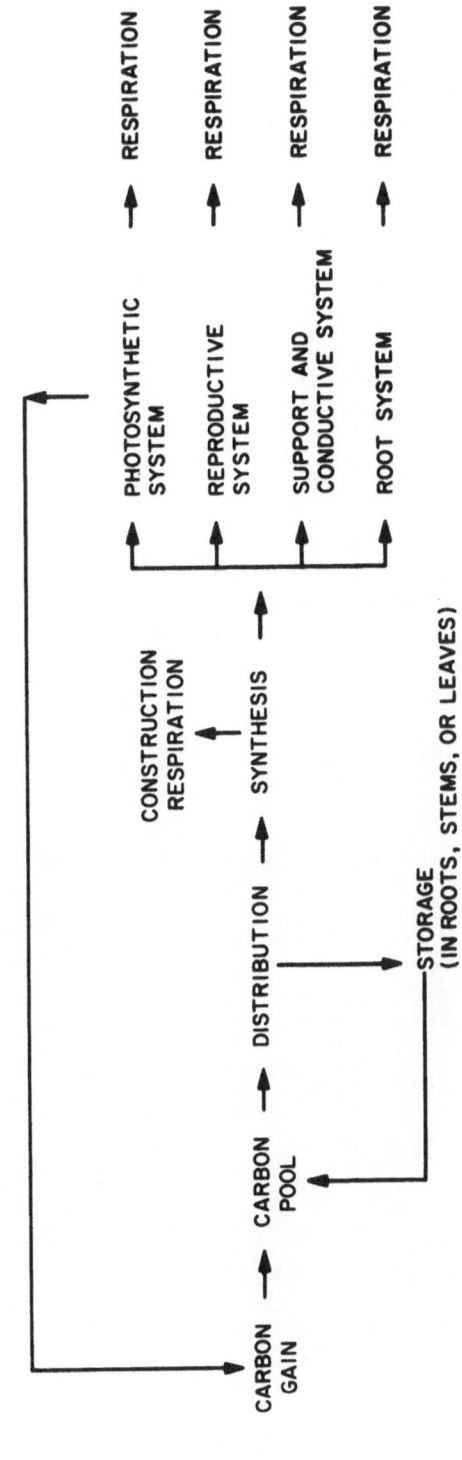

Fig. 6.5. The plant as a system exemplifying a positive feedback process. (Reproduced from Mooney, 1972, with permission of Annual Reviews, Inc.)

7. Social Behavior

The study of social behavior encompasses such questions as why social groups form among individuals of a species, why they are of certain sizes and not others, and what the realtionships are among group members. A theory that has emerged for animal sociality, called sociobiology (Wilson, 1975), is grounded on Darwinian natural selection acting on the individual, and on the modern principles of population dynamics and ergonomics (the study of work, performance, and efficiency). There is almost universal agreement that sociobiology has been highly successful in its application to animal populations, though there is sharp disagreement on its relevance to human society (Wilson, 1978; Alexander, 1979; Sahlins, 1976; Bock, 1980; Lumsden and Wilson, 1981; Singer, 1981).

Sociobiologists, or exponents of any other approach to the study of social behavior in organisms, recognize the importance of feedbacks between organisms within a group, or between individual organisms and the group as a whole. It is the nature of those feedbacks, particularly the ones that lead to the self-reinforcement of social behavior, that are the objects of analysis in this chapter. In this chapter we shall largely restrict discussion to animal populations, though a few examples are also relevant to human society.

7.1 Evolution of r- and K-strategies

Social behavior occurs in many forms and many diverse species, but there are certain preconditions that seem quite common, at least for social behavior in the higher vertebrates. Important among these is a tendency towards "K-*strategies*". "K-*strategist*" is a tag loosely applied to organisms that have relatively long lifespans, low reproductive rates, high investment in care of offspring, and relatively stable population sizes. At the opposite extreme are r-strategists, which have early breeding, high reproductive rates, low parental care, high dispersal ability, and large fluctuations in numbers. The terms r-strategist and K-strategist are simplifications, of course, and in reality organisms display a whole spectrum of strategies between these two extremes. A given species may often display some of the characteristics of an r-selected species, while at the same time having other characteristics typical of K-selection.

Nonetheless, as a broad generalization, the terms are convenient, and there does seem to be at least a tendency towards one extreme or the other, with the occurrence of one typical r- and K-selected characteristic usually implying most of the others as

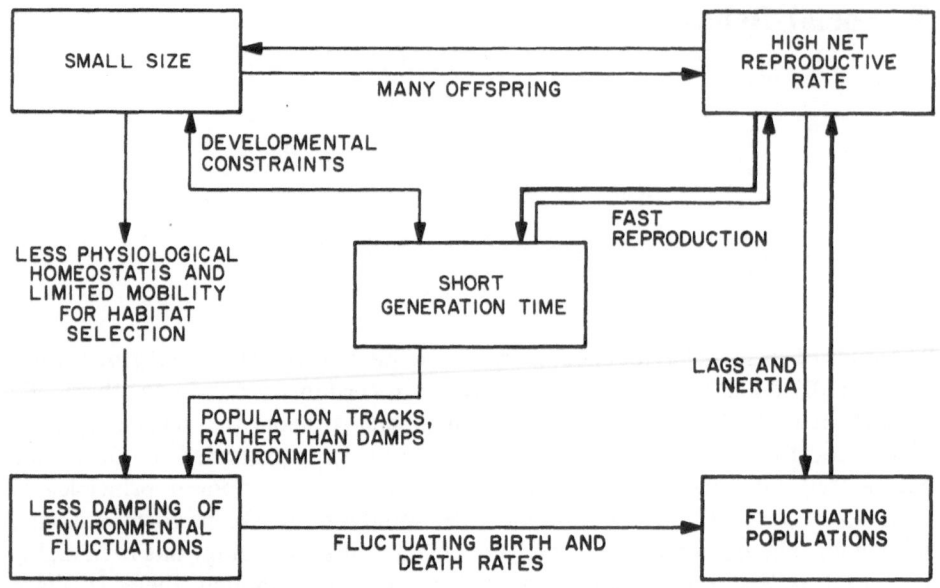

Fig. 7.1. Positive feedback loops reinforcing r-selection. (Reproduced from Horn, 1978, with permission of Blackwell Scientific Publications Limited)

well. Horn (1978) has presented positive feedback diagrams (based on a conceptualization by Southwood, 1977) showing how initial tendencies towards any of the r- or K-selected characteristics tend to reinforce characteristics of the same type and to amplify them by positive feedback.

For example, small animals are often relatively sensitive to environmental fluctuations and may suffer occasional massive die-offs. There will, therefore, be a genetic premium on rapid reproduction to fill up empty space following such disasters. A high reproductive rate leads to smaller sizes and shorter generation times, increasing the vulnerability of the animals to environmental fluctuations (Fig. 7.1).

On the other hand, if an animal is initially relatively large in size, its bulk often gives it some degree of independence from the environment (e. g., better ability to regulate internal temperature), which in turn means that it will be subject to fewer massive population reductions. Since the population is relatively close to the carrying capacity, competition is stiff within the population, and there are few opportunities for young offspring. Hence, reproduction is often deferred and limited so that parental care per offspring is increased. The result is a tendency towards larger size, so there is a feedback amplification (Fig. 7.2).

The evolution of K-strategies laid the groundwork for the evolution of complex social interactions in many species, as will be discussed in the next section.

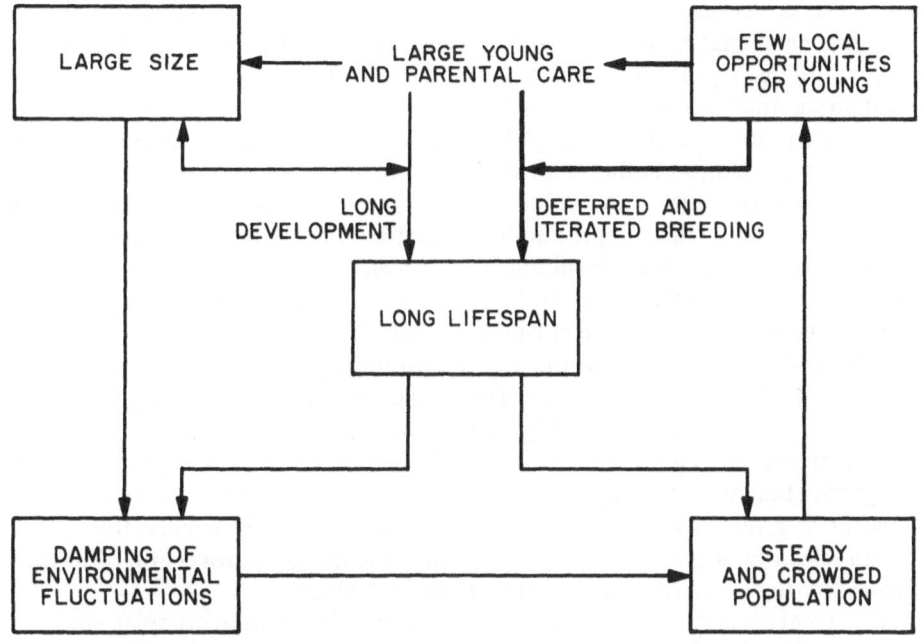

Fig. 7.2. Positive feedback loops reinforcing K-selection. (Reproduced from Horn, 1978, with permission of Blackwell Scientific Publications Limited)

7.2 Development of Social Strategies

A central task of biological science is to explain the origins and continued adaptive value of the social behavior that occurs among a variety of species in several phyla (Wilson, 1975). The study of society in lower organisms is never quite free of reflections on human society, so it is not surprising that discussion of this subject has often had a political flavor, and that metaphors from both Marxian and free-enterprise economics have found their way into the literature on the study of social behavior of animals.

Proponents of a strict interpretation of all evolutionary phenomena in terms of natural selection at the level of the individual tend to emphasize the laissez-faire elements of social behavior, often drawing analogies from Adam Smith's "*Wealth of Nations*" (Ghiselin 1974; Wilson, 1979; Heinrich, 1979). Their view is that, aside from complications arising from kin-selection, natural selection will lead each member of a population to maximize its own fitness. Types of social behavior in which one member of a population "*helps*" another can only be the result of the same sort of "*Invisible Hand*" that causes actions of self-interested humans in a laissez-faire economy to promote the general welfare.

An alternative view, that social strategies have evolved genuine altruistic impulses among the members of a population as a means for the homeostatic self-regulation of the population, was given prominence by Wynne-Edwards (1962). He viewed social strategies, such as the limitation of reproduction through such social phenomena as mating leks, to be adaptive in the evolutionary sense because they help maintain populations close to equilibrium values and tend to prevent large oscillations that would ultimately threaten a population with extinction. While many observations of natural populations seem at first sight to support this view, the theory runs into conflict with the Darwinian idea of natural selection acting exclusively at the level of individuals. Any strategy whereby members of a species altruistically limit their reproductive potential, the counter-argument runs, is unstable since a mutation resulting in a *"cheater"* that failed to obey the social convention on reproductive restraint would increase its relative representation in the population faster than the altruistic genes (Williams, 1966). Hence, acceptance of Wynne-Edwards' ideas would mean postulating a higher order law beyond Darwinian selection. Wynne-Edwards hypothesized that selection might operate on entire groups, selecting in favor of groups in which a greater degree of altruism exists among members. The analog in political theory might be Kropotkin's (1902) suggestion that human society is evolving towards greater mutual cooperation. The concept of group selection has been criticized, but it could conceivably occur if group selection operated at a faster rate then the internal subversion of altruistic groups by the appearance of cheaters (Gilpin, 1975a).

A current view that seems defensible is that development of complex societies is strongly correlated with kin-selection, or, perhaps in some cases, also with reciprocal altruism (in which the individual sacrifices no more than it receives on the average in direct contacts) among unrelated population members. Arguments based on kin selection have been especially attractive since they are interpretable in terms of selection on the level of the individual. Since the purpose of genotypes is the perpetuation of their genes, fitness, from the viewpoint of an individual gene or group of genes in a organism, includes the fitness of all relatives of that organism that might contain the same gene. Close relatives of a given organism are much more likely to share the same gene than other organisms chosen at random from the population. Hence, the enhanced fitness of a close relative is likely to have a positive value to a given genotype. Organism A can increase its inclusive fitness by adopting behavior that improves the fitness of a close relative, B, as long as the consequent decrease in fitness of A is compensated for by the gain in fitness of B corrected for degree of relatedness. Barash (1977) has illustrated the evolution of a network of altruistic interrelationships among close relatives (Fig. 7.3). One can imagine a model for social relationships within kin-groups analogous to the models of mutualistic communities considered in Chap. 8. The model should predict a rapid fall-off of mutualistic behavioral interactions as the degree of relatedness among individuals decreases, since the negative feedback to an individual due to loss of individual fitness should quickly become stronger than the added inclusive fitness of its genotype. The result should be a society of groups with internal mutualism, but competing with each other. The mathematics describing such situations is discussed in Chap. 8.

As mentioned earlier, evolution of a K-strategy can sometimes be the first step towards elaboration of a complex social structure. A possible pathway by which some communally breeding birds, like the Mexican jay, developed their social breeding behavior has been outlined by Brown (1974, 1978). These communal

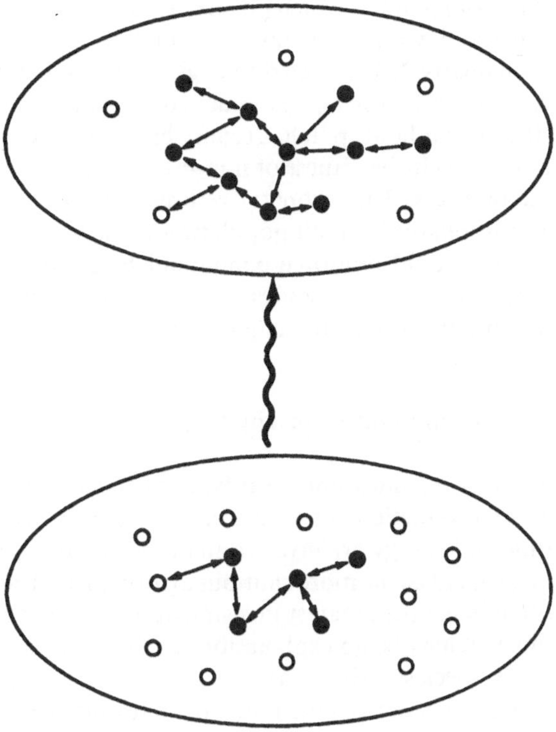

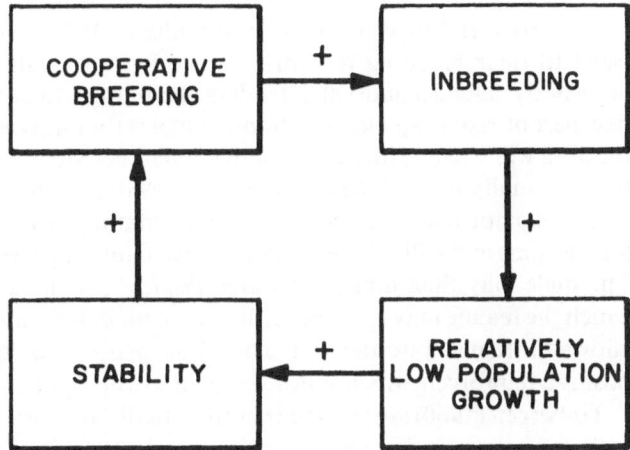

Fig. 7.3. The development through time of a network of mutualistic interactions among close relatives, where arrows represent altruism between population members. If altruism towards kin is advantageous genetically, then it will spread through members of a population; i.e., the system of relationships represented by the bottom diagram changes through time to that represented by the top diagram. (Reproduced from Barash, 1977, with permission of Elsevier North-Holland, Inc.)

Fig. 7.4. Systems diagram showing how communal breeding might be self-reinforcing. Modified from a discussion by Wilson (1975), who did not explicitly include feedback

breeders have lower reproductive rates than their nearest non-communal relatives, partly because of delayed breeding. Rather than dispersing, the yearling offspring stay in the vicinity of their nest and help their parents with later broods.

The evolution of this communal breeding behavior may have been stimulated by certain environmental factors that pushed particular populations beyond a certain threshold point at which positive feedback cycles could develop. The habitat of the Mexican jay is patchy, favoring small, isolated groups. Since the species may have been relatively K-selected to begin with, so that breeding was deferred and parent-offspring relationships extended, there would have been a tendency for offspring to stay in their home patch. Because breeding birds in this patch were likely to be close kin, the inclusive fitness of non-breeders would be raised by their providing help to the breeders. This cooperative breeding would lead both to increased inbreeding and to relatively small population growth, and therefore, greater stability. These factors would reinforce communal breeding (Fig. 7.4). Constraints imposed by patch sizes and closeness of kinship relationships would provide negative feedback, limiting the size of the communal breeding groups.

7.3 Mating and Reproduction

Some of the most universal types of social interaction are those related to mating and reproduction. These are critical activities for any organism since considerable time and energy are invested in offspring. It is natural that the greater its parental investment is, the more cautious an organism should be in choosing a mate. Females often make the greatest investment in time and energy, and it has been noted that this fact may be an explanation for the phenomenon of *"female coyness"* found in many species (Trivers, 1971).

Breaking down reluctance to mate often involves a sequence of interactions between male and female that seem to be mutually reinforcing, leading to a greater and greater degree of intimacy, until mating is consummated. For example, in the case of the green heron, a ritualized interaction is needed to break down the male's strong territorial drive (Suthers and Gallant, 1973). Following migration north-ward to their breeding territories, male herons establish territories, which they vigorously defend against all intruders, including females, and on which they build a new nest or repair an old nest. Soon, females that have subsequently migrated into the area will become interested in the male's activity. At first they will be repulsed, but eventually one will be tolerated on the edge of the territory. After a while, the male behavior changes from hostility to mating display flights. Encouraged, the female joins in the flight, and is later allowed into the territory as far as the nest tree. The male may then make a *"stretch display"*, indicating a readiness to mate, to which the female may respond with a soft call. Pair formation occurs when the male allows the female into the nest. The whole process could break down if the proper positive reinforcements are not received at the right times.

Tinbergen (1960) speculated that the ritualized courtship activities such as head-tossing and mew-calling among herring gulls act both as suppressors of the male's fighting urge and as releasers of sexual responsiveness. The courtship is usually reciprocal, with the female stimulating the male as much as the male stimulates the female. Each stimulation drives the courtship to the next threshold, which can be crossed only with the appropriate next response, until finally the courtship culminates in the mating act. Because a highly complex and species-specific courtship process promotes reproductive isolation (i.e., reduces the risk of hybrids), natural selection has tended to reinforce the development of elaborate rituals.

Another example having strong elements of positive feedback is the courtship of newts, which has even been represented by a mathematical model incorporating this positive feedback (Halliday, 1974; see also McCleery, 1978). During the courtship, which takes place underwater, the male first approaches the female and attempts to take a position in front of her. If the female is interested, she will advance, triggering a retreat display by the male. Further advance of the female may cause the male to turn from the female and to "*creep*" slowly away (this has been termed a "*negative feedback*" phase of the courtship by Halliday since the male cannot see the female during this time and so its degree of "*hope*" diminishes). But if the female has followed the male and touches his tail, the male will deposit spermatophore, which the female then draws into her cloaca.

Further examples of behavioral sequences in mating, as well as for other types of social interactions such as aggression and learning, are presented by Marler and Hamilton (1966). These examples indicate that there is a degree of uncertainty and randomness in the sequences, but that a well-defined organization of alternating male and female actions is evident if the sequences are analyzed carefully. It would be interesting to speculate on what sorts of thresholds have to be breached before courtship behavior can continue towards consummation. Models like Halliday's (1974) model of newt courtship may allow such questions to be explored and answered.

Lorenz, Tinbergen, and other ethologists have speculated on the origin of courtship displays and other ritualized signals (see, e.g., Krebs and Davies, 1981, p. 219). They have conjectured that these signals evolved from minor incidental movements or from displacement activities, those apparently irrelevant gestures (preening, feeding motions, etc.) that tend to occur at moments of indecision. If these actions, whatever their immediate causes, by a courting individual come to be recognized as appropriate signals by the object of the courtship, there will be a selective pressure on the signaler to accentuate the signals. Correspondingly, there will be selective pressure on the receiver of the signals to respond to them. The interaction of the producer and receiver of the signals will form a positive feedback loop that leads to increased stereotyping of the signals into a ritual.

Despite equal representation of genes from both parents in the offspring, the female usually makes an inordinantly greater investment in nourishing and caring for the offspring after they are hatched or born, especially in mammals, where the degree of parental care is often relatively very large. This seemingly unfair situation may have resulted in some instances from a positive feedback loop in natural selection over evolutionary time scales (Barash, 1979). Because the female physically bears the offspring, she initially makes a greater contribution than the male does to their development, and so she has more to lose than the male if the offspring are not successful. The male may actually enhance his inclusive fitness by neglecting his family and attempting to mate with other females, since he can count on the female to protect her heavier investment by providing more care to compensate. When this tendency by the male is rewarded genetically, a vicious circle can develop in evolutionary time, in which the ratio of female to male care of the offspring increases.

Chase (1980) described theoretically how either sexes' contribution to its offspring could have increased through evolutionary time in some species. He used "*reaction curves*" to define the contributions each sex would be willing to make, conditional on a specific contribution from the other sex (Fig. 7.5). For example, in the figure, the female is willing to contribute 45 units of effort when the male's contribution is 10 units (curve F), while the male is willing to contribute only about 10 when the female's effort is 30 (curve M). This state plane diagram is analogous to the state plane diagram of two competing species, with the reaction curves of the former being the analogs of the zero isoclines in the two-species competition diagram. This may seem odd since the reaction curve diagram describes a cooperative activity, the mutual raising of offspring, but ethologists have long been aware that even in cooperative activities there are elements of conflict. For the particular configuration of reaction curves shown in Fig. 7.5, the non-trivial equilibrium point is unstable. Any initial tendency for females or males to unilaterally increase or decrease their effort will cause the ratio of efforts to move away from the equilibrium. This will ultimately result in the sex that initially increased its effort bearing the full reproductive cost. The model is somewhat oversimplified in that neither sex would, in reality, be able to reduce its costs to absolute zero.

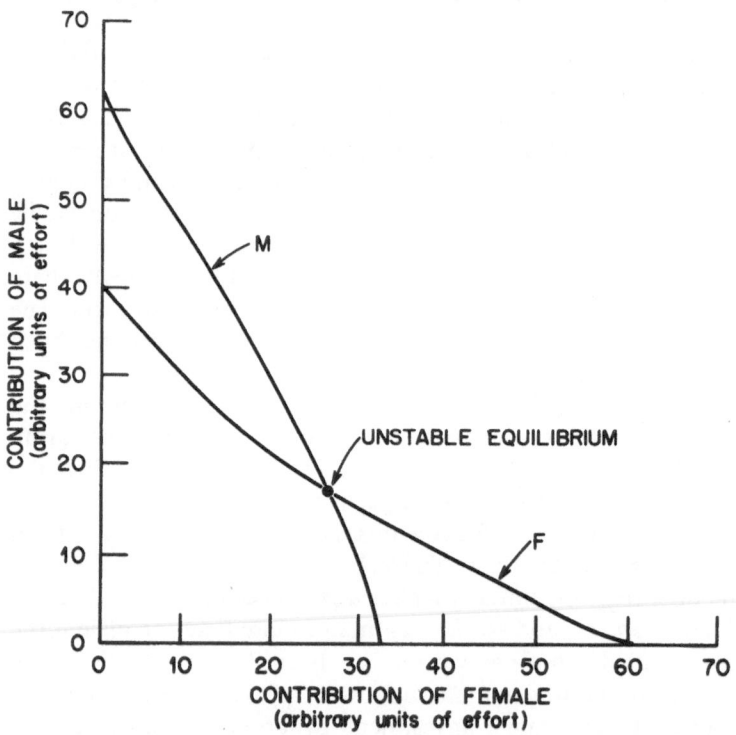

Fig. 7.5. Phase-plane diagram of male and female contributions to offspring. The zero isoclines have the configuration of a two-species competition diagram with an unstable equilibrium point (competitive exclusion), so either the male or the female will end up investing nearly all the time and energy to the offspring. (Reproduced from Chase, 1980, with permission of the University of Chicago Press)

This feedback loop may couple to another positive feedback loop involving female mate selection. If the more successful males are ones that are highly promiscuous and poor providers for their families, the female may be better off in terms of inclusive fitness in choosing the best possible male in terms of his overall fitness, since she can expect little in the way of help in feeding her offspring no matter which male she mates with. This tendency on the part of the female should lead to intensified competition among males for females since there is no limit, in principle, on the number of females a given male can mate with. Because the males must fight for females, increasing sexual dimorphism is expected, and, in fact, is found in many mammal species, especially ones not characterized by monogamy (see Chap. 4).

7.4 Population Models Incorporating Sexual Reproduction

Most models of the dynamics of populations ignore the fact that reproduction occurs sexually. The models assume that the rate of growth of the population is a function of the total population or, in the case of Leslie matrix models, a function of population numbers (sometimes only females) in the adult age classes. This type of simplification may be adequate in the majority of cases, but when the populations being modeled are subject to reductions to low densities and imbalances in the relative numbers of males and females are possible, erroneous conclusions can be drawn from such simple models, and a more detailed approach is necessary. If the population of a sexual species happens to be very low, and the movements of individual organisms are approximately random, the chances of mating and reproduction occurring may be more closely correlated with the product $X_m X_f$, of the male population density, X_m, and female population density, X_f, than with only the female population density. In this case the rates of growth of both male and female subpopulations are proportional to this product. A model based on this assumption would consist of two equations, one for each subpopulation:

$$\frac{dX_m}{dt} = r_m X_m X_f - d_m X_m - ((r_m - d_m)/K_m)\ (X_m + X_f)X_m^2, \tag{7.1a}$$

$$\frac{dX_f}{dt} = r_f X_m X_f - d_f X_f - ((r_f - d_f)/K_f)\ (X_m + X_f)X_f^2. \tag{7.1b}$$

The final terms in each of these equations represent density dependent mortality; these terms are assumed to involve the third power of population density so they can regulate the population numbers (i.e., keep the population from increasing indefinitely). Since this mortality results from crowding, the rates depend on the total population (both males and females), $X_m + X_f$.

Typical zero isoclines for this system are shown in a graph of the X_m, X_f-plane (Fig. 7.6). Notice that there are three equilibrium points, E_0, E_1, and E_2, of which E_0 and E_2 are always stable and E_1 is always unstable. There is a threshold curve, the separatrix (dotted line in Fig. 7.6) that separates trajectories that ultimately reach the origin, E_0, from those that ultimately reach the other stable equilibrium, E_2.

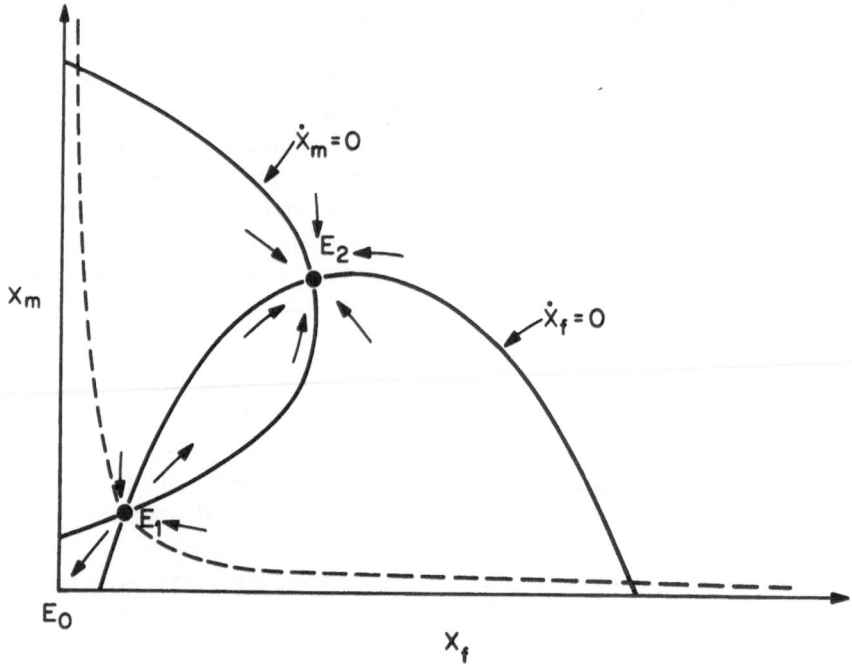

Fig. 7.6. Phase-plane diagram of a sexual species, indicating that both males and females must be present in minimum numbers for the population to persist

This curve represents the threshold that the subpopulations must exceed for the population as a whole to persist. The numerical trends of the subpopulations are mutually reinforcing in the same way as in the case of two mutualistic species (Chap. 8).

The ecological circumstances of certain types of species may make the probabilities of individuals finding mates during their lifetimes very low. This could have promoted the evolution of parthenogenetic reproduction that is found in many parasites. Price (1977) described a positive feedback cycle that may be instrumental in sustaining parthenogenesis in parasites. The environment of obligate parasites can be highly patchy when their hosts are spatially dispersed. Because encounters with possible mates are rare, parthenogenesis is favored. Parthenogenesis results in a lower genetic variability of offspring than does sexual reproduction, which implies a lower fitness in a fluctuating environment (Williams, 1975). The parasitic way of life, with its close attachment to a host that provides a stable, predictable internal environment, reduces the exposure of the parasite to high environmental variability. Therefore, parthenogenesis reinforces reliance on the parasitic mode.

For those organisms endowed by evolution with a sexual mode of reproduction, mating is sometimes a significant problem. Animals such as the bear and the rhinoceros, which lead solitary lives most of the year, may be especially vulnerable to extinction if their population densities are pushed to very low values, because of the difficulty in finding mates. The Sumatra rhinoceros may already be doomed for this reason (Ehrlich and Ehrlich, 1981).

The existence of a threshold is a consequence of the fact that for reproduction to occur, some social interaction must take place, in this case the meeting of males and females. This is a special case of a more general phenomenon known as the Allee effect (Allee, 1938), which codifies the observation that if species whose reproduction activities are based on a social structure are driven below certain levels, the structure may disintegrate and, consequently, reproduction will decline so that the population goes to extinction.

One theory for the extinction of the passenger pigeon is that overhunting led to low population levels and social disintegration (Halliday, 1980). There is no direct evidence for social disintegration as an important factor, but there is no other explanation that is totally compelling. Destruction of the pigeons' beech and oak forest habitat was significant, but at the time of the bird's extinction shortly after the turn of the century, extensive stretches of habitat remained. Catastrophes such as widespread diseases or massive storms also seem unlikely causes of extinction. There is some question as to whether hunting pressure, which was intense during the time the species was abundant, continued at a high level during the rapid decline of the bird. If hunting pressure did remain high, then an effect termed "*depensatory mortality*" could have occurred; that is, the hunting pressure per bird would have increased as the pigeon population decreased. This would have created a positive feedback loop in which the closer the species came to extinction, the greater became the forces pushing it toward extinction. In this case, additional hypotheses such as social disintegration would not be needed to account for extinction.

If hunting pressure did not remain at high levels, however, another explanation for the extinction of the passenger pigeon must be sought. Halliday (1980) has surmised that the bird was a social breeder and that heavy hunting reduced the average colony size below the critical threshold for member replacement, even when the total bird population on the continent was still quite large. Precisely what advantage sociality gave the birds is not known. Goodwin (1967) speculated that it may have helped reduce predation from mammals and other birds, while Murton and Westwood (1976) have suggested that, since the pigeons fed on scattered food sources, colony size may have played an important role in foraging efficiency. Combined with a decrease in habitat, the positive feedback loop of small colony size, reduced foraging efficiency, and low reproductive success may have made extinction of the passenger pigeon inevitable long before the problem was noticed by human observers.

Whether or not the demise of the passenger pigeon can be attributed to disruption of its social stucture, this tragedy should be a warning that extinction of a species can result from alterations in the social structure that are caused by population reduction, and that such alterations may be difficult to observe. However, other detrimental changes in social behavior that are even more difficult to observe can also become entrained in a population. Ratcliffe (1970) has investigated the effects of organochlorine pesticides on eggshell damage in certain British raptors. Presence of such pesticides in the birds seems to stimulate changes in hormonal balance, which in turn affects calcium metabolism and results in thinner eggshells. The eggshells are easily damaged by accident, following which the parent destroys or removes the affected egg. What makes the problem worse is that egg-destroying behavior ·tends to become a self-reinforcing habit, so that parents that have previously destroyed damaged eggs may later destroy healthy eggs. Psychological positive feedback thus produces end effects that would not have been expected from even a perfect understanding of the physiology of the raptors.

7.5 Small Group Dynamics

Social groups larger than immediate families are exceptions in nature. Below we consider a model that was conceived for human populations, though it may also apply to some other primates and perhaps canids. Herbert Simon (1952) has developed a qualitative mathematical model based on a verbal model by G. C. Homans (1950) describing the performance of a task by a small group. The model is characterized by three endogenous variables; $I(t)$, the intensity of interaction among group members, $F(t)$, the level of friendliness, and $A(t)$, the amount of activity within the group. There is also an exogenous variable, $E(t)$, the amount of activity imposed on the group by the external environment.

Hypotheses were advanced concerning the relationships among these variables. In particular, it was suggested that:

1. The interaction intensity, $I(t)$, tends to increase if either $A(t)$ or $F(t)$ does.
2. The friendliness, $F(t)$, tends to increase when $F(t)$ is too low for the current value of interaction, $I(t)$, and to decrease if it is too high for the current value of $I(t)$.
3. The activity, $A(t)$, tends to increase when it is too low for the current level of $F(t)$ or $E(t)$, and to decrease when it is too high for these levels.
4. Both $F(t)$ and $A(t)$ are self-limiting.

We can express these qualitative hypotheses, as Simon (1952) did, as mathematical equations (s.a. Bender, 1978);

$$I(t) = r(A, F), \tag{7.2a}$$

$$\frac{dF(t)}{dt} = s(I, F), \tag{7.2b}$$

$$\frac{dA(t)}{dt} = \Psi(A, F, E). \tag{7.2c}$$

If the level of interaction is initially rather low, then

$$\frac{\partial r}{\partial A} > 0 \quad \frac{\partial r}{\partial F} > 0, \tag{7.3a}$$

$$\frac{\partial s}{\partial I} > 0 \quad \frac{\partial s}{\partial F} < 0, \tag{7.3b}$$

$$\frac{\partial \Psi}{\partial A} < 0 \quad \frac{\partial \Psi}{\partial F} > 0 \quad \frac{\partial \Psi}{\partial E} > 0. \tag{7.3c}$$

Both friendliness and activity lead to increased interaction, which fosters more friendliness, leading to increased activity. Only the self-limitations on activity and friendliness prevent unlimited increases of all three endogenous variables.

7.6 Castes In Insect Societies

Oster and Wilson (1978) proposed a simple mechanism that could generate a range of castes within a colony of social insects. Their idea derives from the phenomenon we noted when discussing growth strategies; that is, a slight initial difference in sizes between organisms can be amplified through time if growth rates increase with size. Suppose newly hatched larvae of a social insect have a normal distribution in weights. If the growth rate increases with weight, then, after some time period, the weight distribution will have broadened, becoming skewed towards the larger size classes. The result will be a relative degree of size polymorphism in the population, although the size distribution will still be unimodal, with the peak occurring at the smaller sized morphs.

According to Oster and Wilson (1978), this weak polymorphism might be accentuated by the existence of certain decision points during the course of development of the larvae. At a critical development time, t_{crit}, larvae that have attained a certain threshold size, s_{crit}, can be sorted out and influenced in such a way as to increase their growth rates even further, while larvae below this threshold would be held back by having their growth rates decreased even further. The result would be a bimodal size distribution; that is, a more sharply defined caste structure. This reinforcement of growth rate differences could be achieved by preferential feeding of the larvae, among other possible methods. Hence a system of castes is developed through a positive feedback mechanism.

Social insects have evolved a system of division of labor involving different castes that cope with different contingencies facing a colony. Commonly, there is at least a worker caste to take care of the nest, gather food, and feed the larvae, while a soldier caste defends the colony against threats from invaders.

One of the key questions concerning the ecology of social insects is what factors determine the ratios of soldiers to workers in a particular colony at any time. Oster and Wilson (1978) have devised a mathematical model to study this problem. They chose three variables to represent their system; energy, E, number of workers, W, and number of soldiers, S. The equations governing these variables are assumed to be (with some changes to demonstrate our analytic approach)

$$\frac{dE}{dt} = \text{foraging} - \text{maintenance} - \text{manufacturing}$$
$$= f_1(W,S) = \beta(W)W - (m_w W + m_s S) \tag{7.4a}$$

87

$$\frac{dW}{dt} = \text{manufacture} - \text{mortality}$$

$$= f_2(E, S, W) = \frac{uE}{c_w} - \eta(S)W \qquad (7.4b)$$

$$\frac{dS}{dt} = \text{manufacture} - \text{mortality}$$

$$= f_3(E, S) = \frac{vE}{c_s} - v(S)S, \qquad (7.4c)$$

with the inequalities

$$\frac{\partial f_1}{\partial W} > 0 \quad \frac{\partial f_1}{\partial S} < 0$$

$$\frac{\partial f_2}{\partial W} < 0 \quad \frac{\partial f_2}{\partial E} > 0 \quad \frac{\partial f_2}{\partial S} > 0 \qquad (7.5)$$

$$\frac{\partial f_3}{\partial E} > 0 \quad \frac{\partial f_3}{\partial S} < 0$$

normally pertaining (see Fig. 7.7) There are two positive feedback loops in this model. The first is a simple loop in which net stored energy and the number of workers reinforce each other. In the second loop if an increased amount of energy is stored, then the amount of energy that can be diverted to soldiers increases, which in turn raises the degress of protection that can be given to workers. This boosts the amount of energy that can be stored, and so forth. There are three negative feedback loops, representing attrition to soldiers and workers, and the net consumption of energy by the non-productive soldiers.

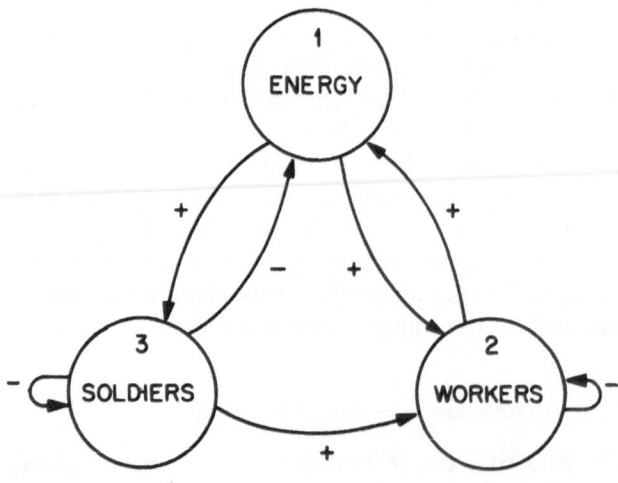

Fig. 7.7. Systems diagram of a social insect colony divided into castes

If the parameters of the model equations can be specified with reasonable accuracy, one can use matrix analysis to determine which strategies of allocation of energy between soldiers and workers were efficacious in producing colony growth. For example, suppose for an initial small colony size the linearized matrix for Eqs. (7.4) takes the form

$$\mathbf{A} = \begin{vmatrix} 0 & a_{12} & -a_{13} \\ a_{21} & -a_{22} & a_{23} \\ a_{31} & 0 & -a_{33} \end{vmatrix}, \tag{7.6}$$

where the a_{ij}'s are all positive; i.e. $a_{ij} = |\partial f_i / \partial X_j|$, where $X_1 \equiv E$, etc.

This is not a non-negative matrix, but a similarity transform,

$$\mathbf{S} = \begin{vmatrix} 1 & 0 & \alpha \\ 0 & 1 & 0 \\ 0 & 0 & 1 \end{vmatrix}, \; \mathbf{S}^{-1} = \begin{vmatrix} 1 & 0 & -\alpha \\ 0 & 1 & 0 \\ 0 & 0 & 1 \end{vmatrix} \tag{7.7}$$

can be used to put $\mathbf{A}$ in the form

$$\mathbf{A}' = \begin{vmatrix} \alpha a_{31} & a_{12} & -\alpha a_{13} - (\alpha^2 + 1) a_{33} \\ a_{21} & -a_{22} & -\alpha a_{21} + a_{23}, \\ a_{31} & 0 & -\alpha a_{31} - a_{33} \end{vmatrix} \tag{7.8}$$

which has only positive off-diagonal elements if a value of α can be found such that $-(\alpha^2 + 1) a_{33} - \alpha a_{13} > 0$. This inequality requires that the effect of the soldier caste on energy must be relatively small.

If α can be chosen such that $\mathbf{A}'$ is essentially non-negative, we can apply the conditions on the principal minors to assess the stability of the system at its equilibrium. In particular, if the sum of the two positive feedback terms, $a'_{12} a'_{23} a'_{31}$ and $a'_{12} a'_{21} a'_{33}$, is outweighed by the term $a'_{31} a'_{13} a'_{22}$, then the system is stable. An unstable system, for $E = 0$ and $D = 0$, means that the colony can grow from an initially small size.

7.7 Dominance Within Groups

As Wilson (1975) pointed out, dominance hierarchies found within groups or populations of some species sharing a single large territory serve a purpose similar to the division of an area into smaller territories of individuals or pairs, as done by certain other species. Both practices tend to reduce the amount of inefficient scramble competition and risky fighting.

For many species, including wasps, chickens, buffaloes, and rhesus monkeys, a linear dominance order exists in such a hierarchy, ranging from the alpha animal on top, which dominates all others in priority to food, mates, etc., to the omega animal at the bottom, which must yield to all others (Chase, 1982). Dominance hierarchies are initially established through fighting, which may simply consist of ritualized contests in some species. Observations show these hierarchies to be highly linear, much more so than would be expected on the basis of prior biological information on the individual animals in the group (Landau, 1965). Landau assumed that ability in contests depends on a number of biological factors, which he also assumed to be distributed in some random fashion within the group. Some few animals will be clearly superior and some few clearly inferior, but Landau showed statistically that most animals would a priori be expected to win, on the average, about as many contests as they lose. This would make it very difficult for the animals to sort out on a linear scale, and in principle, more uncertainty as to relative position in the hierarchy would be predicted than is actually observed.

Chase (1974) attempted to resolve this paradox by postulating that the ability of a particular animal in a contest is not constant through time, but depends on the outcomes of previous contests in which the animal has been involved. An animal that wins a contest increases its self-confidence and is more likely to win the next contest, while one that loses a contest is much more likely to lose the next contest. There may be concomitant hormonal changes in the two animals, increasing the relative degree of future aggressiveness in the winner (Archer, 1975; Leshner, 1975).

Hence, two animals of nearly identical biological abilities may develop opposite mobilities on the dominance scale because of the chance outcome of an initial contest. The result of this positive feedback is that even those animals of roughly equal abilities quickly sort themselves out on a linear scale. This phenomenon also reduces the amount of actual fighting that takes place. Consistent winners exude enough confidence to intimidate those that have been less successful, while the more submissive individuals become even more passive, so that fighting occurs mainly between animals that have had roughly equal amounts of success so that neither is clearly more confident than the other.

7.8 Models of Group Formation and Size

There are three basic ways in which membership in a group can help an individual:

1) the group gives some protection from predators,
2) it improves the ability to defend territories against competitors, and
3) it can aid the individual in foraging and predation.

Often the individual gains some combination of all three advantages. Living in groups also has its disadvantages. In particular, the rewards of food gathering must be shared among the members. Except in the case of social insects, where group cooperation is strongly favored by genetic relationships, life in a group must involve considerable tension, because each individual tries to gain as many of the advantages and as few of the disadvantages as possible for itself. In the same way that coevolution between two species has often resulted in "arms races," evolution of social animals had led to the refinement of intraspecific "cheating" strategies that try to take advantage of the group, and to "anti-cheater" counter-strategies (Darlington, 1980).

Protection against predators can take several forms. For example, the alarm calls of prairie dogs that have spotted danger warn others nearby in the prairie dog town, making it difficult for a predator to stalk a group of these animals. It is also possible that a confusion principle is exploited by some groups of prey. The swarming movements of myriads of fish in a school, or ungulates in a herd, may make it difficult for a predator to focus on any particular individual. Finally, there is strength in numbers. Predatory birds such as hawks and owls are frequently "*mobbed*" by smaller birds. Hamilton (1971) and Vine (1973) discuss the reduction of vulnerability to predation in the case of bird flocking. Groups of large animals, such as the buffalo and elands, will confront even lions when necessary, and can inflict serious wounds on them.

An example of another sort of predator defense has been noted by Darling (1938). He showed that egg laying in large colonies of sea birds is closely synchronized, presumably by a positive feedback loop of mutual stimulation. This has the effect that the period of vulnerability of the offspring of the colony taken as a whole is much shorter than if breeding were spread out. If there are large numbers of birds in the colony, most potential predators will be overwhelmed by the large numbers of vulnerable prey available only for a short period of time. Each offspring will have a very small chance of being taken by a predator. In a small colony, on the other hand, two factors will operate together to increase the vulnerability of the offspring. First, synchronization in breeding is likely to be weaker, and second, since there will be smaller numbers of vulnerable prey, the predators are not as likely to be satiated.

Whereas many prey species form groups to minimize predation, predators form groups to facilitate the capture of large or agile prey. This phenomenon is especially notable on the African grasslands, where lions, hyenas, and African hunting dogs hunt in groups. Some temperate and boreal species, such as wolves, also hunt in groups. A fascinating parallel to these large animals is that the advantage of group predation occurs also on the microscopic level. Bonner (1980), citing Dworkin (1972), notes that "*Myxobacteria operate by the wolf pack principle allowing the group to achieve something denied to separate individuals. The individual bacterial cells feed in a swarm, in this way conquering large prey that they could not digest as separate cells. Their massive, communal excretion of extracellular enzymes could only be accomplished by cooperation.*"

What determines whether or not a group will form and, if it does, what size it will typically have? The tendency towards group formation and growth will be positively reinforced if it is beneficial to each individual. Let us first consider predatory hunting groups. Caraco and Wolf (1975), working with Schaller's (1972) data on lion prides, showed that the mean group size of feeding lions increases linearly with prey biomass, having the mathematical form,

$$\text{lion group size} = 2.5 + 0.02 \times (\text{prey biomass}), \tag{7.9}$$

where biomass is measured in kilograms. The optimal group size is the result of two factors. First, to some extent, the more lions there are in the hunting group, the greater its success rate in capturing prey of a certain size. However, beyond a large enough group size, the addition of more lions does not significantly improve capture efficiency. Second, the more lions in the group, the more ways the prey carcass must be divided. Hence, the group stops growing when the marginal gain in hunting efficiency does not cover the marginal cost of dividing the spoils. If F represents capturable food available per individual to the group, and S represents group size, then the equations describing the interaction of group size and available food would have the general form,

$$\frac{dF}{dt} = f_1(F, S), \qquad (7.10a)$$

$$\frac{dS}{dt} = f_2(F, S). \qquad (7.10b)$$

For relatively small values of S, Eqs. (7.10a) and (7.10b) will behave as a positive feedback system. The variable F in these equations need not, of course, necessarily be interpreted as available food, and any other factor affecting group survival could be substituted, though the specific forms of the equations will then be quite different.

If, the functional forms, of $f_1(F, S)$ and $f_2(F, S)$ can be determined, Eqs. (7.10a, b) can be used to predict the threshold values, F_0 and S_0, below which F and S will simultaneously decrease. An example where knowledge of such thresholds would be of use is the case of herring gulls, which as adults congregate into breeding colonies. (In this case F would not represent capturable food, but some other measure of fitness.) These gulls have been the object of study on some islands of the British Isles (Duncan, 1978), where they are responsible for serious damage to soil cover and vegetation. Mature gulls are recruited to the colony, the rate of recruitment depending on colony density, and only being limited when space becomes scarce. Optimal attraction has been found to occur at densities of from 2 to 10 nests/m^2, although this varies from colony to colony because of topographical and other factors. To control the damage from gulls, large scale culling has been undertaken. But to be successful, culling on a given island must reduce the population below the threshold at which it is attractive to new recruits. Otherwise, recruitment from other areas will soon return the population to its former high levels. Over the long run, of course, even if a given island population is reduced below the attraction threshold so that local extinction on that island occurs, this will only be temporary since recolonization will eventually occur. The initial colonists will be at a greater risk, however, than members of large, established colonies.

While adaptiveness to the environment is the basis of group formation, proximate factors are more immediate causes; for example, chemical pheromones that stimulate grouping. The locust, to take a specific case, exists in three genetically identical phenotypic forms, the *solitaria* type, the *gregaria* type, and the *transiens* type. The *solitaria* type exists at low densities, the swarming *gregaria* type at high densities, and *transiens* at intermediate densities. Transformation from *solitaria* to *transiens* and then to *gregaria* seems to involve a positive feedback relationship between population density and a pheromone. When the population density of the locusts increases in some locale, stimulated perhaps by increased rainfall and improved vegetation, the females lay heavier eggs. The locusts hatched from these eggs tend to be more gregarious than their parents. An airborne pheromone released by immature locusts may act as an attractor. With each breeding cycle, the population density and pheromone concentration increase, leading ultimately to swarming in search of new food supplies. Woodcock and Davis (1978) have described the swarming and eventual disintegration of the swarm through the medium of catastrophe theory. If one is only interested in describing the population buildup phase preceding the swarming, however, it is sufficient to study a positive feedback system consisting of equations for locust density, N, and pheremone concentration, P. For small values of N and P, the two variables will positively reinforce each other.

7.9 The Schooling of Fish

We now turn to the question of school size among fish, where an explicit model has been formulated by Clark and Mangel (1979). Biological observations of tuna (Sharp, 1978) have shown that only a part of the tuna population exists in schools at any given time. Environmental conditions may influence what percentage of fish enter schools. Surface schools of tuna form around "*attractors*", which may be porpoises or even floating debris. Following Clark and Mangel (1979), let $N(t)$ represent the number of background (subsurface) tuna in a region, $Q(t)$ represent the number of tuna in an individual (generic) school, and K be the number of attractors. If the tuna associate with a given attractor at a rate αN, and dissociate at a rate βQ, the school size at any time t is given by

$$\frac{dQ}{dt} = \alpha N - \beta Q. \tag{7.11}$$

The total number of tuna in the region in schools at time t is $S = KQ$. Clark and Mangel use the Schaefer logistic model (Schaefer, 1957),

$$\frac{dN}{dt} = rN(1 - N/\bar{N}) - \theta, \tag{7.12}$$

to describe the subsurface tuna population. In Eq. (7.12) r is the intrinsic growth rate, $\bar{N}$ the environmental carrying capacity, and θ the net transfer rate to the surface schools. In one of their two models the authors let θ be

$$\theta = \alpha KN - \beta S. \tag{7.13}$$

The model for surface and subsurface populations is then (repacing Q by S in Eq. (7.11) and substituting θ into Eq. (7.12)

$$\frac{dS}{dt} = \alpha KN - \beta S, \tag{7.14a}$$

$$\frac{dN}{dt} = rN(1 - N/\bar{N}) - (\alpha KN - \beta S). \tag{7.14b}$$

93

This model has the form of a positive feedback system, the phase plane portrait of which is shown in Fig. 7.8. Study of the model is extremely simple since it can be represented on a two-dimensional state space. However, it is easy to imagine more complex situations where more sophisticated analytic methods are needed. For example, consider competition between two fish species with different schooling proclivities. For added interest, let schools be exposed to additional mortality, $\beta^* S$, due to fishing. The equations for species 1 and 2 are then

$$\frac{dS_1}{dt} = \alpha_1 K_1 N_1 - \beta_1 S_1 - \beta_1^* S_1 - C_{1s} S_1 S_2, \tag{7.15a}$$

$$\frac{dN_1}{dt} = r_1 N_1 (1 - N_1/\bar{N}_1) - (\alpha_1 K_1 N_1 - \beta_1 S_1) - C_{1n} N_1 N_2, \tag{7.15b}$$

$$\frac{dS_2}{dt} = \alpha_2 K_2 N_2 - \beta_2 S_2 - \beta_2^* S_2 - C_{2s} S_1 S_2, \tag{7.15c}$$

$$\frac{dN_2}{dt} = r_2 N_2 (1 - N_2/\bar{N}_2) - (\alpha_2 K_2 N_2 - \beta_2 S_2) - C_{2n} N_1 N_2, \tag{7.15d}$$

where C_{1s}, C_{2s}, C_{1n}, and C_{2n} are competition coefficients between surface and subsurface populations, respectively. The relevant stability matrix at an equilibrium point where the two species coexist has the form

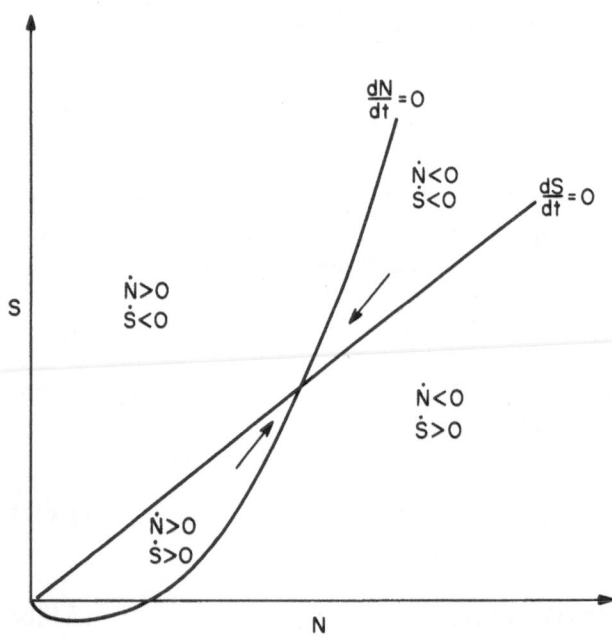

Fig. 7.8. Phase-plane diagram of Eqs. (7.14a, b), showing the tendency towards an equilibrium number of tuna in schools

$$
\mathbf{A} = \begin{vmatrix} -a_{11} & a_{12} & -a_{13} & 0 \\ a_{21} & -a_{22} & 0 & -a_{24} \\ -a_{31} & 0 & -a_{33} & a_{34} \\ 0 & -a_{42} & a_{43} & -a_{44} \end{vmatrix}, \tag{7.16}
$$

which has the form of a positive feedback matrix. Stability of the system at the equilibrium point can be determined by use of the principal minor criteria on matrix $\mathbf{A}$.

7.10 Social Interactions and Game Theory

The mathematical theory of games was originated by Von Neumann and Morgenstern (1944) as a method for analyzing the dynamics of human conflicts. In recent years there has been increased use of game theory in attempting to understand animal behavior. Maynard Smith (1976), Dawkins (1976), and others have speculated that natural selection may have resulted in the evolution of stereotyped strategies for maximizing the profit, or minimizing the loss, to individual members of populations from interactions with other members; for example, in disputes over territories and the sharing of parental investment in offspring. Here we shall give only the briefest resume of game theory and shall introduce a game situation involving positive feedback.

In any encounter (encounter may be interpreted in a fairly loose sense) between two animals, each is assumed to have a choice among a variety of discrete strategies, any one of which can be chosen at that time. The outcome of the encounter for each animal depends on the strategy chosen by each. For example, suppose there are n alternative strategies. Population member 1 chooses strategy I and member 2 chooses strategy J. The profit, or payoff, to member 1 is denoted by $E(I, J)$ and the payoff to member 2 is denoted by $E(J, I)$. The possible outcomes of an encounter can be represented in matrix form

$$
\mathbf{A} = \begin{vmatrix} E(1,1) & E(1,2) & \ldots & E(1,N) \\ E(2,1) & E(2,2) & \ldots & E(2,N) \\ \vdots & \vdots & & \\ & & & \\ E(N,1) & E(N,2) & & E(N,N) \end{vmatrix}, \tag{7.17}
$$

where row I represents the vector of payoffs to strategy I of member 1 and column J is the vector of payoffs to strategy J of member 2.

Strategies that result, on the average, in relatively small payoffs compared with other strategies will be selected against. The fitness, $W(I)$, of a particular strategy, I, can be defined as the summation over the payoffs, $E(I, J)$, that a population member pursuing strategy I will receive, multiplied by the fraction, P_J, of population members pursuing the Jth strategy;

$$W(I) = \sum_{J=1}^{N} E(I, J) P_J. \tag{7.18}$$

This definition assumes equal likelihood of all strategy combinations in the encounters within the population.

There will be an equilibrium when all pure strategies have the same fitness,

$$W(1) = W(2) = \ldots = W(N). \tag{7.19}$$

Unless this is a stable equilibrium, however, the population can move towards a different state of relative occurrence of strategies. One task for the mathematical theory of games is to determine if a point at which Eqs. (7.19) hold is stable and, therefore, an evolutionarily stable strategy (ESS), as defined by Maynard Smith (1979).

Game theory has been used largely to describe competitive or contest encounters. It has also been used to describe situations where cooperative or mutualistic interactions can occur, as in the famous example of the Prisoner's Dilemma. Since a version of the Prisoner's Dilemma game has been proposed as an analogue of the evolution of cooperation (Axelrod and Hamilton, 1981), it is worthwhile considering this game in some detail.

The Prisoner's Dilemma game, as modified by Axelrod and Hamilton (1981) for the purpose of illustrating the evolution of cooperative behavior, can be outlined quite simply. There are two players, each of whom has the choice of the same two strategies of interaction with the other player, "*cooperation*" or "*betrayal*". The payoffs to Player A are shown in Fig. 7.9. Player A receives the biggest payoff when it betrays while Player B cooperates. That is, A takes advantage of the altruism of B, but does not reciprocate; it plays B for a sucker. The next biggest payoff comes when both players cooperate. Punishment is received when both players betray, and severe punishment is received by Player A if it cooperates and Player B betrays. It is assumed neither player knows what the other's choice will be.

It is easy to see that the best strategy for Player A to use is betrayal. In that case if the opponent uses cooperation as a strategy, then Player A is well rewarded. On the other hand, if Player B uses betrayal also, then the punishment to A is only mild and not severe, as it would be if A used cooperation. Player B will no doubt reason the same way as Player A, and as a result, both will betray and thus both will be mildly punished. The logic behind such behavior is unimpeachable, even though the

PLAYER A	PLAYER B	
	COOPERATION	BETRAYAL
COOPERATION	E(1,1) REWARD FOR MUTUAL COOPERATION	E(1,2) PUNISHMENT TO SUCKER
BETRAYAL	E(2,1) REWARD FOR CYNICISM	E(2,2) PUNISHMENT FOR MUTUAL BETRAYAL

Fig. 7.9. Table of payoffs in Prisoner's Dilemma game $E(2, 1) > E(1, 1) > E(2, 2) > E(1, 2)$

consequence is that both players are punished rather than rewarded, as they would be if both cooperated.

If the safest strategy is betrayal, how can mutual cooperation evolve in a population? Nicolis (1980), Axelrod and Hamilton (1981) and Axelrod (1984) show how such evolution is possible by assuming that interactions between players are not *"once-only"*, but can be repeated. Players can alter their strategies on subsequent encounters, depending on the previous strategies of their opponents. Axelrod and Hamilton (1981) show that a strategy called *"tit for tat"*, proposed by Anatol Rapoport, is invariably the most successful; that is, being cooperative on the first encounter and thereafter following cooperation by one's opposite player with cooperation and betrayal with betrayal.

Nicolis (1980) has analyzed a *"tit for tat"* strategy mathematically, assuming there are two players, each with variable probabilities, X_1 and X_2, of cooperating. A phase plane diagram of the iterative game is shown in Fig. 7.10. The game obviously has the form of a positive feedback system with a threshold. If the initial likelihoods of players A and B of being cooperative lie below the separatrix in Fig. 7.10 subsequent iterations of strategies will follow a trajectory towards the pure betrayal strategy $(0, 0)$. If the likelihood of cooperative behavior of the two players initially lies above the separatrix, the subsequent evolution will be towards pure cooperation.

However, this still leaves unanswered the question of how a population that initially consists only of betrayers could reach the threshold necessary to become a society of cooperators. Axelrod and Hamilton (1981) suggest that kin selection may have formed nuclei of cooperating kin groups, since altruism towards kin would have *"paid off"* genetically even if there were no direct reciprocity. Thus, cooperative genes would have at least been present in the population. Computer

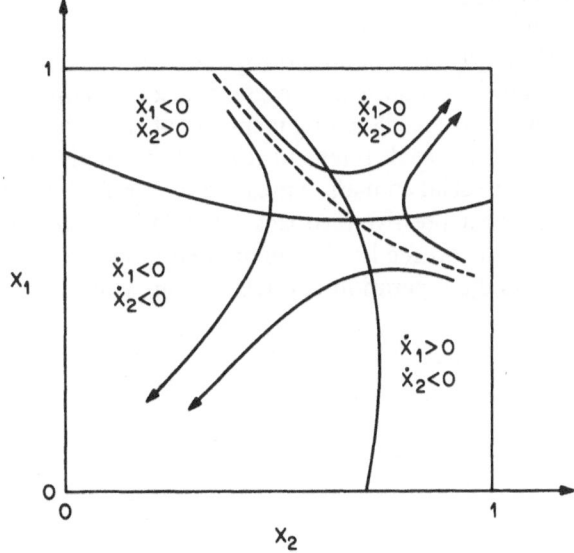

Fig. 7.10. Phase-plane plot of "tit for tat" strategy. X_1 and X_2 represent probabilities of Players A and B, respectively, playing a cooperative strategy. The dotted line is the separatrix or threshold above which cooperative interactions increase until a pure cooperative strategy is reached. (Adapted from Nicolis, 1980, with permission from Springer-Verlag)

simulations by Axelrod (1984) show that, if a large enough subgroup in a population employs the *"tit for tat"* strategy, members of this subgroup will be at an advantage relative to those outside the subgroup. Although the *"tit for tat"* strategy loses in each encounter with a new noncooperative outsider, these losses may be more than compensated for by payoffs from encounters with other members of the subgroup. In this way cooperators in the population would be able to benefit from each other's services, even when cooperators are relatively few in number. The number of cooperators should then be able to increase from a lower threshold number. The *"tit for tat"* strategy will have higher fitness and, over the time span of generations, will increase its proportional representation in the community. This increase in representation will accelerate through positive feedback, because the fitness of the *"tit for tat"* strategy increases as its representation in the community increases.

An important assumption in the above scenario is the ability of a player to recognize the other players as individuals (Colgan, 1983). In this way the player can cooperate with other players that were cooperative in the past and retaliate against those that are not.

7.11 Summary and Conclusions

This chapter has considered only a small sample of the great diversity of social phenomena that have been studied scientifically. But a pattern has begun to emerge from these examples that shows positive feedback to be an important contribution to the fabric of social phenomena. Over evolutionary time scales, trends towards sociality in animals have mutually reinforced other trends such as protected parental care and greater stability of populations. Positive feedback over evolutionary time has also led to ritualization of types of social behavior such as courtship behavior, as initially insignificant gestures were reinforced by sexual selection. Sexual and caste divisions of labor may have intensified through positive feedback.

On the level of population dynamics the coalescing of herds, swarms or schools of animals incorporate positive feedback mechanisms. In social insects the proximate mechanism for allocating members of colonies into various castes may depend on positive feedback intensification of initial size differences. In all populations of obligate social animals depression of population size below certain threshold levels can trigger positive feedbacks that can destroy the population.

Finally, on the level of individual organisms and groups of organisms, mutual trust and cooperation can result from a sequence of beneficial exchanges.

8. Mutualistic and Competitive Systems

The simplest and most obvious positive feedback interaction between species is mutualism. Direct interspecies mutualism can result from a multitude of interactions involving dispersal, shelter, nutrient cycling, energy provision and reproduction (Boucher et al., 1982; Faegri and Van der Pijl, 1966; Heinrich and Raven, 1972; Muscatine and Porter, 1977; Whittaker, 1975; Howe, 1977; Temple, 1977). Mutualistic interactions also arise when mutualists mediate competitive or predator-prey interactions (Wright, 1973; Janzen, 1969; Addicot, 1979; Messina, 1981; Osman and Haugsness, 1981; Heithaus et al., 1980).

Assignment of the term *"mutualism"* to interspecies interaction, even when it involves only a pair of species, can often be ambiguous. The difficulties of defining and analyzing mutualistic interactions within a community of several different species are even greater. Given this complexity, an understanding of the role of mutualistic interactions in the structure and evolution of ecological communities requires knowledge from field observations, experiments, and theory. Until recently mutualism was regarded as relatively unimportant by ecologists. Risch and Boucher (1976) surveyed 12 ecology textbooks and found that competition and predator-prey relationships dominate these books not only in terms of the amount of space allotted but also in the formulation of organizing principles of communities and ecosystems. The limited treatment of mutualism comprises anecdotal examples that are regarded as interesting but eccentric exceptions to the general rules. Recent empirical work demonstrates the ecological significance of mutualism (see Table 8.1). This research, however, has been limited to studies of particular cases. There have been no attempts to develop a unified theoretical basis for the role of mutualistic interactions in ecological communities.

Our purpose, in this chapter, is to provide the beginnings of a unified theoretical foundation that can be used to characterize the dynamics of mutualistic interactions. We will review facts that are known about mutualistic interactions and use these facts as a guide in our theoretical development. We will also examine some general features and community implications of mutualistic interactions, and then explore some hypotheses concerning their evolutionary development.

Let us loosely define mutualism as the interaction between two populations (usually, but not always, different species) that results in a net benefit for one or both populations. Mutualistic interactions between species assume a wide variety of types that can be classified according to the strength of the interaction. Commensalism is the extreme case where the interaction is of direct benefit to one species but the reciprocal effect is negligible (see Fig. 8.1). This type of interaction is common; for example, epiphytes and their supporting plants, invertebrates and fish

(a) MUTUALISM

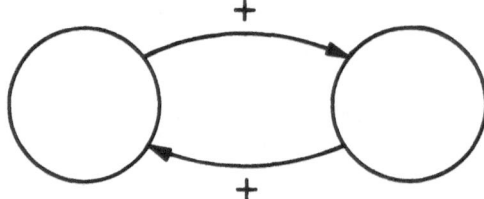

(b) COMMENSALISM

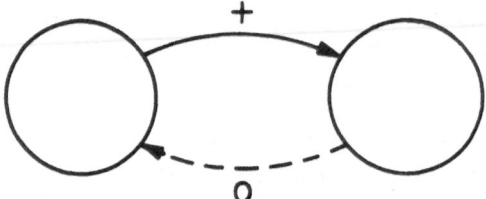

Fig. 8.1. a Mutualism involves the reciprocal exchange of benefits between two populations that are usually regarded as coevolved. **b** When one population benefits but the other is neither benefited, nor harmed, the relationship is called commensalism. This relationship is usually fortuitous but may be highly evolved in certain situations

that share the burrows of the marine worms, insects that live in ant colonies, etc. These relationships offer a domicile, food or nutrients, dispersal of protection to the symbiont. Some commensal relationships actually harm the host, but this harm is insignificant. Thus, there is no selection against the relationship on the part of the host. For the species receiving (or obtaining) the benefit, however, it may be advantageous to increase this benefit through dietary specialization, morphological modifications, and so forth. Good examples of the range of commensalist dependency are found in a case involving birds and large mammals. The association between cattle egrets and cattle is casual, with the number of egrets associated with cattle strongly dependent on the activities of the cattle. The egrets follow cattle that are grazing in the sun, capturing insects that are disturbed by the movement of the cattle (Heatwole, 1965). At the other end of the spectrum, the African tick-birds or ox-peckers are highly specialized and spend their entire life on their partner's back except when disturbed and during nesting season. All the tick-bird's food is derived from its host, and consists of ticks buried in the skin and insects that land on the body. The birds use the back of their hosts as platforms for sunbathing, courtship displays and mating. Their claws are sharp and curved and their tails long and stiff, enabling them to move with ease all over the host's body.

The degree of dependency on another species for benefit is also important in interactions in which both species benefit. At one end of the spectrum there are obligate mutualisms, where neither species can survive without the other. At the other end there are indirect mutualisms involving species that do not interact with each other directly but, nonetheless, benefit each other through an intermediary agent. In between these extremes, lies a range of mutualistic interactions demonstrating various facultative dependencies. Table 8.1 places interactions into these categories.

Table 8.1. Partial catalog of mutualists and the benefits they exchange

Organisms	Benefits provided or exchanged	Ref.
Obligate Mutualism		
Endosymbionts		
Lichens (fungus–algae)	Favorable environment–sugars	Ahmadjian (1966)
Coelenterates–zooxanthellae	Favorable environment–sugars	McLaughlin and Zahl (1966)
Animals–bacteria	Favorable environment–proteins	Howard (1966), Cavanaugh (1983)
Cells–organelles	Favorable environment–food, organization	Margulis (1970)
Higher plants–bacteria	Favorable environment–nitrates	Lange (1966)
Insects–fungi, bacteria	Favorable environment–various organic compounds	Koch (1966)
Algae–protozoa	Favorable environment–nutrients	Karakashian and Karakashian (1965)
Coelenterate–fish		
Sea-anemone–damselfish	Protection–food, habitat	Verwey (1930)
Pollination Systems		
Figs–wasp	Larval food–pollination	Galil and Eisikowitch (1971), Weibes (1979), Janzen (1979)
Yucca–moth	Larval food–pollination	Powell and Mackie (1966)
Obligately outcrossed plants–insects	Nectar–pollination	Faegri and van der Pijl (1966)
Ant–plant associations		
Ant plants–ants	Domicile–nutrients, germination	Rickson (1979), Janzen (1974), Kleinfeldt (1978)
Various trees (*Acacia, Cecropia, Barteria, Ochroma*)–ants	Domicile, food–protection, "allelopathy"	Janzen (1966, 1967, 1969, 1972), Hocking (1975)
Insect–insect		
Ant–aphid, treehopper	Dispersal, reproduction, protection–food	Way (1963), Fritz (1982)
Facultative Mutualism		
*Vascular plants–mycorrhiza	Carbon compounds–nutrients	Meyer (1974), Haselwandter and Read (1982)

* Obligate in the tropics, and nutrient poor soils

Table 8.1 (continued)

Organisms	Benefits provided or exchanged	Ref.
Vascular plants–bacteria	Carbon compounds–nitrogen	Lange (1966)
Marine cleaner associations	Food–parasite removal	Feder (1966), Limbaugh (1961)
Anemones, Epibionts–various invertebrates	Protection–locomotion, food	Ross (1971), Glynn (1981), Street (1975), Vance (1978)
Flowering plants–pollinators	Food–reproduction	Faegri and van der Pijl (1966), Plowright and Hartling (1981), Rickson (1979)
Anemones–fish	Protection–food	Roughgarden (1975)
Fruiting plants–dispersers (largely ants, birds)	Food–germination, dispersal	van der Pijl (1969), Temple (1977), Culver and Beattie (1978), Howe (1977), Vanderwall and Balda (1977), Janson (1983), Berg (1972), McKey (1975), Howe and Vande Kerchove (1979)
Insects–fungus	Food–nutrient, culture	Batra and Batra (1969), Batra (1966), Graham (1967), Starmer (1981)
Insects (largely ants)–plants	Food–protection	Bently (1976, 1977), Tilman (1978), Room (1972), Stephenson (1982), Keeler (1979, 1981a, 1981b), Schemske (1980, 1982), Risch and Rickson (1981)
Ants–butterfly larvae	Food–increased survival	Ross (1966), Pierce and Mead (1981)
Phoretic relationships	Dispersal	Colwell (1973), Wilson (1980)
Plants with diaspores–ants	food (elaiosomes)–nutrients, germination, protection from predation, dispersal	Heithaus et al. (1980), Beattie and Culver (1981), O'Dowd and Hay (1980)
Algae–insect	Attachment–domicile, food	Brock (1960)

Table 8.1 (continued)

Organisms	Benefits provided or exchanged	Ref.
Indirect Mutualism		
Beneficial Predation		
Isopods–mangrove		Simberloff et al. (1978)
Leaf miners–holly		Owen (1978a)
Plants–aphids		Owen (1978b)
Plants–grazers		Stenseth (1978), Dyer and Bokhari (1976)
Parasite–Host		
Spider–wasp		Valerio (1974)
Cowbirds–*Orapendula*		Smith (1968)
Pathogens–humans		Desowitz (1977)
Indirect competition		
Coexisting ant populations		Davidson (1980)
Salamander–midge		Dodson (1970)
Zooplankton		Neill (1974)
Plant Defense Guilds		
Grapes and blackberries		Doutt and Nakata (1973)
Coexisting crucifers		Takvanainen and Root (1972)
Palatable–unpalatable grasses		McNaughton (1978)
Other examples		Astatt and O'Dowd (1976)
Keystone Mutualist		
Flowering plants–pollinators		Waser and Real (1979), Thompson (1979), Bobisud and Neuhaus (1975)
Fruiting plants–dispersers		Gilbert (1980), Howe (1977)

8.1 Dynamics of Mutualistic Communities

The simplest model of mutualism is obtained by modifying the two species Lotka-Volterra equations of competition so that the interspecies interactions express benefit rather than harm, to either species where $\alpha_{12}, \alpha_{21} > 0$

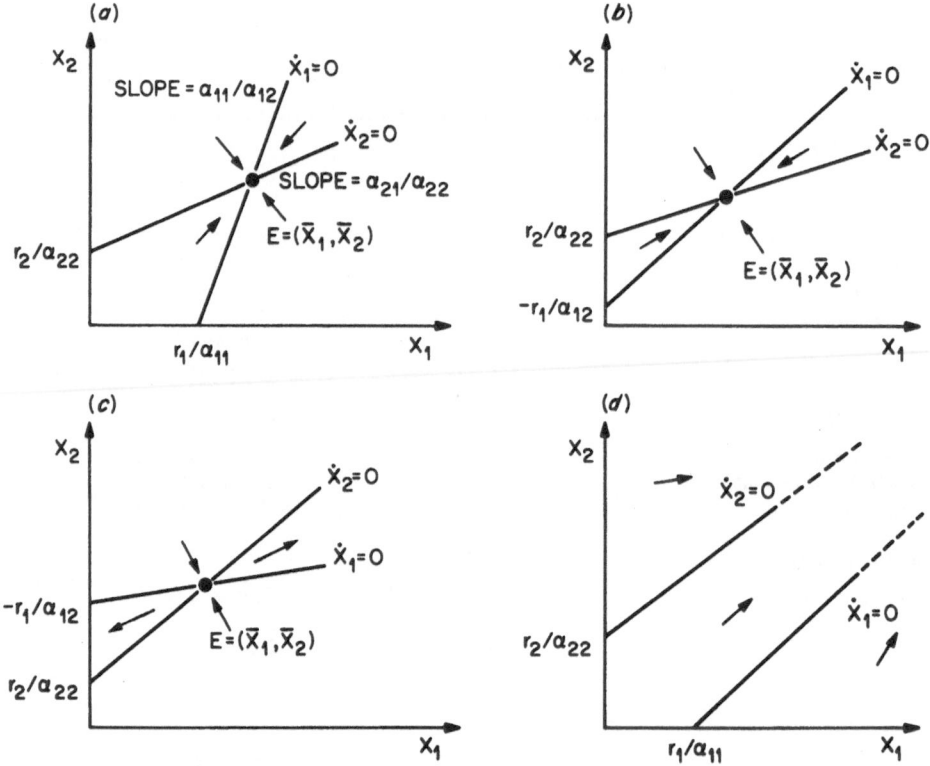

Fig. 8.2a–d. Several dynamical possibilities of two species mutualistic Lotka-Volterra equations. **a**, **b**, and **d** depict parameter ranges in which both species coexist at the stable equilibrium E. In **a** and **d** both species are facultative mutualists since they would persist in absence of interaction. In **d** weak self-regulation and strong interaction lead to large populations. In **b** and **c** species 1 is an obligate mutualist since it has a nonpositive carrying capacity and will disappear when mutualist is absent. **c** shows an unstable equilibrium

$$\frac{dX_1}{dt} = X_1(r_1 - \alpha_{11} X_1 + \alpha_{12} X_2)$$

$$\frac{dX_2}{dt} = X_2(r_2 + \alpha_{21} X_1 - \alpha_{22} X_2).$$

(8.1)

This model and a discussion of its salient features have been presented by Pianka (1978), Vandermeer and Boucher (1978), Goh (1979), and Travis and Post (1979). Many of the features of more complicated models of two-species mutualism are exemplified by this simple (though admittedly unrealistic) model. The equilibrium populations $\bar{X}_1$, and $\bar{X}_2$ for this two-species mutualistic system are shown in Fig. 8.2. Note that $\bar{X}_1 > r_1/\alpha_{11}$, and $\bar{X}_2 > r_2/\alpha_{22}$, so that both equilibrium populations are larger than they would be in the absence of the mutualistic interaction. This is a characteristic of positive feedback systems.

Vandermeer and Boucher (1978) examined all possible cases of stability and persistence of two mutualistic populations, using the graphical analysis of the

Lotka-Volterra equations. We will impose some additional constraints and present a slightly different analysis.

First we constrain the equation parameters so that the equilibrium populations of the two species are positive ($\bar{X}_1 > 0$, $\bar{X}_2 > 0$). This is ensured by Cramer's rule if and only if

$$\det \begin{vmatrix} -r_1 & \alpha_{12} \\ -r_2 & -\alpha_{22} \end{vmatrix} > 0 \quad \text{or} \quad \frac{r_2}{\alpha_{22}} > \frac{-r_1}{\alpha_{12}}$$

and

$$\det \begin{vmatrix} -\alpha_{11} & -r_1 \\ \alpha_{21} & -r_2 \end{vmatrix} > 0 \quad \text{or} \quad \frac{r_1}{\alpha_{11}} > \frac{-r_2}{\alpha_{21}}.$$

(8.2)

For Eqs. (8.1) to be stable to small perturbations from equilibrium, it is necessary that slight movements away from the equilibrium position set up forces tending to restore equilibrium. The mathematical conditions that must be satisfied for a model system to be stable to small perturbations from equilibrium are well known and can be stated in terms of the *"community matrix"*. The elements of the community matrix are the coefficients of the linearized system. Equations (8.1), when linearized at the equilibrium, become

$$\frac{dY_1}{dt} = -\alpha_{11}\bar{X}_1 Y_1 + \alpha_{12}\bar{X}_1 Y_2,$$

$$\frac{dY_2}{dt} = \alpha_{21}\bar{X}_2 Y_1 - \alpha_{22}\bar{X}_2 Y_2$$

(8.3)

where $Y_i = X_i - \bar{X}_i$. The equilibrium point is stable if and only if all the eigenvalues of the community matrix

$$\mathbf{S} = \begin{vmatrix} -\alpha_{11}\bar{X}_1 & \alpha_{12}\bar{X}_1 \\ \alpha_{21}\bar{X}_2 & -\alpha_{22}\bar{X}_2 \end{vmatrix},$$

(8.4)

have negative real parts. It can be shown (Travis and Post, 1979) that this is equivalent to the condition

$$\alpha_{11}\alpha_{22} > \alpha_{12}\alpha_{21}.$$

(8.5)

Condition (8.5) may also be easily derived from a graphical analysis of two species mutualism (Fig. 8.2). Goh (1979) showed that condition (8.5) is necessary and sufficient for global stability of Eqs. (8.1).

Inequality (8.5) can be intrepreted as meaning that stability is ensured if and only if the product of the species self-regulation is stronger than the product of the interspecific benefits. Since stabilizing effects will not arise from mutualistic interspecific interactions, stability must be provided by self-regulation. A surprising observation is that stability about the equilibrium point of the Lotka-Volterra equations describing two competing species is also given by condition (8.5). This is related to the fact that, like the model of two mutualists, the model of two competing species constitutes a positive feedback system. We will return to this subject later.

The zero isoclines of the Lotka-Volterra equations describing mutualistic equations have positive slopes so that the equilibrium sizes are larger than the respective carrying capacities. This is an essential property of mutualistic interactions. However, the Lotka-Volterra equations can sometimes yield unrealistically large equilibrium population sizes, even when the mutualistic interaction is small. If the product of the mutual benefits $\alpha_{21}\alpha_{12}$ is approximately equal to the product of the self-limitation terms $\alpha_{11}\alpha_{22}$, then the isoclines are nearly parallel and the equilibrium population sizes approach infinity (Fig. 8.2d). To address this problem we must adopt more complicated equations.

Rather than limit our discussion to a particular system of equations, we introduce a general pair of ordinary differential equations that describe the population dynamics of two mutualistic species:

$$\frac{dX_i}{dt} = X_i f_i(X_1, X_2) \quad (i = 1, 2). \tag{8.6}$$

As before, X_i represents the population density of species i. The per capita growth rate of species i is represented by f_i, which is some function of the species population densities. For Eq. (8.6) to represent the population dynamics of mutualistic species it is necessary to make an assumption regarding the nature of the dependence of the per capita growth rate of a species on the population density of the other species. Assume that when the community is at equilibrium, each species has a beneficial effect on the other. We define the interaction coefficient

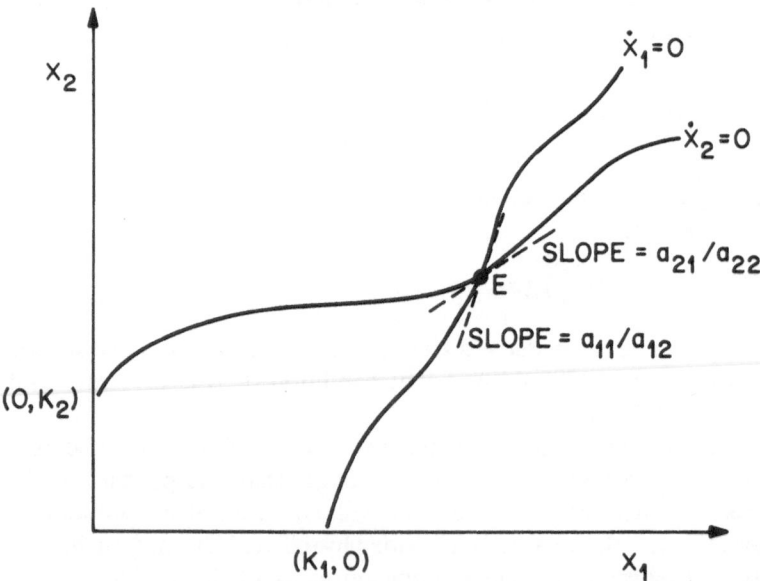

Fig. 8.3. Dashed lines indicate tangent lines to the isoclines at the equilibrium point E. The two facultative mutualists coexist at the stable equilibrium since $a_{11}/a_{12} > a_{21}/a_{22}$

$$a_{ij} = \frac{\partial f_i(X_1, X_2)}{\partial X_j}. \tag{8.7}$$

The species interact mutualistically if $a_{ij} > 0$ for $i \neq j$. Here attention is restricted to a small region near a feasible equilibrium $\bar{X} = (\bar{X}_1, \bar{X}_2)$, where the equilibrium point is assumed to lie in the positive quadrant of the state space plane. The one or more feasible equilibrium points of the system (8.6) can be found by solving the algebraic equations

$$f_i(\bar{X}_1, \bar{X}_2) = 0, \quad (i = 1, 2). \tag{8.8}$$

Local stability of the equilibrium of this general nonlinear two species system may be determined graphically. In this case, we examine the slopes of the tangent lines to the isoclines at the equilibrium point (Fig. 8.3). The tangent line to the species isocline has the slope a_{11}/a_{12}, while the tangent line to the species 2 isocline has the slope a_{21}/a_{22}. The slope of the species 1 isocline tangent must exceed the slope of the species 2 tangent for the equilibrium to be locally stable. That is,

$$\frac{a_{11}}{a_{12}} > \frac{a_{21}}{a_{22}}, \tag{8.9}$$

or

$$a_{11}a_{22} > a_{12}a_{21}. \tag{8.10}$$

This condition is analogous to the stability condition (8.5) for the Lotka-Volterra equation describing mutualism [Eq. (8.1)].

8.2 Limits to Mutual Benefaction

To keep the equilibrium population size of mutualistic species bounded, the per capita growth rate of the species must be zero or negative for large population sizes. For a two species community, this may be stated as

$$f_1(X_1, X_2) \leq 0 \quad \text{if} \quad X_1 > L_1, \tag{8.11a}$$

$$f_2(X_1, X_2) \leq 0 \quad \text{if} \quad X_2 > L_2. \tag{8.11b}$$

If conditions (8.11a, b) are assumed to hold, it can be shown (Hirsch and Smale, 1974) that every trajectory beginning in $X_1 > 0$, $X_2 > 0$ tends toward an equilibrium in the region $0 < X_1 < L_1$, $0 < X_2 < L_2$. If, in addition, both species are self-regulating, $\partial f_1/\partial X_1 < 0$, $\partial f_2/\partial X_2 < 0$, then there is only one stable equilibrium toward which all trajectories approach (see also Albrecht et al., 1974).

The constant L_1 may be thought of as an upper bound for the species one isocline (Fig. 8.4). As the isocline $dX_1/dt = 0$ approaches the boundary $X_1 = L_1$, the rate of change of population 1 with respect to population 2 approaches zero. That is

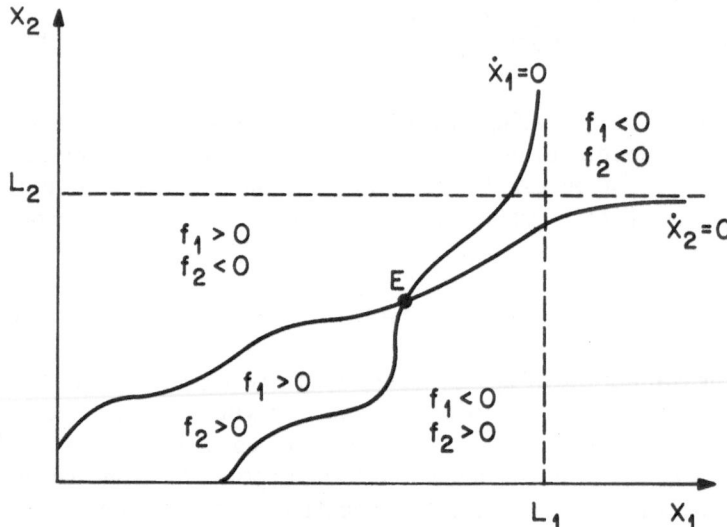

Fig. 8.4. Isoclines of a two-species mutualistic system in which the population sizes are ultimately limited. L_1 is an upper bound of the population size of species 1. L_2 is an upper bound for species 2. The equilibrium E pictured here is stable

$$\frac{\partial X_1}{\partial X_2} = \frac{\partial f_1/\partial X_2}{\partial f_1/\partial X_1} \Rightarrow 0 \quad \text{as} \quad X_1 \Rightarrow L_1, \tag{8.12a}$$

$$\frac{\partial X_2}{\partial X_1} = \frac{\partial f_2/\partial X_1}{\partial f_2/\partial X_2} \Rightarrow 0 \quad \text{as} \quad X_2 \Rightarrow L_2. \tag{8.12b}$$

In the mutualistic Lotka-Volterra equations these ratios are positive constants, so that an increase in interaction coefficients could theoretically cause equilibrium population sizes to become infinite. From (8.12a) we observe that the species isocline may approach its asymptote through a combination of two different causes. The first of these is that the benefit one species, say species 1, receives through interaction with the other $(\partial f_1/\partial X_2)$ may approach zero as its population density approaches its limit (X_1 approaches L_1). The second way in which the species 1 isocline may approach its asymptote is for species 1 self-interaction $(\partial f_1/\partial X_1)$ to become large as X_1, approaches L_1.

Host-symbiont models of the first method, where the per capita benefit decreases as population size increases, have been suggested by Whittaker (1975) and May (1976a, b). These models are stabilized by the addition of a saturation factor that limits the benefit, per host individual, that the host receives from interaction with the symbiont (provided the equilibrium is feasible). In the particular formulations the above authors develop, the host receives the maximum per capita benefit when the symbiont population is small. The benefit decreases monotonically as the symbiont population increases (Table 8.2). Dean (1983) presents a model for mutualists in which the carrying capacity of one species depends on the other species abundance but the increase in carrying capacity decreases as the population size

Table 8.2. Equations expressing saturation of mutualistic benefit

Author	$f_2(X_1, X_2)$	$\partial f_2/\partial X_1$
Whittaker (1975)	$r_2\left(\dfrac{-K-X_2-aX_1C}{K}\right)$ where $C = D/(D+X_2)$	$-\dfrac{ar_2}{K(1+X_2/D)}$
May (1976)	$-d+\dfrac{IX_1}{X_1X_2+X_2E+EF}$	$\dfrac{I(X_2E+EF)}{(X_1X_2+X_2E+EF)^2}$
Dean (1983)	$\dfrac{r_2(k_2-X_2)}{k_2}$ where $k_2 = K_2\left\{1-\exp\left(\dfrac{-b_2X_1+c_2}{K_2}\right)\right\}$	$\dfrac{X_2b_2\exp\left(\dfrac{-b_2X_1+C_2}{K_2}\right)}{k_2^2}$

In Whittaker's (1975) and May's (1976b) models, subscript 1 refers to the symbiont population, 2 refers to the host population. The coefficient a represents the effect of species 1 on the population of species 2. K and r_2 are the carrying capacity and intrinsic rate of increase, respectively of population 2. C is a saturation factor that limits the benefit per host individual, from interaction with the symbiont. D is a scaling factor. The coefficient d represents an intrinsic death rate of the host population. The benefit species 2 receives from interaction with species 1 saturates at I. E and F are factors that scale the amount of benefit the host receives when the symbiont populations are plentiful and few, respectively. In Dean's (1983) model of two mutualists, K_2 is the maximum value of k_2, and b_2, c_2 are constants

increases. This results in a diminishing return on the benefit received sufficient to ensure stability, because eventually self-limitation exceeds the mutualistic benefits received.

The host-symbiont models of Whittaker (1975), May (1976b), and Dean (1983) illustrate another feature of mutualistic models with nonlinear per capita growth rate functions that may be important in many natural mutualistic systems. In their models, the host isocline is curved in such a way that it crosses the symbiont isocline twice, resulting in two feasible equilibria. One equilibrium, point at relatively low population densities for both species, is unstable. The other, at higher population densities, is stable. This produces a threshold effect; that is, it is necessary for the population levels of both the host and symbiont to exceed critical levels for the positive feedback to result in growth to the stable equilibrium. If the population levels of one or both species populations are low, then the symbiont will decline toward zero (Fig. 8.5a). If the host is also obligatory, then it too will disappear (Fig. 8.5b) if either population is low.

The second case, where self-interaction strength increases with population size, will occur when species 1 completely consumes a resource necessary for its own growth, or when the population is limited by some exogenous factor. Heithaus et al. (1980) have provided an example of enhanced stability of two mutualistic species populations by a predator. The mutualists are ants and violets. The violets produce seeds that bear energy rich elaiosomes that the ants use as a food source. The ants in turn deposit the seeds in their nests, a safe site for germination, reducing predation,

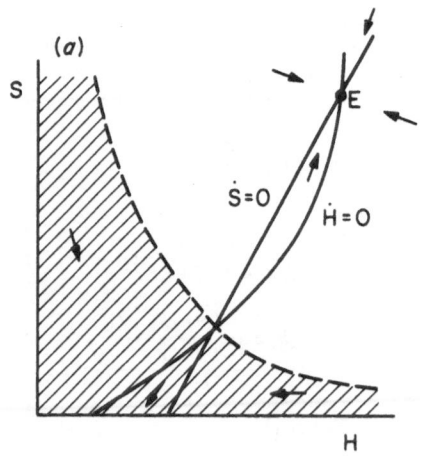

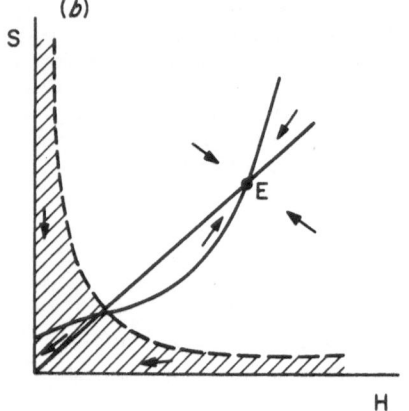

Fig. 8.5a and b. Perturbation of the host and symbiont populations into the hatched region will cause the symbiont population to disappear in **a** and both populations to disappear in **b**. The dotted lines are separatrices

and increasing both germination stimuli and the supply of nutrients to the seeds (Culver and Beattie, 1978). Because each of these species can live in the absence of the other, their carrying capacities may be assumed to be positive. If their interaction is modeled by the Lotka-Volterra equations (8.1) stability would be ensured if and only if the isoclines appear similar to those in Fig. 8.2a. Small rodents, however, are serious predators of the violet seeds. Introducing the rodent predators, the three species system may be represented by the following equations

$$\frac{dX_1}{dt} = X_1(r_1 - \alpha_{11}X_1 + \alpha_{12}X_2),$$

$$\frac{dX_2}{dt} = X_2(r_1 - \alpha_{22}X_2 + \alpha_{21}X_1 - \alpha_{23}X_3), \tag{8.13}$$

$$\frac{dX_3}{dt} = X_3(-r_3 + \alpha_{32}X_2).$$

The Routh-Hurwitz stability conditions (see May 1973b) can be applied to this system in the neighborhood of equilibrium $(\bar{X}_1, \bar{X}_2, \bar{X}_3)$. From these conditions, stability is ensured if and only if

$$\alpha_{12}\alpha_{21} < \alpha_{11}\alpha_{22}\left[1+\frac{\alpha_{32}\alpha_{23}\bar{X}_2\bar{X}_3}{(\alpha_{11}\bar{X}_1)^2+\alpha_{11}\alpha_{22}\bar{X}_1\bar{X}_2}\right].$$

Since we assume that the equilibrium population sizes $\bar{X}_1$, $\bar{X}_2$, $\bar{X}_3$ are positive, we see that because of the presence of the predator, the parameters α_{21} and α_{12} can assume larger values while stability is preserved. The allowable values of α_{12} and α_{21}, with and without rodent predation on the violet seeds, are depicted in Fig. 8.6. The rodent population exerts a negative feedback on the violet population that enhances the effect of its own self-regulatory term (α_{22}). Other examples of mutualisms stabilized by rate-limiting resources and inhibitory resources are given by Meyer et al. (1975).

8.3 Multi-species Mutualism

There are several examples of ecological communities that consist of more than two mutualistic species. Smith (1968) studied mutualistic and commensal relationships between *Orapendulas*, cowbirds and several hymenopteran species. L.E. Gilbert (1980) described several multispecies tropical and temperate communities that are

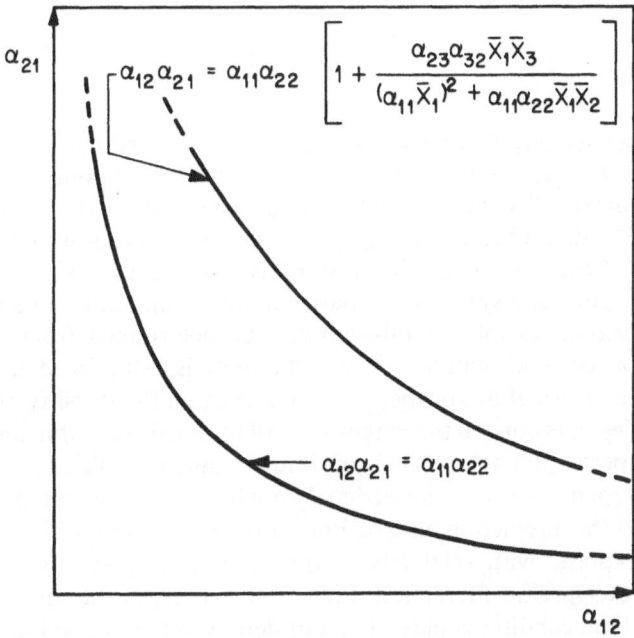

Fig. 8.6. The region under the lower hyperbola describes the allowable parameter values of mutualistic benefit (a_{12}, a_{21}) for stability of two mutualists. The region under the upper hyperbola defines the allowable values for mutualistic benefit when one of the mutualists is subject to predation. (From Post et al., 1981, with permission of Academic Press, Inc.)

111

characterized by a "*keystone mutualist*" and associated mutualistic "*mobile links.*" These are largely seed dispersal or flower pollination systems involving many species that may be regarded as mutualistic. In order to develop a mathematical framework for constructing models of such systems, we must extend our results to more than two populations. We will proceed by considering a general system of differential equations that describe the population growth of a community of n interacting mutualistic species. Letting

$$\frac{dX_i}{dt} = X_i g_i(X_1, X_2, \ldots, X_n) \quad (i = 1, 2, \ldots, n), \tag{8.14}$$

where X_i, and g_i are defined as before, except that the per capita rate of increase g_i now depends on the population density of all n species. The equilibrium point or points of Eqs. (8.14) may be found by solving the algebraic equations,

$$g_i(\bar{X}_1, \bar{X}_2, \ldots, \bar{X}_n) = 0 \quad (i = 1, 2, \ldots, n). \tag{8.15}$$

We assume that (8.15) has at least one positive solution with all $X_i > 0$. For Eqs. (8.14) to represent the population dynamics of a mutualistic community, it is necessary that, when the community is at equilibrium, the jth species ($j \neq i$) have a beneficial effect on the per capita growth rate of the ith species. A precise mathematical statement of this condition is that the elements of the interaction matrix $\mathbf{A}$,

$$a_{ij} = \frac{\partial g_i(\bar{X}_1, \bar{X}_2, \ldots, \bar{X}_n)}{\partial X_j}, \tag{8.16}$$

satisfy $a_{ij} \geq 0$ for $i \neq j$. The model mutualistic community (8.14) will be stable in the neighborhood of the equilibrium $\bar{X}$ if and only if all of the eigenvalues of the matrix $\mathbf{DA}$ have negative real parts. $\mathbf{D}$ is the diagonal matrix defined by $\mathbf{D} = \text{diag} (\bar{X}_1, \bar{X}_2, \ldots \bar{X}_n)$ and $\mathbf{A}$ is the interaction matrix composed of elements defined by (8.16). This stability condition, however, may be simplified for mutualistic systems. A characteristic of mutualistic communities is that stability near a feasible equilibrium can be determined from the matrix $\mathbf{A}$ alone. This property of mutualistic communities is established in Appendix B. As a consequence of this property, we observe that the stability of a mutualistic community depends only on the magnitudes of interactions within and between the component species and not on the equilibrium values ($\bar{X}_1, \bar{X}_2, \ldots \bar{X}_n$). The exact nature of this dependence is not immediately obvious since it is stated in terms of the eigenvalues of the interaction matrix. For mutualistic communities, however, we can explicitly express, with relatively simple, biologically interpretable conditions, how the interspecific interactions affect the stability of the community. We will summarize these conditions here. The full details are contained in Appendix B.

The stability of a mutualistic community near an equilibrium is dependent on the strength of the diagonal or intraspecific regulation terms in the population matrix. If these terms are negative and strong compared to the mutualistic benefit a species receives then the mutualistic community will remain near equilibrium. The precise relationships between the interaction matrix elements may be expressed by the condition known as quasi-diagonal dominance. This states that a necessary and sufficient condition for a mutualistic community to be stable in the neighborhood of a feasible equilibrium is that there exist positive constants $d_1, d_2, \ldots, d_n$ such that

$$-d_i a_{ii} > \sum_{\substack{j=1 \\ j \neq i}}^{n} d_j a_{ij}, \quad i = 1, \ldots, n. \tag{8.17}$$

An equivalent condition is that there exist another set of positive constants $d_1, d_2, \ldots, d_n$ such that

$$-d_i a_{ii} > \sum_{\substack{j=1 \\ j \neq i}}^{n} d_j a_{ji}, \quad i = 1, \ldots, n. \tag{8.18}$$

The conditions (8.17) or (8.18) are known as quasi-diagonal dominance. Although conditions (8.17) and (8.18) indicate how stability is maintained in mutualistic communities, they are of little help in determining whether a particular mutualistic community will be stable. The problem presented by these conditions is that it can be difficult to establish whether or not a vector $\mathbf{d} = (d_1, d_2, \ldots, d_n)$ exists such that (8.17) or (8.18) is satisfied. Fortunately, there exists an equivalent condition that is verifiable in a finite number of arithmetical steps (see Chap. 2 and Appendix B). Necessary and sufficient conditions for a mutualistic community of n species to be stable in a neighborhood of a feasible equilibrium can be stated in terms of the principal minors of the interaction matrix

$$\mathbf{A} = \begin{vmatrix} a_{11} & a_{12} \cdots a_{1n} \\ a_{21} & a_{22} \cdots a_{2n} \\ \cdot \\ \cdot \\ \cdot \\ a_{n1} & a_{n2} \cdots a_{nn} \end{vmatrix}.$$

The principal minors of this matrix are

$$D_{11} = \det |a_{11}|,$$

$$D_{22} = \det \begin{vmatrix} a_{11} & a_{12} \\ a_{21} & a_{22} \end{vmatrix},$$

$$D_{33} = \det \begin{vmatrix} a_{11} & a_{12} & a_{13} \\ a_{21} & a_{22} & a_{23} \\ a_{31} & a_{32} & a_{33} \end{vmatrix},$$

$$\cdot$$
$$\cdot$$
$$\cdot$$

$$D_{nn} = \det |\mathbf{A}|.$$

113

The equilibrium point of the model mutualistic community [Eqs. (8.14)] is stable if and only if

$$(-1)^k D_{kk} > 0 \quad (k = 1, 2, \ldots, n).$$ (8.19)

For the two species system described by Eqs. (8.1), condition (8.19) reduces to condition (8.10).

8.4 Models of the Evolution of Mutualism

Simple cost-benefit models for the evolution of commensalism and mutualism were developed by Roughgarden (1975), who considered the association of damselfish with sea anemones (a protection association) and by Keeler (1981b) who summarized the literature concerning ant-extrafloral nectary associations. Keeler's model also suggests several hypotheses concerning ant-extrafloral nectary associations that may be evaluated with field observations and experiments. In both of these models, the costs and benefits of a guest either associating or not associating with a host are evaluated in terms of the fitness of the guest. Mutualism is assumed to evolve if the fitness of the mutualistic guest genotypes is higher than that of nonmutualistic guest genotypes. These models lead to conditions under which mutualistic associations should form and specify the extent to which the associations should be obligatory or facultative.

These models, however, consider evolutionary changes only in one of the partners. Many mutualistic interactions required coevolution; that is evolutionary changes in both partners. To develop coevolutionary models of mutalism, we need to employ dynamic models of population interactions.

Vandermeer and Boucher (1978) point out that stable coexistence (neighborhood stability of the feasible equilibrium) is of limited value in deciding the fate of a species engaged in a mutualistic interaction. From Lotka-Volterra equations they argue that persistence is the biologically interesting feature of mutualism and may be unrelated to stable coexistence. Mutualism and commensalism can tip the balance deciding whether or not a species may enter and persist in a community of organisms or an otherwise unfavorable environment. Hallam (1980) shows that interaction with a mutualist can permit coexistence in a two-competitor Lotka-Volterra system where one competitor would have been excluded in the absence of the mutualist.

The analysis that we have already presented is capable of determining whether or not a mutualistic species will persist. A population can increase when rare, escaping extinction, if the community matrix is unstable when evaluated at the equilibrium point where this population is zero; i.e., if the point is a repellor. This is equivalent to there being at least one positive eigenvalue and it means that the solution to these equations will move away from the equilibrium in the direction of increasing the population size of the rare species (Allen, 1977; Schaffer, 1977). This type of analysis was used in Chaps. 6 and 7 and will be exploited further in the following chapters.

114

One way for persistence of a species to be favored in the face of environmental perturbations is for the equilibrium population density to be increased. In mutualistic communities this is readily accomplished by positive feedback between species. Analysis of how all population sizes change when the equilibrium population size of a single population in a mutualistic community is altered shows that, if a species is capable of increasing its per capita growth rate, then its equilibrium population density, as well as the density of the other species with which it interacts mutualistically, will increase. Rigorous proof of this intuitive assertion is provided in Appendix E.

There are two evolutionary mechanisms by which increased persistence can be achieved. First, a species per capita growth rate may increase by more efficient use of resources. For a species in a mutualistic association with another, this may increase the positive effect of the benefit it derives from the other species. Such a development can readily selected for through natural selection. If this in turn increases the benefits that the evolving species may return to its mutualist, then the relationship will strengthen. If on the other hand, this involves increased exploitation of the other mutualist, then the negative effects may outweigh the positive effects and the relationship will develop into a parasitism. Meyer (1966) views the relationship between higher plants and mycorrhiza as reflecting such a dynamic balance between positive and negative effects. Soluble carbohydrates produced by the plant are utilized by the mycorrhizal fungus as a source of energy. The plant obtains nutrients and water from the fungus. The mycorrhizal fungus is capable of secreting auxins that intervene in the metabolism of the plant, causing soluble carbohydrates to flow from the main root to the infected roots. If this carbohydrate drain is too large compared to the nutrient gains, then the interaction is more properly viewed as a parasitism.

Second, a species equilibrium density may increase, if this increase supplies more benefit to its mutualistic partners. For example, consider the development of the obligate mutualism between ants and *Acacias* (Janzen, 1967). Early in the relationship ants probably casually visited weakly developed extrafloral nectaries or Beltian bodies of the *Acacias* but did not provide protection in return. If a mutation occurred such that descendents of a queen eliminated competing plants and animal predators of the *Acacia* then the tree would have accelerated its growth and provided more extrafloral nectaries and Beltian bodies for the ant colonies to exploit. Thus through indirect positive feedback the equilibrium density of the ants would have increased.

However, not all the ant colonies of one tree will exhibit protective behavior at once. Mutations that increase aggresive behavior will occur in only a small number of the ant colonies. The tree will benefit from this occurrence, but this benefit will feed back on the entire and population, nonprotector and protectors alike. Under this circumstance it is possible for cheating ants to take advantage of the mutualists without providing any benefits. In fact, Janzen (1975) describes an ant species *(Pseudomyrmex nigropilosa)* that is such a parasite of the ant-*Acacia* mutualism. We shall now develop a possible solution to this difficulty.

The ability to determine the stability and instability of mutualistic populations is necessary if one wishes to explore the conditions under which mutualism will evolve from a commensal association. Let us consider a hypothetical example which may

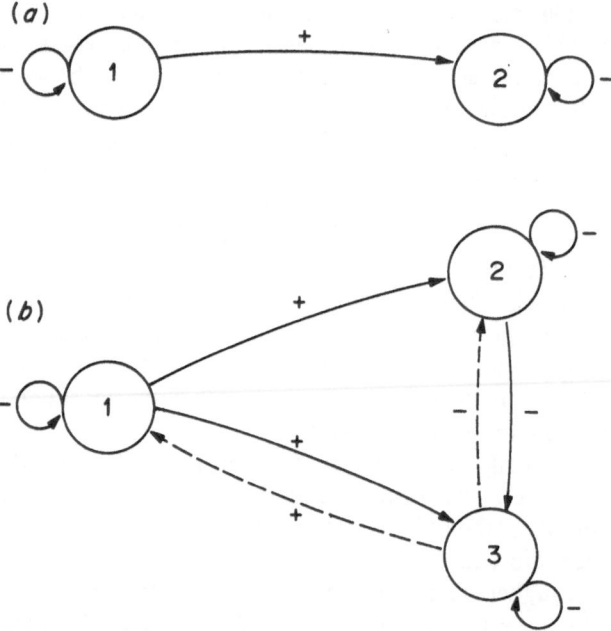

Fig. 8.7.a Depicts an ancestral commensal relationship between host species 1 and guest species 2, and **b** shows the three-population system when a new phenotype, population 3, invades. Population 3 is similar to population 2 except that it is capable of providing benefit to species 1. The dashed arrows indicate interactions that are nearly zero when population 3 is small, as would be the case when invading. (From Post et al., 1981, with permission of Academic Press, Inc.)

relate to the ant-*Acacia* system. Figure 8.7 depicts an evolutionary ancestral commensal relationship between two species. Into this community a new phenotype of species 2 arises (population 3) that has population parameters identical to those of population 2, except that it provides some benefit for the host population 1. Because population 3 is nearly identical to population 2, we expect these two populations to compete.

If the community interaction matrix, evaluated at $(\bar{X}_1, \bar{X}_2, 0)$ is unstable, then population 3 will increase when introduced into the system. However, this possibility is unlikely. Assume for a moment that the benefit population 1 receives from population 3 imposes no cost to the fitness of 3. The self-damping term, $-a_{33}$, will be small when the population size of 3 is small, but the negative effect of population 2 on 3 will be equivalent to 2's self-damping term resulting in the two populations having identical population constraints. If there is a cost to population 3 for providing the benefit to population 1 then population 3 will always be at a disadvantage compared to population 2. The positive feedback between populations 1 and 3 does not benefit 3 more than 2 unless another ingredient is added to the system. If the relationship between the host and the guest involves a one-to-one interaction between individuals, then mutualistic behavior or benefit extended by the guest is directed back only on itself. This is possible when individual mutualist pairs are spatially or behaviorly isolated from one another, which effectively divides the host population into two subpopulations. One subpopulation consists of individuals associated with commensal guests; the other subpopulation involves individuals with mutualistic guests (Fig. 8.8). This community is a positive feedback system. Given particular model equations and ranges of parameter values, we could use the criteria presented earlier for determining the stability of the equilibrium of the four-population system (i.e., the conditions for a stable mixed population of commensal and mutualistic phenotypes). If the four-population equilibrium is unstable, and if the original ancestral commensal relationship is unstable to invasion by a mutualistic phenotype, we expect the system to collapse to a stable two species mutualistic relationship by the competitive exclusion of the commensal phenotype. Actual trajectories of the populations, however, require a full global analysis of a particular system of equations.

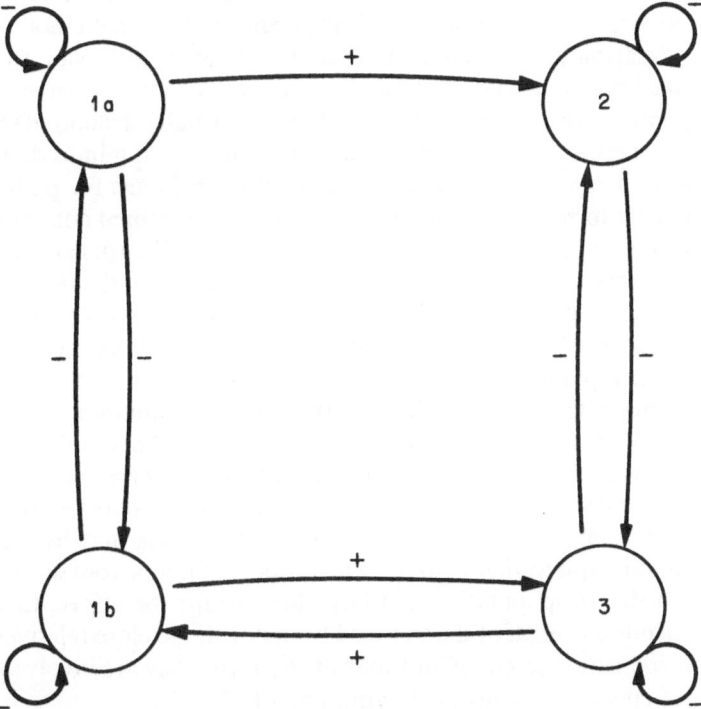

Fig. 8.8. Spatial or behavioral isolation of individual mutualistic or commensal pairs has the effect of subdividing the host population into two subpopulations: those with commensal guests (1a) and those with mutualistic guests (1b). (From Post et al., 1981, with permission of Academic Press, Inc.)

8.5 Isolation and Obligate Mutualism

In the preceding section we showed that it is necessary to introduce spatial or behavioral isolation of pairs of mutualistic individuals for natural selection to operate on differences in fitness between commensal and mutualistic phenotypes. To determine if such isolation at the individual level actually occurs in natural mutualistic systems, we surveyed the literature and tabulated all the obligate mutualistic interactions we could locate (Table 8.1). Obligate mutualists were chosen because in these cases these is no question that both species involved receive benefits. If isolation is absolutely essential, then it should be a feature of all highly coevolved mutualistic associations.

Examining Table 8.1, we see that all of the examples in the endosymbiotic category support the hypothesis of isolation almost by definition. The symbionts actually live their entire life cycle within the body of their host. The sea anemone-damselfish association is an example of behavioral isolation of mutualist pairs (this is an implicit assumption of Roughgarden's cost-benefit model). In strictly obligate associations, the damselfish exhibits territorial behavior toward other fish encroaching on its sea anemone. In pollination systems where nectar is provided as a reward for dispersal of pollen, the flower is usually structured so that nectar robbing (i.e., nectar extraction without pollination) is minimized. In other pollination systems, where larval food is extracted in exchange for pollination, the isolation appears to be more complicated (K. Keller, personal communication, 1980). For example, yuccas are perennial and the moth pollinators are annual. An adult moth pollinates a yucca and deposits eggs in the ovary of the flower. The larvae mature, emerge from the pod, and drop to the base of the plant to overwinter. In the spring, the yuccas bloom and the moth adults emerge. It is extremely likely than the moth will then pollinate the plant it fed on previously. It is not known if similar events happen with the fig species and their wasp pollinators.

The remaining obligate associations involve ants and either various plants or aphids. It is possible to reconcile these examples with our hypothesis if we accept a broader definition of the individual. For example, consider the ant-*Acacia* obligate mutualism. All of the occupants of a single *Acacia* may be descendants of a single founding queen that colonized a young seedling or root sprout. If this is the case, then the ant population is a large kin group; the efforts that the individual ant expends on behalf of the tree feed back directly to close relatives. The haplo-diploid reproductive system of ants may explain why they are involved in a broad range of mutualistic relationships (Hamilton, 1967).

A similar argument can be made for the ant-aphid mutualism. Aphids reproduce parthenogenetically so that a single plant is usually occupied by many genotypically identical individuals. Janzen (1977) refers to such populations as "*evolutionary individuals.*" It still remains to be determined if there is a one-to-one relationship between aphid evolutionary individuals and individual ant colonies.

This test of the isolation hypothesis for obligate mutualism is currently inconclusive for two reasons. First, we need to stretch the definition of individual. Whether this is valid remains to be determined. Second, not enough is known about the natural history of some of the obligate mutualists to decide if isolation does occur.

Many of the mutualisms listed under the facultative category of Table 8.1 are also characterized by the isolation of mutualist pairs from conspecifics. These include the plant-bacteria associations, best exemplified by the legume-rhyzobium combination, marine cleaner associations, combinations involving various marine invertebrates and anemones, mycorrhizae and others. There are, however, important exceptions. These include plant pollination and seed dispersal mutualisms as well as other facultative associations, particularly those listed in Table 8.1 as indirect. We will consider some of the indirect mutualistic relationships in the next section. The origin of these may have been fortuitous, but a further strengthening of these relationships by positive feedback mechanisms could also have occurred.

Pollination and dispersal relationships, however, are frequently so highly coevolved that they cannot be regarded as casual or fortuitous (Gilbert, 1975; MyKey, 1975) though they defy precise ecological classification. In one sense they can be viewed as mutually exploitative. Plants have complicated floral modifications that force pollinators to carry pollen. Many times these modifications are particularly adapted to take advantage of particular behavioral features of specific pollinators. Pollinators will visit flowers only if offered sufficient rewards such as nectar, pollen (for food), protection or brood places. Similarly, mobile animals will only eat those portions or fruit that they find attractive. It is usually only inadvertently that an animal predator fails to entirely consume all of the seeds or propagules that it transports. For example, squirrels usually do not find all the acorns they have cached, and birds excrete or regurgitate viable seeds that are undigestible remains of fruit. If, on the other hand, we examine the effect of numbers of individuals of one species on the other, disregarding a specific resource analysis, the interactions of pollinators and dispersers of plants appear mutualistic.

The various pathways through which mutualism may evolve can be classified (see Fig. 8.9). The first important distinction is whether or not the original relationship between the species was symbiotic or non-symbiotic. The symbiotic pathways begin with two species that live close to one another due to ecological preferences. This would logically imply that these species are quite different in some aspect of their natural history so that this close association results in commensalism rather than competition (Briand and Yodzis, 1980). The non-symbiotic pathways to mutualism, while requiring that the two species have some interaction, do not require an intimate association due to common environmental preferences. These pathways begin with direct interactions where one species exploits the other for food, or they compete with each other for resources.

The symbiotic pathways begin with host-symbiont relationships in which the symbiont receives some benefit such as shelter, transportation or nourishment from its host, while not injuring the host. Usually, the symbiont becomes a parasite of the host (Fig. 8.9) by feeding on the body of the host and doing the host a certain amount of harm (Buchsbaum, 1948). From here the relationship can develop in two more or less distinct directions depending on whether the host or the symbiont is responsible for the changes. The host may become tolerant of the parasite and may exploit some feature of the parasite for its own benefit. In this case both organisms remain largely distinct *("heterosymbiotic")* and exploit each other for complementary resources. Mycorrhizal associations are of this type. If, on the other

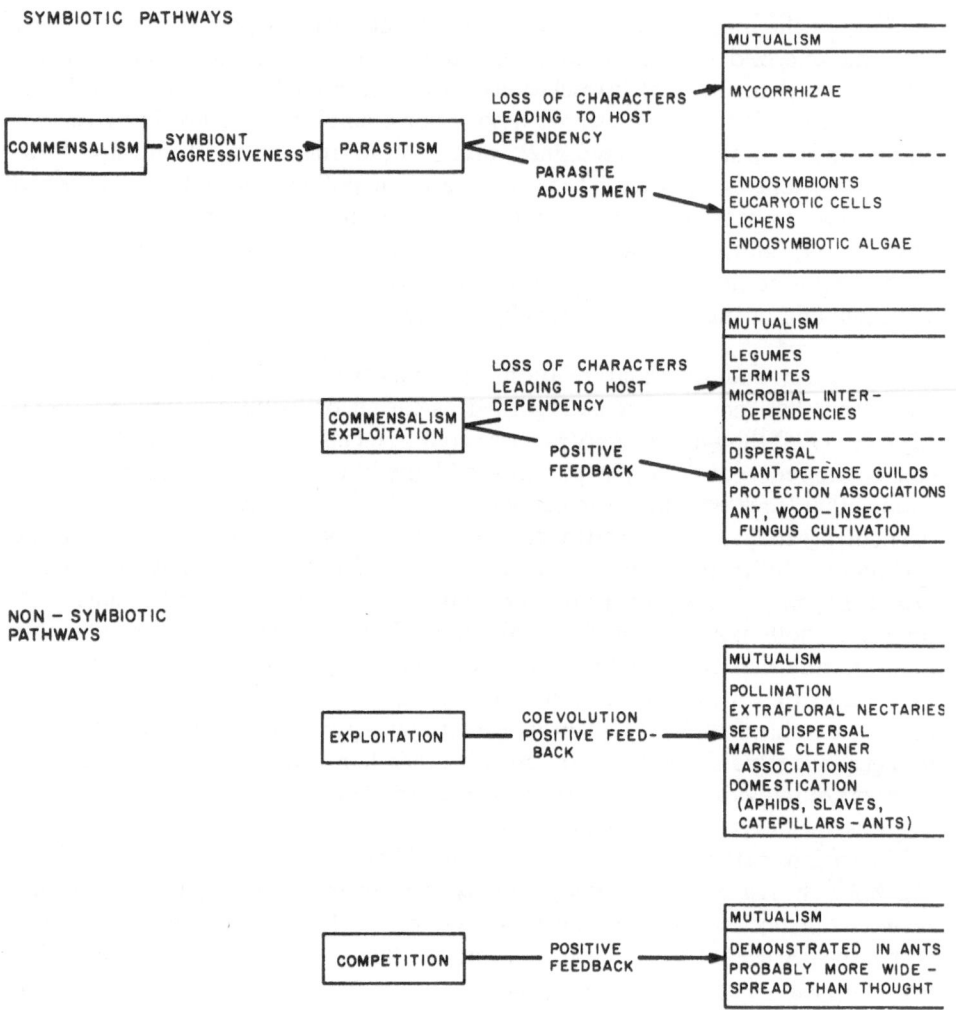

Fig. 8.9. Diversity of evolutionary pathways resulting in mutualism between species

hand, the parasite adjusts to be less virulent to the host, it may eventually be of some service to the host. The parasitism then usually becomes "*homosymbiotic*," and both organisms function as if they were a single individual. Examples of this trend are insects and their endosymbiotic partners, lichens, and coelenterates with their photosynthetic zooxanthellae.

Symbiotic mutualisms also appear to evolve directly from the commensal association between the host and symbiont without passing through a parasitic stage (Fig. 8.9). If the association is a frequent occurrence from the host's perspective, then the host may become dependent on some benefit that the symbiont is capable of providing. In this fashion termites rely on intestinal protozoans to develop their cellulose eating habits. In other cases, it may be to the symbiont's advantage to provide a benefit to the host in order to increase its own fitness. This positive feedback can lead to a strengthening of the relationship through evolutionary changes as discussed by Roughgarden (1975) for marine protection associations and Wilson (1980) for special dispersal relationships and bark-beetle fungus cultivation.

Non-symbiotic evolution of mutualistic associations is largely a process of coevolution of predator-prey or exploitation systems. In many cases, for example nectar-reward pollination systems and extrafloral nectary-ant associations, the mutualism is the result of a coevolutionary strengthening of mutual exploitation in which positive feedback provides the dynamic impetus. When the positive feedback loop is due to the actions of only one of the species, the interaction resembles domestication. For example, ants that tend aphids have selected aphids that respond to their protection and manipulation of the adults and their care of the aphid's eggs. Such aphids become incapable of existence independent of the ants, so that even though the aphids are exploited by the ants, they also benefit from the ant's attentions. The other nonsymbiotic pathway in Fig. 8.9 to mutualism through competition is discussed in the next section.

The foregoing classification demonstrates the diversity of evolutionary pathways that result in the tremendous variety of mutualistic associations. While this classification is far from perfect, it serves as a beginning to organize the many examples of mutualistic associations in order to draw similarities and generalizations from the details of physiological, morphological and behavioral adaptations. This particular organization emphasizes the importance of positive feedback in the evolution of mutualism.

8.6 Limited Competition

A community consisting of populations of mutualistic species can be modeled as a positive feedback system, thereby allowing powerful mathematical tools to be used in analyzing the dynamics of the community. However, in most biotic communities, mutualistic populations also interact with nonmutualistic populations to a degree that cannot be conveniently ignored. It thus becomes an important question to find out if communities with exploitative and competitive relationships as well as mutualistic relationships can be modeled as positive feedback systems. The answer to this question is a qualified yes. We can apply the results to a larger class of ecological communities than strictly mutualistic ones, but not all communities.

Consider two competing populations depicted in Fig. 8.10. Population 1 has a negative effect on population 2. If there is an increase in the negative effect that population 1 has on population 2 (e.g., by heightened aggressive behavior or by increased production of secondary compounds), then the equilibrium size of population 2 will decrease. This reduces the negative effect population 2 has on population 1 and population 1 increases. Thus, the effect population 1 has on population 2 feeds back positively on itself. The converse holds if population 2 is increased. Thus two competing populations form a simple positive feedback system (discussed also in Chap. 2). Mathematically, the community matrix for two

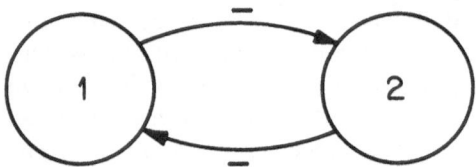

Fig. 8.10. Two competing populations – a simple positive feedback system

competing species is similar to the community matrix for two mutualistic species, and similar matrices have identical stability properties. Matrix $\mathbf{A^*}$ is similar to matrix $\mathbf{A}$ if there exists a nonsingular matrix $\mathbf{S}$ such that

$$\mathbf{A^*} = \mathbf{SAS}^{-1}. \tag{8.20}$$

The matrix

$$\mathbf{S} = \begin{vmatrix} 1 & 0 \\ 0 & -1 \end{vmatrix} \tag{8.21}$$

transforms the matrix representing two competing populations,

$$\mathbf{A} = \begin{vmatrix} -a_{11} & -a_{12} \\ -a_{21} & -a_{22} \end{vmatrix}, \tag{8.22}$$

to the interaction matrix

$$\mathbf{A^*} = \begin{vmatrix} -a_{11} & a_{12} \\ a_{21} & -a_{22} \end{vmatrix} = \begin{vmatrix} 1 & 0 \\ 0 & -1 \end{vmatrix} \begin{vmatrix} -a_{11} & -a_{12} \\ -a_{21} & -a_{22} \end{vmatrix} \begin{vmatrix} 1 & 0 \\ 0 & -1 \end{vmatrix}, \tag{8.23}$$

for two mutualistic populations and has the same stability properties as $\mathbf{A}$. Thus we can study the community matrix of the corresponding mutualistic system to determine whether or not the equilibrium of the two competing populations is stable.

Two competing populations form a positive feedback system because the path from a species to the other (consisting of the links of the effect of one population on the other near equilibrium) and back again consists of an even number of negative effects, resulting in a net positive effect. This concept can be generalized to include any number of interacting populations. If the effect of all pathways in a community which loop from a population, through any other populations and back again is positive (i.e., contains an even number of negative effects) then the community is a positive feedback system. This can be determined by examining the corresponding graph of the community matrix. Another method is to divide the community into two subcommunities with the following properties:

1) the interactions between the populations within a subcommunity are all positive or zero in effect (commensal, mutualistic or no interaction),

2) the interactions between populations between different subcommunities are all negative or zero (amensal, competitive or no interaction).

Mathematically, this is equivalent to performing identical row an column permutations on the community matrix to bring it into the form

$$A = \begin{vmatrix} A_{11} & A_{12} \\ A_{21} & A_{22} \end{vmatrix},$$ (8.24)

where the submatrices A_{ij} have the property that A_{11}, A_{22} are square matrices each with the dimension corresponding to the number of populations in the respective subcommunities. They also have the sign conditions $A_{11}, A_{22} \geq 0$ (ignoring the diagonal elements) and $A_{12}, A_{21} \leq 0$. The corresponding matrix for which our stability conditions are applicable is obtained by the similarity transformation (8.20), where

$$S = \begin{vmatrix} I_1 & 0 \\ 0 & -I_2 \end{vmatrix}$$ (8.25)

and I_1 and I_2 are identity matrices with dimensions corresponding to A_{11}, A_{22} respectively.

At first glance, this class of communities may seem small, but many ecologically relevant communities may satisfy conditions similar to (8.17). For example, the competitor-competitor-mutualist system considered by Rai et al. (1983) is a special case of limited competition. They modeled a system consisting of a pair of competitors X_1 and X_2, and a pair of mutualists, X_1 and U with the equations

$$\frac{dU}{dt} = Uh(U, X_1)$$

$$\frac{dX_1}{dt} = \alpha X_1 [g_1(U, X_1) - q_1(U, X_1, X_2)]$$ (8.26)

$$\frac{dX_2}{dt} = X_2 [g_2(X_2) - q_2(X_1, X_2)].$$

This same basic model can be used to describe interactions between flies, beetles, and mites (Springett, 1968), beetles, ants and membracids (Messina, 1981), hermit crabs and hydroids (Wright, 1973), and leaf-cutting ants and fungi (Quinlan and Cherret, 1978). The interaction matrix A, given by

$$A = \begin{bmatrix} \dfrac{\partial h}{\partial U} & \dfrac{\partial h}{\partial X_1} & 0 \\ \alpha \dfrac{\partial}{\partial U}[g_1 - q_1] & \alpha \dfrac{\partial}{\partial X_1}[g_1 - q_1] & -\dfrac{\partial q_1}{\partial X_2} \\ 0 & -\dfrac{\partial q_2}{\partial X_1} & \dfrac{\partial}{\partial X_2}[g_2 - q_2] \end{bmatrix} = \begin{bmatrix} - & + & 0 \\ + & - & - \\ 0 & - & - \end{bmatrix},$$ (8.27)

has the sign pattern in the category represented by (8.24) and thus represents a case of limited competition.

The ability to apply similarity transforms to the interaction matrix has a significance that goes beyond pure mathematics. Consider two three-species communities (Fig. 8.11). In the first community (Fig. 8.11a) there are competitive interactions between species 1 and 2 and between species 2 and 3, but a mutualistic relationship between species 1 and 3. The community matrix of this system is transformable into a positive feedback system. The second community (Fig. 8.11b) differs in that species 1 and 3 are competitors and not mutualistic. The community matrix of this system is not transformable into a positive feedback system. The difference between these systems embodies an important idea, *"an enemy of an enemy is a friend."* By interacting mutualistically with species 3, species 1 in the system in Fig. 8.11a is reinforcing the positive relationship that already exists between these species as mutual competitors of species 2. The net effect of species 1 on species 3 is always positive and vice versa; hence the two species are mutualists, both directly and indirectly. In the system in Fig. 8.11b, however, the net effects

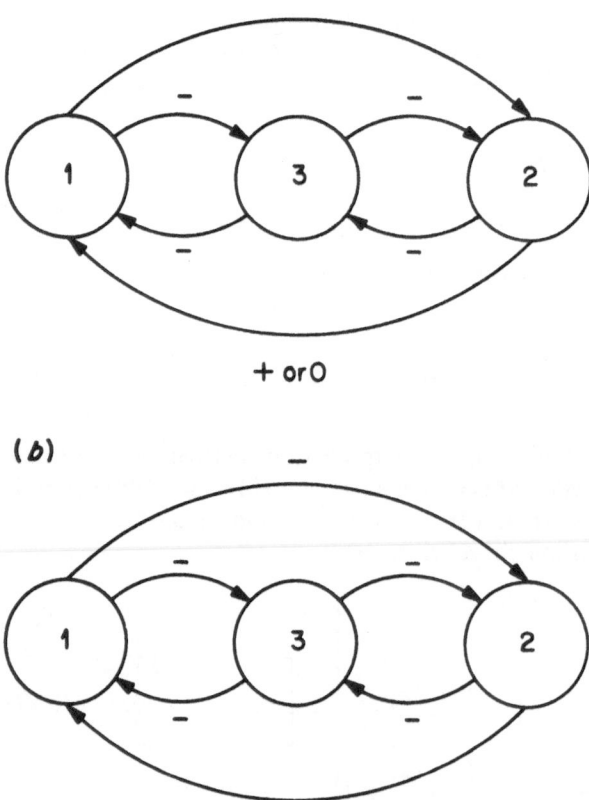

Fig. 8.11. **a** A positive feedback system. **b** Competition between populations 1 and 2 cause this not to be a positive feedback system

124

between species 1 and 3 are ambiguous, because they cannot be represented as pure positive feedback but only as a combination of positive and negative feedbacks.

This idea may be more helpful in classifying relationships in larger systems. For example, consider the study of coexisting ant popualtions by Davidson (1980). Davidson determined that a community of competing ant populations was food limited, and measured the magnitudes of both exploitative and interference interactions by observing the ants' diets and behavioral interactions at bait locations. When estimates of the magnitudes of the competition coefficients are arranged into an interaction matrix, ignoring the smaller elements that Davidson measured ($\alpha_{ij} < 0.2$), the community matrix approximates a community with limited competition. Figure 8.12 depicts the strong interactions among the ant species. From this figure and the corresponding sign matrix, we predict that the abundances of *Pogonomyrmex rugosus* should be positively correlated with *Pheidole xerophila* and *Novemessor cockerelli*. All of these species should be negatively correlated with *Poponomyrmex desertorum*. These predictions, obtained by Davidson (1980) using different methods, are supported by field manipulations. In fact, Davidson (1980) found that *N. cockerelli* and *P. rugosus* are usually nearest neighbors of each other in the study area, suggesting net interspecific facilitation. These two species are, unambiguosly, indirect mutualists.

8.7 Summary and Conclusions

The salient features of the population dynamics of mutualistic species can be described by fairly simple mathematical models. Obligatory and facultative associations can be modeled with equilibrium sizes of mutualistic species larger than the carrying capacities of the populations in absence of interactions. The stability of mutualistic interactions must be derived from stronger density-dependent self-limitation of the component populations than the positive feedback obtained from the exchanged benefits. This is guaranteed if there is a saturation of the benefits received from a mutualist partner. This also introduces a threshold phenomenon where a mutualism can form if population densities are large enough, but will lead to extinction of obligate mutualists if population densities are too low. The results for two-species interactions can be extended to multiple-species mutualistic interactions. Multiple-species models demonstrate the same stability characteristics as two-species models, namely, equilibrium stability is guaranteed if and only if for each species, density-dependent self-regulation is stronger than the effect of mutualistic benefits exchanged. General multispecies equations can be used to develop models of the evolution of mutualism from a commensal association. This model leads to the hypothesis that spatial or behavioral isolation of pairs of interacting individuals is necessary for the evolution of mutualism. A broad-ranging view of the literature describing obligate mutualistic interactions generally lends support to this hypothesis.

The stability results can be extended to communities of mutualistic and competitive populations provided the community can be subdivided into two subcommunities with mutualistic interactions between populations within each subcommunity and competitive interactions between populations in different subcommunities.

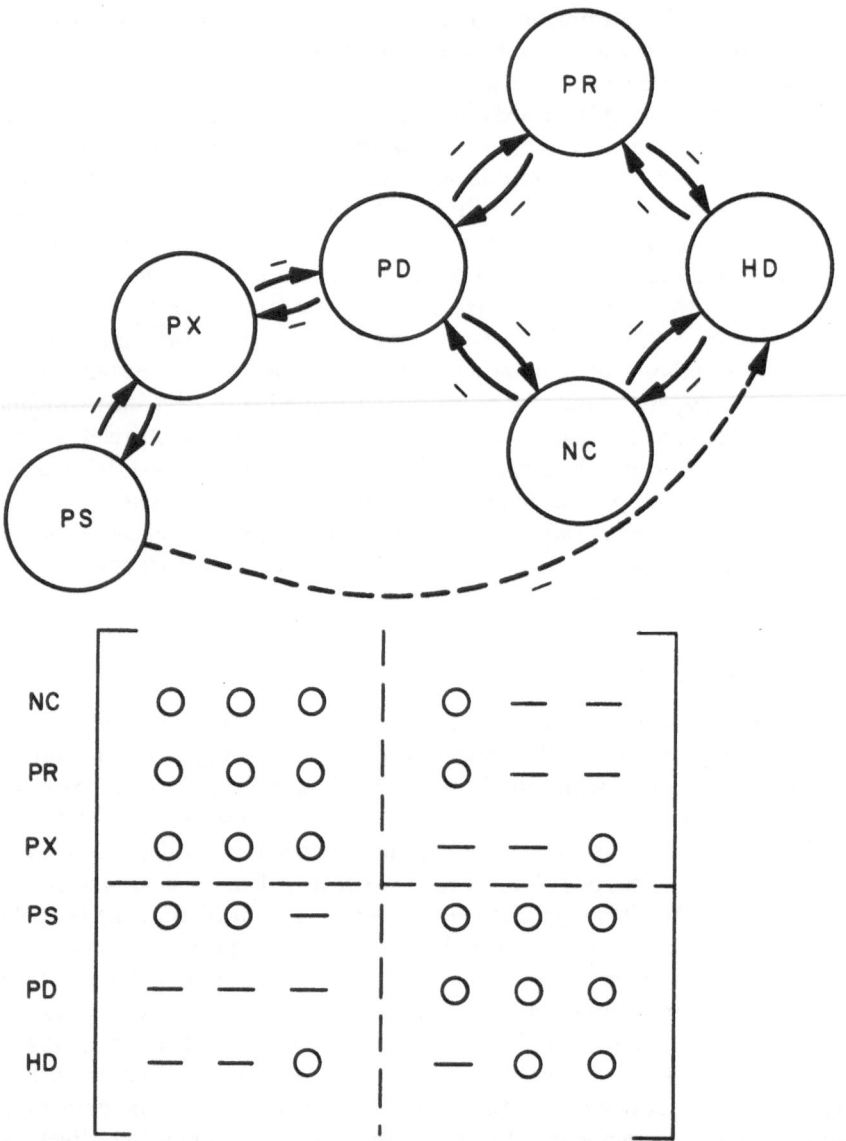

Fig. 8.12. *NC = Novomessor cockerelli, PR = Pogonomyremex rugosus, PD = Pheidole desertorum, PX = Pheidole xerophila, PS = Pheidole sitarches.* This representation of the major competitive interactions in an ant community shows a community with limited competition except for the interaction between *PS* and *HD* (dashed). This interaction introduces a negative loop of length 5, which is probably not significant compared to the strength of the feedback loops between *PR, PD, NC,* and *HD* (after Davidson, 1980)

9. Age-Structured Populations

9.1 Age-Structure

Life, in part, is a process of aging; individuals are born, they mature, and at some point they die. The aging process has a direct effect on individual probabilities of survival and reproductive activity, as well as on total population size. For example, fecundity is zero until the age of reproductive maturity; it then increases and remains high through adulthood, finally declining in old age. Mortality, on the other hand, is usually high in the very young and very old individuals, and lower for adults. Any detailed description of population dynamics must take the age-dependent nature of these variables into account.

Chapter 8 was devoted to the study of systems of mutualistic and/or competitive species. The equations that we used to describe these systems had one common feature: they all assumed that population growth depends continuously on the time variable. In fact, all equations used in Chap. 8 were extensions of the simple Verhulst-Pearl logistic equation,

$$\frac{dN}{dt} = rN - aN^2 . \tag{9.1}$$

This equation, which was introduced in 1838 by the French mathematician Verhulst, is one of the most popular in the field of biomathematics. Its popularity stems, in part, from the fact that it provides a satisfactory description to a pattern of population growth that is frequently observed in nature (Fig. 9.1). One criticism of the model, however, is that it treats the entire population as a single unit without regard to age structure. Since most life history parameters (e.g., fecundity, mortality, body weight, interspecific competition) are dependent on age, the age-structure of the population should be taken into account in any realistic model of its growth. Another difficulty is that, while the Verhulst-Pearl equation is capable of predicting gross population growth, it cannot predict the proportion of the population expected to be found in a given age class. In an attempt to overcome such difficulties, Bernadelli (1941), Lewis (1942), and Leslie (1945, 1948) introduced a discrete model for predicting age-structured population growth, which bears Leslie's name.

In a discrete model of population growth, rather than time being treated continuously as with the logistic model (9.1), the continuum of time is divided into discrete intervals. To illustrate this idea, consider a population in which individuals survive over many years, but breed during only one season of the year. In any given

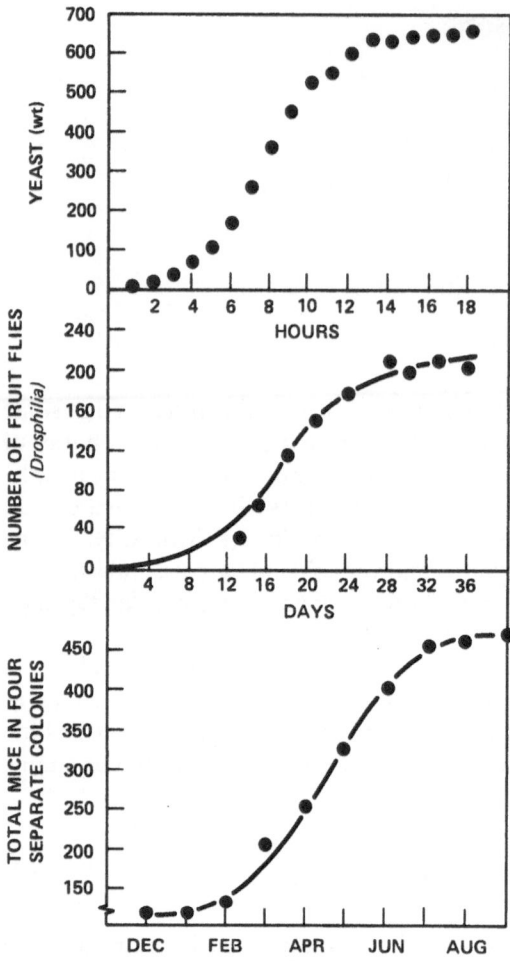

Fig. 9.1a–c. Three sigmoid growth curves for **a** yeast cells, **b** fruit flies (*Drosophila*), and **c** mice colonies. The top two curves closely approach the mathematical logistic curve. (Reproduced from Dinsmore, 1970, with permission of the University of Puerto Rico Press)

year t, the population can be divided into age-classes composed of individuals born $1, 2, 3, \ldots$ years previously (see Fig. 9.2). Assume that the fecundities and probabilities of survival from one breeding season to the next are different for each of these age-classes. The problem faced by Bernadelli (1941), Lewis (1942), and Leslie (1945, 1948) was to predict the number of individuals in each age class at the beginning of the breeding season in year $t+1$, given the number of individuals in each age class at the beginning of the breeding season in year t and the fecundity and probability of survival for each age class.

In the next section, we will discuss a solution to this problem. While Leslie's treatment of this problem has been discussed in detail in many textbooks, we present a more intuitive development based on the ideas of positive feedback and the theory of positive matrices. After the treatment of Leslie matrices, we introduce compensatory Leslie matrix models, in which fecundity and mortality are not only

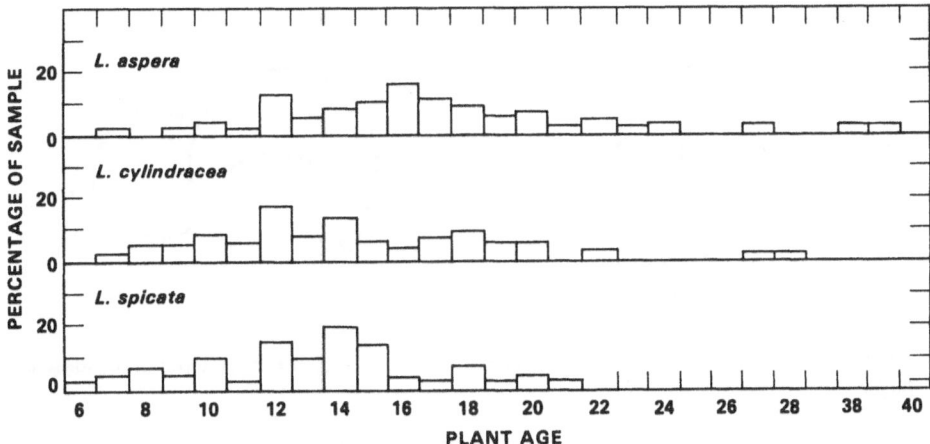

Fig. 9.2. The age structures (in years) of three *Liatris* plant species. (Reproduced from Levin, 1973, with permission of Allen Press, Inc.)

age-dependent, but are density-dependent as well. We then present extensions of the Leslie matrix concept to multi-species systems in which positive feedback dominates, as in the systems of Chap. 8. Age structure will be seen to have a significant effect on the dynamics of competing species. Finally we close this chapter with an application of age-dependent models to the problem of the evolution of life-history strategies.

9.2 Leslie Matrices

In its simplist terms, an age-structured population is a positive feedback system between the young and the adults of a population (see Fig. 9.3). The young, over a period of time, mature, become adults and thus augment the adult age class. Adults augment the juvenile age class through the bearing of offspring. Age-structure in a population causes the positive feedback in the system to be delayed in time. After adults augment the juvenile age class by having offspring, it may be many years before the young mature, enter the adult age class and complete the positive feedback cycle. Much of the diversity in natural ecosystems results from this delay in positive feedback, since the ecological niche of an age-structured population is thus allowed to have a temporal component as well as a spatial one.

Consider, for the moment, the population discussed in the previous section. Its individuals are assumed to survive over many years, but reproduction takes place only during the breeding season. The population can be divided into a finite number, say n, of age groups. Let $X_1(t)$ denote the number of individuals in the first age group at the beginning of the breeding season in year t. Let $X_2(t)$ denote the number of individuals in the second age group, etc. Assume that $f_i \geq 0$ is the effective fecundity per unit time of an individual in age class i and p_i is the probability of

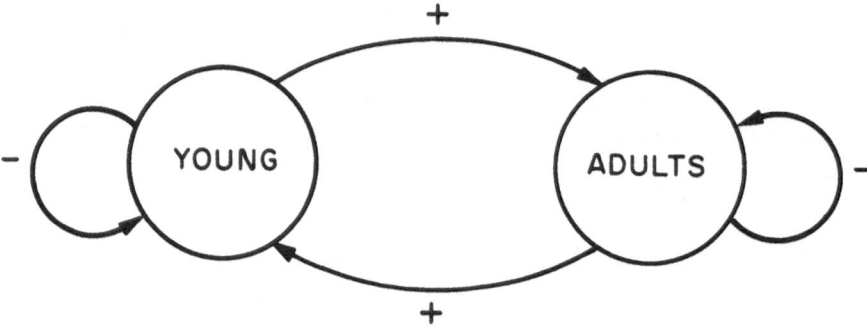

Fig. 9.3. Diagram of an age-structured population. The feedback loop between the young and the adults is positive

survival from age class i to age class $i+1$. The term "*effective*" is used because we only count females born in the time interval from t to $t+1$ that survive to $t+1$. The newborn females that do not survive to $t+1$ are not included in the effective fecundity rates. It simplifies the discussion to consider only the females of the population. The same arguments would also apply if one considered both sexes, provided the ratio of males to females remains constant and that age-specific mortality is the same for both sexes.

Given the number of females in each age class at time t, the effective fecundities f_i and the propabilities of survival p_i, we should be able to determine the number of females in each age class at time $t+1$. Because individuals can only enter the youngest age class by being born into it, and the total number of females in age class 1 at time $t+1$ is the sum of surviving female offspring born at time t in all other age classes,

$$X_1(t+1)=f_1 X_1(t)+f_2 X_2(t)+ \ldots +f_n X_n(t). \tag{9.2}$$

The number of females in the older age classes at time $t+1$ is determined by the number surviving from the immediately preceding age class. Thus,

$$X_2(t+1)=p_1 X_1(t),$$
$$X_3(t+1)=p_2 X_2(t),$$
$$\cdot$$
$$\cdot \tag{9.3}$$
$$\cdot$$
$$X_n(t+1)=p_{n-1} X_{n-1}(t).$$

Equations (9.2) and (9.3) form a positive feedback system in which the older age classes amplify the first age class through reproduction, and each age class (except the first) is amplified through survival from the previous age class (see Fig. 9.4).

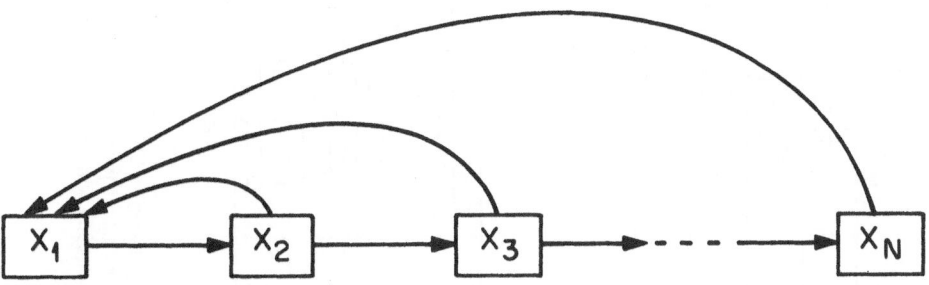

Fig. 9.4. Diagram of a multi-age-structured population. The feedback loops are positive

These equations were first employed by the English economist, Cannan (1895), and by the demographers, Bowley (1924) and Whelpton (1936). It was not until later that Bernadelli (1941), Lewis (1942), and Leslie (1945) recognized that the process could be conveniently formulated in matrix terms (s. Cohen, 1979).

To formulate Eqs. (9.2) and (9.3) in matrix form, let $\mathbf{X}(t)$ $=[X_1(t), X_2(t), \ldots, X_n(t)]$ be a vector representing the number of individuals in each of the n age groups. Define the matrix $\mathbf{A}$ (commonly known as a Leslie matrix) as follows:

$$\mathbf{A} = \begin{vmatrix} f_1 & f_2 & f_3 & \cdots\cdots & f_n \\ p_1 & 0 & 0 & \cdots\cdots & 0 \\ 0 & p_2 & 0 & \cdots\cdots & 0 \\ . & . & . & & \\ . & . & . & & \\ . & . & . & & \\ 0 & 0 & 0 & \cdots p_{n-1} & 0 \end{vmatrix} \tag{9.4}$$

where f_i and p_i are as defined above. The matrix formulation of Eqs. (9.2) and (9.3) is then given by

$$\mathbf{X}(t+1) = \mathbf{A}\mathbf{X}(t). \tag{9.5}$$

For age-structured populations whose fertility and mortality rates remain constant in time [as is the case with Eqs. (9.2, 9.3)], the strong ergodic theorem (Seneta, 1973) characterizes growth of the population. This theorem states that, regardless of the initial age distribution of the population, an age distribution is rapidly reached in which the rate of population increase in each age class is the same as the rate of increase of the total population (see Fig. 9.5). Such an age distribution is known as a stable age distribution. In a stable age distribution, even though total population size is expanding, the fraction of the total population found in each age class remains fixed. It can be shown that the stable age-structure and growth rate are

131

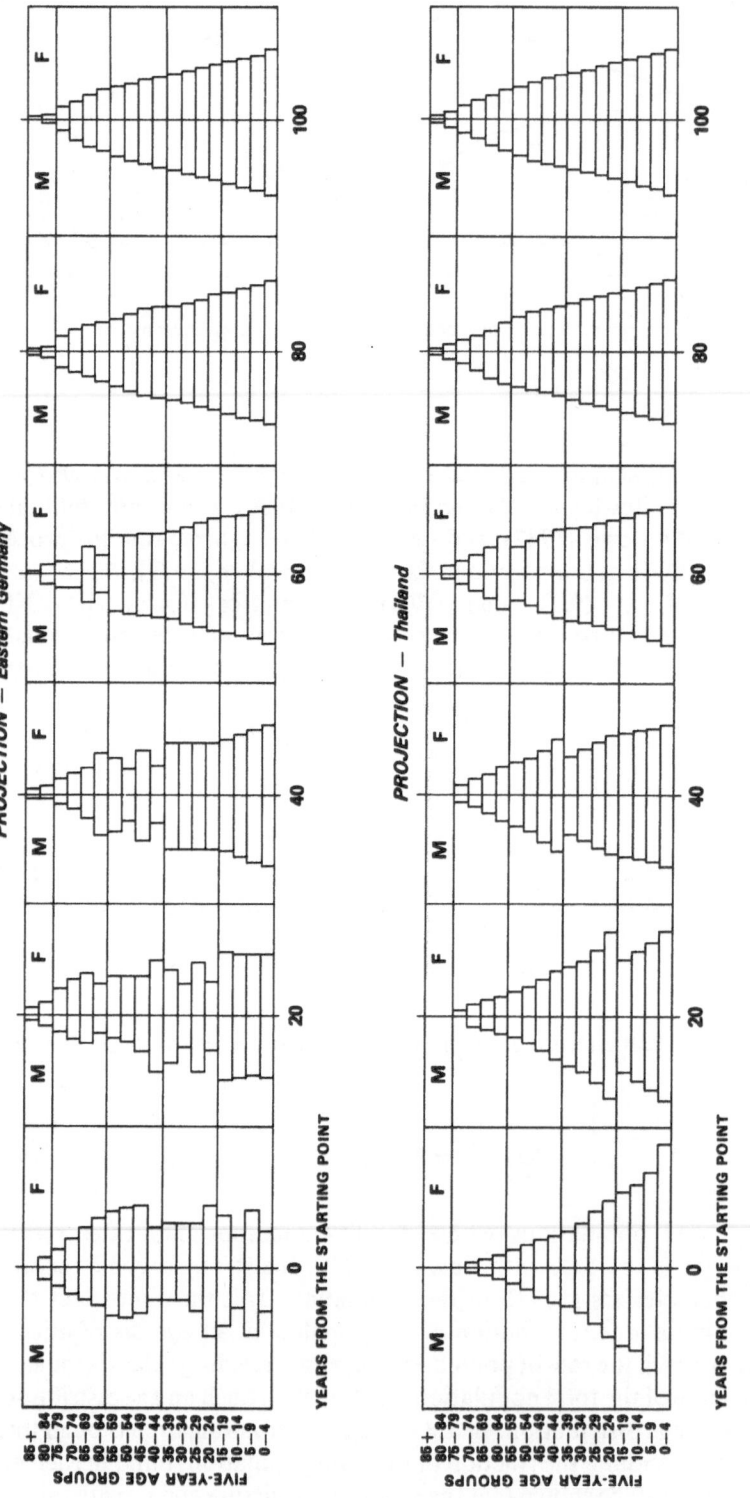

Fig. 9.5. Two sets of projections computed on the basis of the population of East Germany in 1957 and of an estimate of the populations of Thailand in 1955, respectively. The age-distributions are divided into 5-year age classes and M = male and F = female. Hypothetical vital rates used in both projections assume an expectation of life at birth for both sexes of 60.4 years and a gross reproduction rate of 1.50. (Reprinted from Bourgeois-Pichat, 1969, with permission from United Nations)

132

independent of the initial state of the population, and are determined only by values of the demographic parameters.

A relationship exists between the rate of growth of the population and the eigenvalues of the Leslie matrix $\mathbf{A}$. (We recall that a number λ is an eigenvalue of a matrix $\mathbf{A}$ if there exists a nonzero vector $\mathbf{X}$ such that $\mathbf{AX} = \lambda\mathbf{X}$. It is well-known that an $n \times n$ matrix has n eigenvalues.) A stable age distribution can exist only if the ratio of population sizes in a given age class remains constant from year to year; that is, only if there exists a constant $\lambda > 0$ such that for large t, $X_1(t+1)/X_1(t) = \lambda$, $X_2(t+1)/X_2(t) = \lambda$, etc. This can be condensed to vector notation by saying that if the population is in a stable age distribution, there must exist a $\lambda > 0$ such that for large t, $\mathbf{X}(t+1) = \lambda\mathbf{X}(t)$, where $\mathbf{X}(t)$ is the vector of population sizes. It follows from Eq. (9.5) that λ must be an eigenvalue for the Leslie matrix $\mathbf{A}$, that is,

$$\mathbf{A}(t+1) = \mathbf{AX}(t) = \lambda\mathbf{X}(t). \tag{9.6}$$

Sykes (1969) has shown that it is a corollary of the Perron-Frobenius Theorem (s. Appendix B) that the λ in Eq. (9.6) is, in fact, the maximal eigenvalue of the matrix $\mathbf{A}$. Thus, the maximal eigenvalue of the Leslie matrix gives the intrinsic rate of growth of an age-structured population, while the normalized eigenvector corresponding to this eigenvalue determines that fraction of the total population found in each age class. If $\lambda < 1$, the population decreases toward zero (extinction), while if $\lambda > 1$, positive feedback causes the population to increase exponentially without bound. Only in the unlikely case that λ is exactly equal to one does the population avoid ultimately approaching zero or infinity.

In summary, if the population growth dynamics of a species can be described by the Leslie matrix model (9.5), then the population growth rate and stable age distribution can be determined through an eigenvalue analysis. As an application of these concepts, Hess et al. (1975) developed the population projection matrix (Leslie matrix) for the marine fish species winter flounder *(Pseudopleuronectes americanus)*. The age-specific fecundity was multiplied by the proportion of females (0.7) to obtain the f_i for each age class. The element p_1 was calculated by the method of Vaughan and Saila (1976) to achieve an equilibrium population. The resulting population projection matrix and a plot of its eigenvalues are presented in Fig. 9.6. The maximal eigenvalue has a value of 1.0, indicating, as assumed, that the population is in equilibrium.

In general, the maximal eigenvalue of a matrix is difficult to compute. However, since the Leslie matrix is positive, simple criteria can be used to determine the magnitude of the largest eigenvalue. The maximal eigenvalue will be greater than one (implying that the population is growing exponentially without bound) if and only if $(-1)^n \det(\mathbf{A} - \mathbf{I}) < 0$ (see Appendix B). Applying this criterion to the example in Fig. 9.4 with three age classes, we obtain

$$(-1)^n \det(\mathbf{A} - \mathbf{I}) = (-1)^3 \begin{vmatrix} -1 & f_2 & f_3 \\ p_1 & -1 & 0 \\ 0 & p_2 & -1 \end{vmatrix} < 0 \tag{9.7}$$

Element of the population projection matrix for the winter flounder

Age (i)	Eggs per adult (f_i)	Survivorship s_i
0	0	0.0000157
1	0	0.1454
2	0	0.33
3	183.000	0.33
4	310.100	0.33
5	445.900	0.33
6	569.100	0.33
7	679.000	0.33
8	774.900	0.33
9	851.900	0.33
10	910.700	0.33
11	956.900	0.33
12	966.800	0

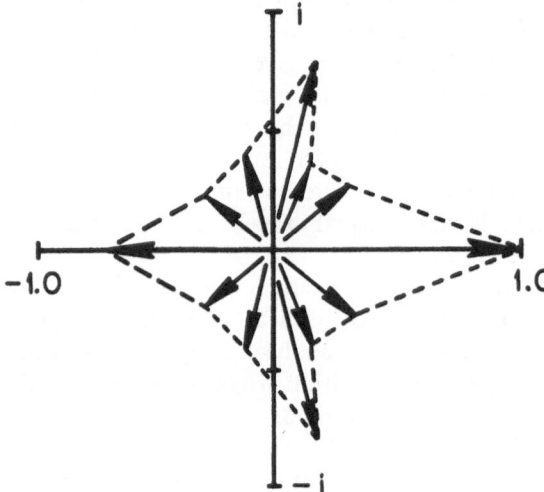

WINTER FLOUNDER

Fig. 9.6. Table of parameter values for the equilibrium Leslie matrix for the winter flounder, and the distribution of eigenvalues in the complex λ-plane. (Reproduced from Horst, 1977, with permission of Pergamon Press)

or $p_1 f_2 + p_1 p_2 f_3 > 1$. For the general Leslie matrix model (9.2, 9.3), the criterion is

$$(-1)^n \det (\mathbf{A} - \mathbf{I}) = (-1)^n \begin{vmatrix} f_1 - 1 & f_2 & f_3 & \cdots & f_n \\ p_1 & -1 & 0 & \cdots & 0 \\ 0 & p_2 & -1 & \cdots & 0 \\ \cdot & \cdot & & & \\ \cdot & \cdot & & & \\ \cdot & \cdot & & & \\ 0 & 0 & \cdots & p_{n-1} & -1 \end{vmatrix} < 0 \quad (9.8)$$

or

$$f_1 + p_1 f_2 + p_1 p_2 f_3 + \ldots + p_1 p_2 \ldots p_{n-1} f_n > 1. \tag{9.9}$$

The quantity

$$R \equiv f_1 + p_1 f_2 + p_1 p_2 f_3 + \ldots + p_1 p_2 \ldots p_{n-1} f_n \tag{9.10}$$

is known as the net reproductive rate and represents the expected number of offspring that a newborn will have in its lifetime. If we let $l_k = p_1 p_2 \ldots p_{k-1}$ denote the probability of a newborn surviving to age class k, and define $l_1 = 1$, then R can be written

$$R = \sum_{k=1}^{n} l_k f_k. \tag{9.11}$$

Thus, condition (9.9) can be rewritten as $R > 1$, with the obvious biological interpretation that if the expected number of offspring that a newborn will have during its lifetime is greater than one, the population will grow exponentially without bound.

9.3 Compensatory Leslie Matrices

It is generally recognized that all living organisms participate in a competitive struggle for existence. At low population densities, intraspecific competitive pressures are slight and survival and reproductive rates are at a maximum, but at high population densities, competition for resources could become intense and have a negative effect on both survival and reproduction. Thus, whenever population sizes deviate from equilibrium in either direction, competitive pressure or its absence may provide a tendency for eventual return to average size. As Krebs (1972) points out... *"Natural populations do not usually become extinct or increase to infinity."* This concept is what has been loosely termed in the literature as *"the balance of nature."*

The concept of compensation has been developed to describe this situation. Compensation refers to the tendency of populations to experience 1) a decrease in survival and reproductive rates as they grow in density, thus establishing an upper limit to population size, and 2) an increase in survival and reproductive rates as population density declines, thus establishing a lower limit to population size under all but the most usual conditions.

The compensatory concept has found its most extensive application in stock-recruitment models in which total population size is represented by a single variable. The two general types of models most commonly used in the fisheries literature to describe the density-dependent nature survival and fecundity are Beverton-Holt-type models (Beverton and Holt, 1957) and Ricker-type models (Ricker, 1954). Although these appelations actually refer to specific models, they are often used to designate models that do not and do, respectively, allow for overcompensation.

Overcompensation occurs when an increase in the numbers of individuals in age-class i causes the number of individuals surviving to age-class $i+1$ (or the number of offspring of age-class i) to decrease. This is a much stronger density-dependent effect than mere compensation, where an increase in the number of age-class i causes a decrease in the survival rate or fecundity rate but not in the number of survivors or offspring. In Fig. 9.7, survivorship to age-class $i+1$ is plotted as a function of population of age-class i in both the a) Beverton-Holt-type and b) Ricker-type models. Similar curves could be plotted for fecundity.

In most applications of the Leslie matrix model (9.5) to natural populations in a nonequilibrium state, the largest eigenvalue of the projection matrix is larger than one, reflecting the fact that the populations are growing at an exponential rate. Obviously, such a state cannot exist over an extended period of time. In order to make Leslie matrix model predictions conform more with natural observations, density-dependent survival and reproductive rates have been introduced into these models (Leggett, 1977; Krebs, 1972). As an example, consider the density-dependent or compensatory Leslie matrix model with three age classes:

$$\mathbf{Z}(t+1)=\mathbf{L}\{\mathbf{Z}(t)\}f(t), \tag{9.12}$$

where $\mathbf{Z}(t)\equiv(X_1, X_2, X_3)$ and

$$\mathbf{L}=\begin{vmatrix} 0 & f_2 & f_3 \\ p_1 & 0 & 0 \\ 0 & p_2 & 0 \end{vmatrix}. \tag{9.13}$$

Let f_2, f_3, and p_2 be constants while p_1, the survival fraction from the first to the second age-class, is the density-dependent function

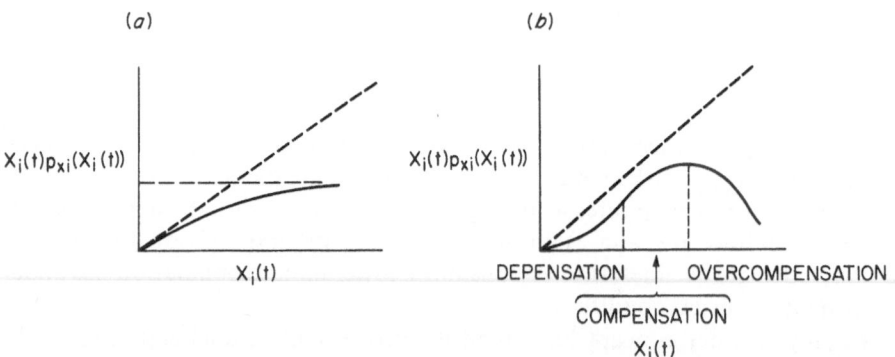

Fig. 9.7a and b. Survivorship to age-class $i+1$ plotted as a function of population size of age-class i. The dashed diagonal line shows this relationship for the traditional linear Leslie matrix model. a Shows the Beverton-Holt-type density-dependent relationship that is always compensatory. b Shows a Ricker-type density dependence that is only compensatory when the equilibrium population size is not in the region of depensation or overcompensation. (Reprinted from Travis et al., 1980, with permission of Academic Press, Inc.)

$$p_1 = \frac{r}{1 + \alpha X_1}.$$

In this model it is assumed that the young-of-the-year are subject to density-dependent mortality resulting from competition for food and other resources with members of the same age-class. The model has only one equilibrium point with positive population densities, $\bar{Z} = (\bar{X}_1, \bar{X}_2, \bar{X}_3)$, where

$$\bar{X}_1 = \frac{(f_2 + p_2 f_3)r - 1}{\alpha},$$

$$\bar{X}_2 = \frac{r\bar{X}_1}{(f_2 + p_2 f_3)r},$$

$$\bar{X}_3 = p_2 \bar{X}_2.$$

When $\mathbf{L}$ is linearized about $\bar{Z}$, the following perturbation matrix is obtained:

$$\mathbf{A} = \begin{vmatrix} 0 & f_2 & f_3 \\ r/(1 + \alpha \bar{X}_1)^2 & 0 & 0 \\ 0 & p_2 & 0 \end{vmatrix}. \tag{9.14}$$

Since this is a positive matrix, stability of the model (9.12) is determined by the condition $(-1)^3 \det (\mathbf{A} - \mathbf{I}) > 0$ or

$$1 > \frac{f_2 r}{(1 + \alpha \bar{X}_1)^2} + \frac{p_2 f_3 r}{(1 + \alpha \bar{X}_1)^2}. \tag{9.15}$$

9.4 Interacting Populations

Interactions between species tend to be age-specific. For example, vertebrate carnivores such as wolves feed preferentially on the very young and very old and, hence, weak members of their large prey. Competing insects that have several life stages usually have age-specific interactions. The larvae of one species compete with the larvae of another species for resources, but not with adults of their own species or adults of their competitor. Despite such facts, relatively little has been done to extend Leslie matrix methods to multi-species communities. Part of the reason may be that the generalized Leslie matrices representing multi-species systems are less tractable mathematically than single-species Leslie matrices. What work has been done using multispecies models with age classes involves computer simulations (Pennycuick et al., 1968) or particular forms of two species, two age-class equations with limited competitive interactions (Hassell and Comins, 1976).

We shall show that certain extensions of the Leslie matrix concept to multi-species systems are tractable analytically. Very simple criteria for determining the stability of the system exist for matrices representing mutualistic or competitive interactions, as well as their subcategories commensal and amensal interactions. We first consider a general compensatory model allowing for age-specific interactions between two species.

Consider the following matrix model for two interacting populations

$$\mathbf{Z}(t+1) = \mathbf{L}\{\mathbf{Z}(t)\}\mathbf{Z}(t). \tag{9.16}$$

Here $\mathbf{Z}(t)$ is the age-class population vector for the two species; $\mathbf{Z}(t) = (X_1, X_2, \ldots, X_m, Y_1, Y_2, \ldots, Y_n)$ where X_i and Y_j are the population densities of the ith age class of species X and the jth age-class of species Y, respectively. The total numbers of age classes are m for species X and n for species Y. The matrix $\mathbf{L}$ has the form

$$\mathbf{L} = \begin{vmatrix} \mathbf{L}_x & 0 \\ 0 & \mathbf{L}_y \end{vmatrix}, \tag{9.17}$$

where submatrix $\mathbf{L}_x$ has the form

$$\mathbf{L}_x = \begin{vmatrix} f_{x1} & f_{x2} & f_{x3} & \cdots & f_{x,m-1} & f_{xm} \\ p_{x1} & 0 & 0 & \cdots & 0 & 0 \\ 0 & p_{x2} & 0 & \cdots & 0 & 0 \\ . & . & . & . & . & . \\ . & . & . & . & . & . \\ . & . & . & . & . & . \\ 0 & 0 & 0 & & p_{x,m-1} & p_{xm} \end{vmatrix}. \tag{9.18}$$

The element f_{xj} represents fecundity of the ith age-class and p_{xi} represents the survival rate of the ith age-class of species X. A similar expression pertains for submatrix $\mathbf{L}_y$. In expression (9.18), the survival rate p_{xm} is not explicitly set equal to zero, allowing for two possible interpretations of the oldest age-class. The survival rate p_{xm} will equal zero in the case where all individuals in age-class m die after one time unit. In the case where age-class m represents a lumping of several adult age-classes, p_{xm} will be nonzero, representing the probability of a member of age-class m surviving to remain in age-class m after one time interval.

In a general age-specific model, each nonzero element in the matrix $\mathbf{L}$ can, in principle, be density-dependent; that is, each element can be a function of the population sizes of the various age-classes of the particular species or its competitor. Density dependence in both the fecundities and survival rates of age-structured single-population models has been considered by several authors (e. g., Leslie, 1948; Smouse and Weiss, 1975; Guckenheimer et al., 1977). In the mathematical analysis that follows we make three assumptions about the density-dependence of the elements of $\mathbf{L}$. First, for all age classes i, j, the fecundities, f_{xi} and f_{yj}, and the survival fractions, p_{xi} and p_{yj}, for the two species may not depend negatively on populations of age-classes other than i within the same species. Secondly, for all i and j, the quantities $f_{xi}X_i, f_{yj}Y_j, p_{xi}X_i, p_{yj}Y_j$ may not depend negatively on X_i, Y_j, X_i and Y_j, respectively. Thus, "overcompensation" is omitted from our model. Finally, only nonpositive dependence of the fecundities and survival fractions on any age-class population of the competing species is permitted.

These three assumptions limit the generality of the mathematical analysis, but still permit us to consider a wide range of interesting cases. For example, the first assumption will always hold if, as is often the case, survival through the first age-class of life is density-dependent and individuals from the first age-class do not affect survival of and are not affected by individuals in other age-classes of the same species. This assumption does not preclude a type of interspecific "*competition*" that arises from the feeding of adults of one species on the eggs, larvae or juveniles of another, as long as the positive effect of this predation on the survival of the predators can be ignored. For this positive effect to be negligible, the food intake of each species through predation on the other species should be only a small fraction of the predator's diet. This is often the case in freshwater fish populations.

Our methods are applicable to models of the Beverton-Holt stock-recruitment type regardless of the equilibrium population sizes, since the slope of the survivorship curve is always positive (Fig. 9.7a) and overcompensation therefore does not occur. In the following section, we analyze an example of a model where this type of density-dependence occurs in the first age-class survival. Our results also apply to Ricker-type models whenever the equilibrium population size is low enough that the slope of the survivorship curve (Fig. 9.7b) at equilibrium is positive.

Let there be an equilibrium point $\bar{\mathbf{Z}} = \bar{X}_1, \bar{X}_2, \ldots, \bar{X}_m, \bar{Y}_1, \bar{Y}_2, \ldots, \bar{Y}_n)$, where all $\bar{X}_i, \bar{Y}_j > 0$, defined at the point at which $\mathbf{Z}(t+1) = \mathbf{Z}(t)$. The behavior of the two populations very close to the equilibrium point $\bar{\mathbf{Z}}$ is described by the linearized matrix equation

$$\tilde{\mathbf{Z}}(t+1) = \mathbf{A}\tilde{\mathbf{Z}}(t), \tag{9.19}$$

where $\tilde{\mathbf{Z}}(t)$ represents the vector of deviations from the equilibrium point and

$$\mathbf{A} = \begin{vmatrix} \bar{\mathbf{L}}_x & \mathbf{B}_{xy} \\ \mathbf{B}_{yx} & \bar{\mathbf{L}}_y \end{vmatrix}. \tag{9.20}$$

The submatrix $\bar{\mathbf{L}}_x$ has the form

$$\bar{\mathbf{L}}_x = \begin{vmatrix} g_{x1} & g_{x2} & g_{x3} & \cdots & g_{x,m-1} & g_{xm} \\ q_{x1} & 0 & 0 & \cdots & 0 & 0 \\ 0 & q_{x2} & 0 & \cdots & 0 & 0 \\ \cdot & & \cdot & \cdot & \cdot & \cdot \\ \cdot & & & & & \\ \cdot & & & & \cdot\cdot & \\ 0 & 0 & 0 & \cdots & q_{x,m-1} & q_{xm} \end{vmatrix}, \tag{9.21}$$

where

$$q_{xi} = p_{xi}(\bar{X}_i, \bar{Y}_1, \bar{Y}_2, \ldots, \bar{Y}_n) + \bar{X}_i \left. \frac{\partial p_{xi}}{\partial X_i} \right|_{\mathbf{Z}}$$

and

$$g_{xi} = f_{xi}(\bar{X}_i, \bar{Y}_1, \bar{Y}_2, \ldots, \bar{Y}_n) + \bar{X}_i \left. \frac{\partial f_{xi}}{\partial X_i} \right|_{\mathbf{Z}}.$$

The terms g_{xi} and q_{xi} represent the change (at equilibrium) in the fecundity or number of survivors, respectively, of age-class i with respect to the number of individuals in age-class i. Because we assume that none of these depend on the number of individuals in other age-classes of the same species and also that there is no density overcompensation near equilibrium, all of the elements of the submatrix $\mathbf{L}_x$ are nonnegative. The submatix $\mathbf{L}_y$ has similar form. The submatrix $\mathbf{B}_{xy}$ has the form

$$\mathbf{B}_{xy} = \begin{vmatrix} \sum_{i=1}^{m} \bar{X}_i \frac{\partial f_{xi}}{\partial Y_1} & \sum_{i=1}^{m} \bar{X}_i \frac{\partial f_{xi}}{\partial Y_2} & \cdots & \sum_{i=1}^{m} \bar{X}_i \frac{\partial f_{xi}}{\partial Y_n} \\ \bar{X}_1 \frac{\partial p_{x1}}{\partial Y_1} & \bar{X}_1 \frac{\partial p_{x1}}{\partial Y_2} & \cdots & \bar{X}_1 \frac{\partial p_{x1}}{\partial Y_n} \\ \cdot & \cdot & & \cdot \\ \cdot & \cdot & & \cdot \\ \cdot & \cdot & & \cdot \\ \bar{X}_{m-1} \frac{\partial p_{x,m-1}}{\partial Y_1} & \bar{X}_{m-1} \frac{\partial p_{x,m-1}}{\partial Y_2} & \cdots & \bar{X}_{m-1} \frac{\partial p_{x,m-1}}{\partial Y_n} \end{vmatrix}. \quad (9.22)$$

$\mathbf{B}_{yx}$ has a similar form.

When a competitive relationship exists between the species, all of the elements of $\mathbf{B}_{xy}$ and $\mathbf{B}_{yx}$ are less than or equal to zero. If the relationship is amensal, either $\mathbf{B}_{xy}$ or $\mathbf{B}_{yx}$ equals zero and the other is negative. When the two species are mutualistic, all elements of $\mathbf{B}_{xy}$, $\mathbf{B}_{yx}$ are greater than or equal to zero. If the species are commensal, either $\mathbf{B}_{xy}$ or $\mathbf{B}_{yx}$ equals zero and the other is positive. The theory developed in the following sections applies to all of these cases. This theory, unfortunately, cannot be directly applied to most predator-prey interactions.

9.5 Coexistence of Two Interacting Populations

Most theoretical considerations of competition have been based on Lotka-Volterra differential-equation models,

$$\frac{dX(t)}{dt} = r_1 \left(1 - \frac{X(t)}{K_1} - c_{12} Y(t) \right) X(t) \quad (9.23)$$

and

$$\frac{dY(t)}{dt} = r_2 \left(1 - \frac{Y(t)}{K_2} - c_{21} X(t) \right) Y(t). \quad (9.24)$$

In such models, all members of a population, whatever their ages, are lumped into one variable. Depending on the values of $K_1, K_2, c_{12},$ and c_{21}, and on the initial values of X and Y, mutual coexistence or the exclusion of one or the other species can occur. In particular, mutual coexistence is possible if, and only if, the strength of the intraspecific competition exceeds that of the interspecific competition; that is, if, and only if,

$$1/(K_1 K_2) > c_{12} c_{21}. \qquad (9.25)$$

We shall now develop the corresponding conditions for competing age-structured populations (see Travis et al., 1980). Consider a generalized Leslie matrix model for two competing species, each divided, for convenience, into only two age classes:

$$\mathbf{Z}(t+1) = \mathbf{L}\{\mathbf{Z}(t)\}\mathbf{Z}(t), \qquad (9.26)$$

where $\mathbf{Z}(t) = (X_1, X_2, Y_1, Y_2)$ and

$$\mathbf{L} = \begin{vmatrix} 0 & f_{x2} & 0 & 0 \\ p_{x1} & p_{x2} & 0 & 0 \\ 0 & 0 & 0 & f_{y2} \\ 0 & 0 & p_{y1} & p_{y2} \end{vmatrix}. \qquad (9.27)$$

Let $f_{x2}, f_{y2} p_{x2}$, and p_{y2} be constants while p_{x1} and p_{y1}, the survival fractions from the first to the second age-classes, are density-dependent functions:

$$p_{x1} = \frac{r_x}{1 + \alpha_x X_1 + \beta_{x1} Y_1 + \beta_{x2} Y_2} \qquad (9.28)$$

and

$$p_{y1} = \frac{r_y}{1 + \alpha_y Y_1 + \beta_{y1} X_1 + \beta_{y2} X_2}. \qquad (9.29)$$

This example was deliberately chosen for simplicity, but it incorporates features that have a degree of generality. The young-of-the-year age-class is pictured as the one being most susceptible to mortality resulting from density-dependent factors. Survival diminishes as a result of competition for food and other resources between members of the same age-class in both species. The density of older organisms also has a depressive effect on the survival of young-of-the-year of the competing species. This could result from competition for resources and also from predation, as is often the case in freshwater fish. We assume that, if predation is involved, the resultant food intake constitutes only a small part of the predator's dient, and so has no significant positive effect on f_{x2}, f_{y2}, p_{x2}, or p_{y2}.

The present model has only one equilibrium point at which the two species coexist, $\bar{\mathbf{Z}} = (\bar{X}_1, \bar{X}_2, \bar{Y}_1, \bar{Y}_2)$, where

$$\bar{X}_2 = \frac{\alpha_y f_{y2}(1 - p_{y2})(r_x f_{x2} - 1 + p_{x2}) - (1 - p_{x2})(\beta_{x2} + f_{y2}\beta_{x1})(r_y f_{y2} - 1 + p_{y2})}{(1 - p_{x2})(1 - p_{y2})[\alpha_x f_{x2}\alpha_y f_{y2} - (\beta_{x2} + f_{y2}\beta_{x1})(\beta_{y2} + f_{x2}\beta_{y1})]},$$

$$\qquad (9.30)$$

$$\bar{X}_1 = f_{x2}\bar{X}_2, \qquad (9.31)$$

$$\bar{Y}_2 = \frac{r_x f_{x2} - 1 + p_{x2} - (1 - p_{x2})\alpha_x f_{x2}\bar{X}_2}{(1 - p_{x2})(\beta_{x2} + \beta_{x1} f_{y2})}, \qquad (9.32)$$

and

$$\bar{Y}_1 = f_{y2}\,\bar{Y}_2.\tag{9.33}$$

When **L** is linearized about **Z**, the following matrix **A** is obtained:

$$\mathbf{A} = \begin{vmatrix} 0 & f_{x2} & 0 & 0 \\ s_x & p_{x2} & -h_{x1} & -h_{x2} \\ 0 & 0 & 0 & f_{y2} \\ -h_{y1} & -h_{y2} & s_y & p_{y2} \end{vmatrix},\tag{9.34}$$

where

$$s_x = \gamma_x r_x (1 + \beta_{x1}\,\bar{Y}_1 + \beta_{x2}\,\bar{Y}_2),$$

$$s_y = \gamma_y r_y (1 + \beta_{y1}\,\bar{X}_1 + \beta_{y2}\,\bar{X}_2),$$

$$h_{x1} = \gamma_x r_x \bar{X}_1 \beta_{x1},$$

$$h_{x2} = \gamma_x r_x \bar{X}_1 \beta_{x2},$$

$$h_{y1} = \gamma_y r_y \bar{Y}_1 \beta_{y1},$$

$$h_{y2} = \gamma_y r_y \bar{Y}_1 \beta_{y2},$$

$$\gamma_x = 1/(1 + \alpha_x \bar{X}_1 + \beta_{x1}\,\bar{Y}_1 + \beta_{x2}\,\bar{Y}_2)^2,$$

$$\gamma_y = 1/(1 + \alpha_y \bar{Y}_1 + \beta_{y1}\,\bar{X}_1 + \beta_{y2}\,\bar{Y}_2)^2,$$

By the theorem on stability of discrete models (see Appendix C), $\bar{\mathbf{Z}}$ is a stable equilibrium point (i.e., $\lambda < 1$) if, and only if, all the principal minors of the matrix,

$$\mathbf{SAS}^{-1} - \mathbf{I} = \begin{vmatrix} -1 & f_{x2} & 0 & 0 \\ s_x & -1 + p_{x2} & h_{x1} & h_{x2} \\ 0 & 0 & -1 & f_{y2} \\ h_{y1} & h_{y2} & s_y & -1 + p_{y2} \end{vmatrix}\tag{9.35}$$

are positive. Here **S** is the similarity transformation defined by

$$\mathbf{S} = \begin{vmatrix} 1 & 0 & 0 & 0 \\ 0 & 1 & 0 & 0 \\ 0 & 0 & -1 & 0 \\ 0 & 0 & 0 & -1 \end{vmatrix}.\tag{9.36}$$

It can be shown that this condition yields the following criteria:

$$1 > R_x,\tag{9.37}$$

$$1 > R_y,\tag{9.38}$$

and

$$(1-R_x)(1-R_y) > (h_{x1}f_{y2}+h_{x2})(h_{y1}f_{x2}+h_{y2}), \tag{9.39}$$

where $R_x \equiv p_{x2}+s_xf_{x2}$ and $R_y \equiv p_{y2}+s_yf_{y2}$.

Conditions (9.37) and (9.38) state that the rate of change, R_x and R_y, of the net reproductive rates of species X and Y near equilibrium cannot be large enough for the perturbed population density to move away from the equilibrium point. Condition (9.39) represents limitations on the strengths of the competitive interactions. The terms $(1-R_x)$ and $(1-R_y)$ can be thought of as the strengths of species self-interaction. The terms $(h_{x1}f_{y2}+h_{x2})$ and $(h_{y1}f_{x2}+h_{y2})$ can be interpreted as the effect of one species on the other at equilibrium. Stability about $\bar{Z}$, and hence coexistence, requires that the strength of self-interaction be stronger than that of interspecific competition, a result that is analogous to Eq. (9.25).

To understand the constraints that condition (9.39) imposes on the rate of change of the species net reproductive rates and competitive interactions strengths, we rewrite the inequality (9.39) as

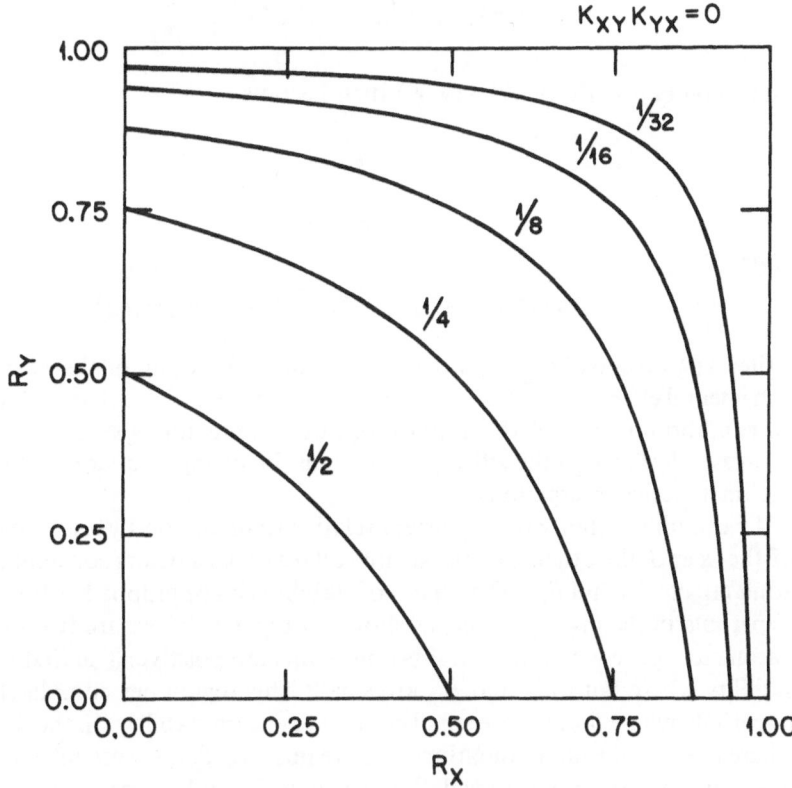

Fig. 9.8. A general stability diagram for models of the form of Eq. (9.27). If R_x and R_y lie below the appropriate $K_{xy}K_{yx}$ boundary, the equilibrium will be stable. (Reproduced from Travis et al., 1980, with permission of Academic Press)

$$(1-R_x)(1-R_y) > K_{xy}K_{yx}, \tag{9.40}$$

Where $K_{xy} \equiv h_{x1}f_{y2}+h_{x2}$ and $K_{yx} \equiv h_{y1}f_{x2}+h_{y2}$. This describes a region under the hyperbola in the positive quadrant of the R_x, R_y plane (Fig. 9.8). In this figure we see that, if $K_{xy}K_{yx}=0$, condition (9.39) is implied by conditions (9.37) and (9.38). As the strength of interspecific interaction, measured by the product $K_{xy}K_{yx}$, increases, the region of stability decreases, and R_x and R_y must be small to ensure stability ($\lambda < 1$). If $K_{xy}K_{yx} \geq 1$, there is no region of stability. The same conclusions hold for models of two competing species with any number of age classes. More age classes merely add more terms to R_x, R_y, K_{xy}, and K_{yx}. The interpretations remain the same. The stability criteria [Eqs. (9.37)—(9.39)] may be expressed as functions of the equation parameters by substitution of the equilibrium values given by Eqs. (9.30)—(9.33) for $\bar{X}_1, \bar{X}_2, \bar{Y}_1$, and $\bar{Y}_2$. The algebraic manipulations required to accomplish this task are simplified by the expressions

and

$$1 + \alpha_x \bar{X}_1 + \beta_{x1}\bar{Y}_1 + \beta_{x2}\bar{Y}_2 = \frac{r_x f_{x2}}{1-p_{x2}} \tag{9.41}$$

$$1 + \alpha_y \bar{Y}_1 + \beta_{y1}\bar{X}_2 + \beta_{y2}\bar{X}_2 = \frac{r_y f_{y2}}{1-p_{y2}}. \tag{9.42}$$

The stability criteria (9.37)—(9.39) then become

$$\alpha_x > 0, \tag{9.43}$$

$$\alpha_y > 0, \tag{9.44}$$

and

$$\alpha_x \alpha_y > (1/f_{x2}f_{y2})(f_{x2}\beta_{y1}+\beta_{y2})(f_{y2}\beta_{x1}+\beta_{x2}). \tag{9.45}$$

Criteria (9.43) and (9.44) show that, for stability, the young-of-the-year must have a detrimental effect on their own survival rate. This requires each population to reach an equilibrium, even in the absence of other interacting species. Equation (9.45) expresses the fact that the self-regulation must be strong compared to the strength of the interspecies interactions.

To examine stability of the system relative to predation by adults on the young-of-the-year of the other species, we hold all other parameters constant and vary the adult effects β_{x2} and β_{y2}. The region of stability can be graphed as the area under a hyperbola in the β_{x2}, β_{y2} plane, as shown in Fig. 9.9. There are two regions in this parameter space where the equilibrium points are positive. The first is a region of stability where both β_{x2} and β_{y2} are small. This region lies completely under the hyperbola where β_{x2}, β_{y2} are small enough to ensure stability. In the second region, where the equilibrium population sizes are positive, β_{x2}, β_{y2} are relatively large and inequality (9.45) is not satisfied. Similar analysis can be carried out for all pairs of interaction parameters with similar conclusions. This pattern of stable and unstable behavior shown in Fig. 9.9 is similar to results found by Hassell and Comins (1976) for a related model that we consider in the following section.

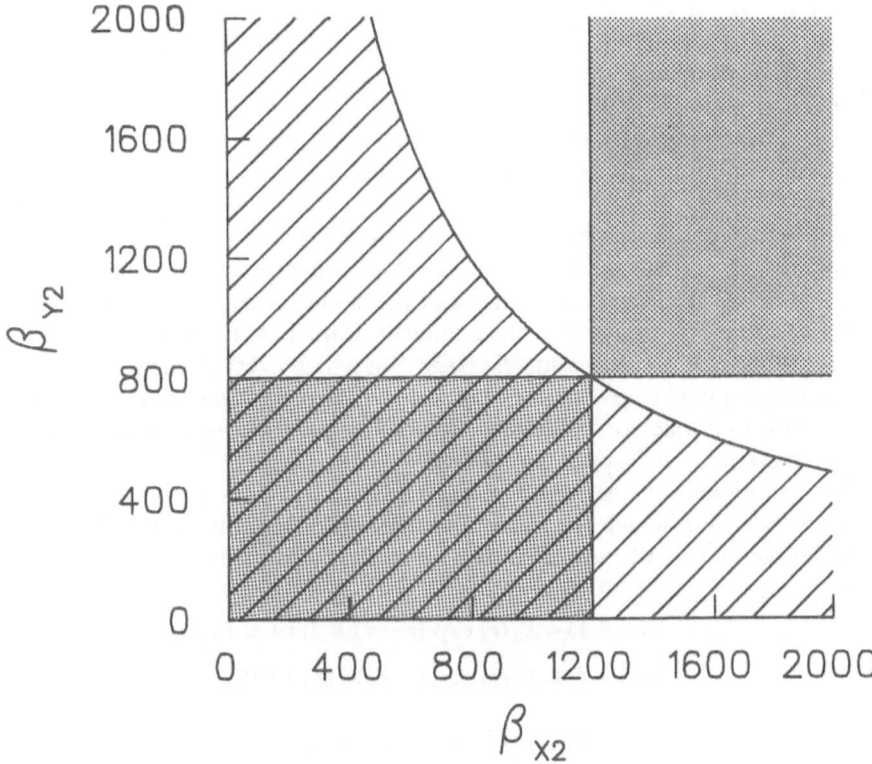

Fig. 9.9. Local stability of the equilibrium point $(\bar{X}_1, \bar{X}_2, \bar{Y}_1, \bar{Y}_2)$ of Eq. (9.26) for a variety of values of β_{y2} and β_{x2}. The other parameter values are $f_{x2}=20,000$, $r_x=0.3$, $r_y=0.3$, $p_{x2}=p_{y2}=0.99$, $\beta_{x1}=\beta_{y1}=0.0$, $\alpha_x=0.04$, $\alpha_y=0.06$. The hatched region under the hyperbola represents parameter values where Eq. (9.26) is stable. The stipled region is where the equilibrium populations are all positive. Only for values of β_{x2} and β_{y2} lying in the region which is both hatched and stipled will Eq. (9.26) have positive equilibria which are stable to all perturbations. (Reprinted from Travis, et al. 1980 with permission from Academic Press)

9.6 Other Compensatory Models

Hassell and Comins (1976) investigated a discrete-time model for two-species competition. Although they did not explicitly use the compensatory Leslie matrix formulation in their presentation, their model is another example of this class of models. They approached the problem of local stability in the traditional way by linearizing the equations about the equilibrium points and obtained characteristic equations from the linearized matrix. This procedure, however, resulted in analytic criteria for stability only in a very limited number of cases. The significance of the present work is that local stability conditions can easily be derived from the theory of positive matrices for two-species competition (or mutualism) models of any number of age-classes and complexity. These stability criteria can be derived as long as the elements of the **A**-matrix satisfy certain sign requirements. These sign requirements will be satisfied as a rule.

145

The derivation of analytic formulas for the stability of competing species in a Leslie matrix model should permit the effects of age structure on competitive interactions to be investigated. For example, one could compare inequality (9.25) with the inequalities (9.37), (9.38), and (9.39), although in this case the comparison is somewhat compromised by the fact that the former inequality is based on a continuous-time model while the latter equalities are based on a discrete-time model.

The assumptions made above regarding density-dependence of the matrix elements will not always hold in practice, because overcompensation and competition between different age-classes of the same species can occur. Violation of these assumptions means that the similarity transform (9.36) cannot convert the consequent matrix $\mathbf{A}$ into a matrix $\mathbf{SAS}^{-1}$ in which every element is nonnegative. However, in many cases, we may be able to find other similarity transformations that result in nonnegative matrices. In such cases, the results from the theory of positive matrices will apply completely.

For example, consider the two-age-class. two-species model of Hassell and Comins (1976) in which the intra- and inter-specific competition involves larvae with larvae and adults with adults. The model is given by

$$X_1(t+1) = X_2(t) \exp\left[r - a'(X_2(t) + \alpha' Y_2(t))\right]$$

$$X_2(t+1) = X_1(t) \exp\left[-a(X_1(t) + \alpha Y_1(t))\right]$$

$$\hspace{8cm} (9.46)$$

$$Y_1(t+1) = Y_2(t) \exp\left[s - c'(Y_2(t) + \beta' X_2(t))\right]$$

$$Y_2(t+1) = Y_1(t) \exp\left[-c(Y_1(t) + \beta X_1(t))\right].$$

Here X_1 and Y_1 represent the larval or juvenile stages, and X_2 and Y_2 the adults. The generations and stages are discrete and nonoverlapping, so that adults and larvae are not present at the same time. The time interval from t to $t+1$ represents the duration of the adult and juvenile stages. The parameters r and s are the intrinsic rates of increase of two species and the terms a, a', c, and c' define "*threshold*" densities above which competition becomes severe. The relevant stability matrix of the system linearized about the equilibrium $(\bar{X}_1, \bar{X}_2, \bar{Y}_1, \bar{Y}_2)$ is

$$\mathbf{A} = \begin{vmatrix} 0 & a_{12} & 0 & a_{14} \\ a_{21} & 0 & a_{23} & 0 \\ 0 & a_{32} & 0 & a_{34} \\ a_{41} & 0 & a_{43} & 0 \end{vmatrix},$$

where

$$a_{12} = (1 - a'\bar{X}_2) \exp\left[r - a'(\bar{X}_2 + \alpha'\bar{Y}_2)\right]$$

$$a_{14} = -a'\alpha\bar{X}_2 \exp\left[r - a'(\bar{X}_2 + \alpha'\bar{Y}_2)\right]$$

$$a_{21} = (1 - a\bar{X}_1) \exp\left[-a(\bar{X}_1 + \alpha'\bar{Y}_1)\right]$$

$$a_{23} = -a\alpha\bar{X}_1 \exp\left[-a(\bar{X}_1 + \alpha\bar{Y}_1)\right]$$

$$a_{32} = -c'\beta'\bar{Y}_2 \exp\left[s - c'(\bar{Y}_2 + \beta'\bar{X}_2)\right]$$

$$a_{34} = (1 - c'\bar{Y}_2) \exp\left[s - c'(\bar{Y}_2 + \beta'\bar{X}_2)\right]$$

$$a_{41} = -c\beta\bar{Y}_2 \exp\left[-c(\bar{Y}_1 + \beta\bar{X}_1)\right]$$

$$a_{43} = (1 - c\bar{Y}_1) \exp\left[-c(\bar{Y}_1 + \beta\bar{X}_1)\right].$$

When there is no density overcompensation (i.e., $a_{12}, a_{21}, a_{34}, a_{43} > 0$), then the sign pattern of the stability matrix is

$$\mathbf{A} = \begin{vmatrix} 0 & + & 0 & - \\ + & 0 & - & 0 \\ 0 & - & 0 & + \\ - & 0 & + & 0 \end{vmatrix}$$

and the similarity transformation defined using the matrix

$$\mathbf{S} = \mathbf{S}^{-1} = \begin{vmatrix} 1 & 0 & 0 & 0 \\ 0 & 1 & 0 & 0 \\ 0 & 0 & -1 & 0 \\ 0 & 0 & 0 & -1 \end{vmatrix}$$

will yield a nonnegative matrix $\mathbf{SAS}^{-1}$. If, on the other hand, a, a', c and c' are large enough to produce overcompensatory behavior in every age-class (i.e., $a_{12}, a_{21}, a_{34}, a_{43} < 0$), the sign pattern is

$$\mathbf{A} = \begin{vmatrix} 0 & - & 0 & - \\ - & 0 & - & 0 \\ 0 & - & 0 & - \\ - & 0 & - & 0 \end{vmatrix},$$

and the similarity transformation defined using the matrix

$$\mathbf{S} = \mathbf{S}^{-1} = \begin{vmatrix} 1 & 0 & 0 & 0 \\ 0 & -1 & 0 & 0 \\ 0 & 0 & 1 & 0 \\ 0 & 0 & 0 & -1 \end{vmatrix}$$

will yield a nonnegative matrix.

Even when our results cannot be applied to particular models by devising appropriate similarity transformations, they may provide at least limited information on the models. For example, Perkowski (1979) investigated two-age-class, two-species models where in addition to interspecific competition there was intraspecific competition in which adults competed with larvae (through cannibalism). He found that this cannibalism intensifies species self-regulation, increasing the probability of stable coexistence of the two species. Although Perkowski's model violates our matrix element assumption so that necessary and sufficient conditions cannot be prescribed for stability, sufficient conditions can be developed if one ignores cannibalism in the stability calculations.

Under certain circumstances the applications of our theory can be extended to more than two interacting species. For example, consider the case of three species, X, Y, and Z. If X and Y compete, Y and Z compete, and X and Z are either mutualistic or noninteracting, the resultant **A**-matrix has a form such that the results from positive matrix theory can be applied to any models having configurations of species in which 1) a friend of a friend is a friend, 2) an enemy of a friend is an enemy, and 3) an enemy of an enemy is a friend (Travis and Post, 1979).

9.7 Life-History Strategies

The probabilities of survival and reproductive success depend on the interactions between an organism and its environment. Natural selection acts to adjust these interactions in such a way as to maximize fitness. Medawar (1946, 1952) noted the close connection between natural selection and the age structure of a population. He pointed out that natural selection works most effectively on genes that exert their influence early in an individual's life. In this final section of the present chapter, we present a brief and elementary introduction to the evolution of life history parameters in an age-structured population. A more extensive treatment of the same subject has been given by Charlesworth (1980).

Any organism has limited resources of time and energy at its disposal. These resources must be divided between various life functions in such a way as to maximize fitness. For convenience, the various life-history functions are generally divided into three broad categories: growth, survival, and reproduction. Because the primary goal of a living organism is to maximize the representation of its progeny in future generations, reproduction must be considered the most important life-history function. The value of devoting resources to growth and survival can only be in their enhancement of reproduction at later stages in an individuals life history. For example, maintenance activities such as building tissue, competition for resources, defense of territories, and dispersal to new geographic locations can be expected to reduce age-specific mortality rates. Devoting greater resources to growth can be expected to result in a larger body size which in turn contributes to both inhanced survival and greater rates of reproduction.

The theory of life history tactics is concerned with the identification of those age-specific rates of reproduction and risks of death that maximize the contribution of progeny to future generations under a given set of environmental conditions. Many factors are correlated with reproductive and survival rates. Variation in age at maturity, number and size of young, reproductive effort, and degree of parental care all affect the contribution of progeny to future generations. Other non-specific factors, such as density of the population in relation to resources, trophic and successional position of the population, and predictability of the environment also appear to be important in the selection of life history strategies (Wilbur et al., 1974; Boyce, 1979).

There are many indications that the various life functions must compete for the limited resources of time and energy. As a specific example of the costs and benefits involved in the division of energy between various life-functions consider the presence of secondary substances in plants. Secondary substances are chemical compounds that, although ubiquitous among plants, have no known metabolic function. Although there is some debate regarding the adaptive significance of secondary compounds (Seigler and Price, 1976), they are generally believed to serve an antiherbivore function. However, plants must pay metabolic cost for the production of these chemical defenses. These costs usually manifest themselves in measurable differences in growth rates or reproductive output. Cates (1975) found that the palatable form of *Asarum caudatum* (Aristolochiaceae), which lacked secondary compounds, had higher growth rates and produced more seeds than did a polymorphic form of that crop species with a chemical-defense system. It is well known that crop plants selected for high yield are often more susceptible to insects and disease. Tester (1977) showed that in the absence of herbivores, soybeans that are insect-resistant produced fewer seeds than ones susceptible to insects.

Insight into another aspect of defense investment is provided by a study by Rehr et al. (1973). They found that defense-chemicals present in non-ant-acacias, were absent from acacia species protected by obligate mutualism (see Chap. 8). They interpreted this to mean that the added metabolic costs of producing secondary products in acacias protected by ants provided selective pressure for their loss.

9.8 Intrinsic Rate of Increase

The intrinsic rate of increase, r, of a population is obviously dependent on the age-specific rates of reproduction and survival. Lotka was the first to derive an analytic expression relating these variables (Lotka, 1925). This equation, known as Lotka's equation or the characteristic equation, has been central to most subsequent theoretical work concerning life history tactics. Let p_x denote the probability of surviving to age x and f_x the instantaneous birth-rate at age x. Assume a stable age distribution and consider the number $N_0(t)$ of individuals born at a given time t. To be newborn at time t, an individual must have had parents who where born at time $t - k$ and survived to age k and given birth. Thus,

$$N_0(t) = \sum_{k=0}^{n} N_0(t-k)p_k f_k. \tag{9.47}$$

If we assume that the number of newborn progeny grows exponentially with time, that is, $N_0(t) = k \exp(rt)$ then

$$k \exp(rt) = \sum_{k=0}^{n} k \exp[r(t-k)]p_k f_k \qquad (9.48)$$

or

$$1 = \sum_{k=0}^{n} \exp(-rk)p_k f_k. \qquad (9.49)$$

It has been generally assumed in the literature (Gadgil and Bossert, 1970; Fagen, 1972; Goodman, 1974; Schaffer, 1974; Charlesworth and Leon, 1976; Michod, 1979) that an optimal life history strategy is one that maximizes the intrinsic rate of increase, r, of the population. Under this interpretation, the intrinsic rate of increase can be a convenient measure of individual fitness. Charlesworth (1973) has provided some justification for this assumption. He showed that in an age-structured diploid, random-mating population, any nonrecessive mutant that decreases r will be eliminated by selection, while a mutant that increases r will be favored.

9.9 Reproductive Strategies

Maynard Smith (Maynard Smith and Price, 1976; Maynard Smith, 1976) introduced the concept of an evolutionary stable strategy (ESS) as a tool for determining which phenotypes, within a range of possible phenotypes, will become established in a population by natural selection. An evolutionarily stable strategy is a phenotype that, when common, eliminates rare mutant phenotypes. As an example of an application of this concept to age-structured populations, we consider selection of age-specific reproductive efforts, $E_i, 1 \leq i \leq n$. Reproductive effort is visualized as the fraction of total resources of time and energy available to an individual at age i that is devoted to reproduction. It is assumed to take values between 0 and 1. The fecundity at age i, $f_i(E_i)$, is assumed to be an increasing function of reproductive effort E_i, while the probability of survival at age i, $p_i(E_i)$, is a decreasing function of E_i.

Assumptions about the exact form of the dependence of fecundity and probability of survival on reproductive effort appear to be crucial. Four general forms of dependence have been discussed in the literature (Gadgil and Bossert, 1970; Schaffer, 1974; Leon, 1976). For fecundity these are 1) linear; 2) concave; 3) convex; and 4) sigmoidal (see Fig. 9.10). The same four categories can be defined for probability of survival, except that the graphs will be decreasing functions of reproductive effort, E. At the present time, no empirical measurements of the actual relationship between fecundity (or survival) and reproductive effort have been made (Stearns, 1976). There is, however, indirect evidence (Tinkle et al., 1970; Gadgil and Solbrig, 1972; and Schaffer and Tamarin, 1973) that under various environmental conditions each of the curves shown in Fig. 9.10 can actually be realized.

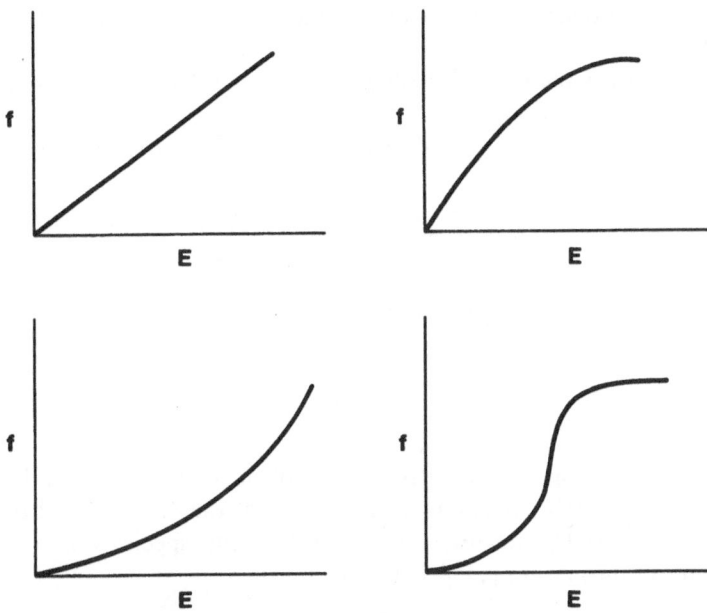

Fig. 9.10. Possible relationships between fecundity and reproductive effort

As we have seen before, the intrinsic rate of increase r, can be defined by the equation

$$1 = \sum_{k=0}^{n} \exp(-rk) p_k(E_k, E_{k-1}, \ldots, E_0) f_k(E_k). \tag{9.50}$$

(Here we have made explicit the assumption that p_k and f_k are functions of the reproductive efforts E_i.) An optimal life history strategy, in the context of the present example, would be a particular choice of age-specific reproductive efforts $E_i, 1 \leq i \leq n$, that maximizes r. However, suppose that $p_i(D, E_i)$ and $f_i(D, E_i)$ are functions of both density and reproductive effort. The Eq. (9.50) becomes

$$1 = \sum_{k=0}^{n} \exp(-rk) p_k(D, E_k, E_{k-1}, \ldots, E_0) f_k(D, E_k). \tag{9.51}$$

Let us further assume that for a given phenotype $\bar{\mathbf{E}} = (\bar{E}_0, \bar{E}_1, \ldots, \bar{E}_{n-1}, 1)$ the population is in an ecological equilibrium with density $\bar{\mathbf{D}} = (\bar{D}_0, \ldots, \bar{D}_n)$. Note that reproductive effort during the last age class is total, i.e., $\bar{E}_n = 1$. Because the population is in equilibrium, the intrinsic rate of increase r is equal to zero and Eq. (9.51) becomes $R(\bar{E}, \bar{D}) = 1$ where $R = \sum p_k(\bar{\mathbf{D}}, \bar{\mathbf{E}}) f_k(\bar{\mathbf{D}}, \bar{\mathbf{E}})$ is the net reproductive rate of the population. Now for the phenotype $\bar{\mathbf{E}}$ to be an evolutionarily stable strategy, and mutant phenotype $\hat{E}$ must be at a selective disadvantage when rare. Thus $R(\hat{\mathbf{E}}, \bar{\mathbf{D}}) < R(\bar{\mathbf{E}}, \bar{\mathbf{D}})$ for any $\hat{E}$ is a necessary and sufficient condition for $\bar{\mathbf{E}}$ to be an evolutionarily stable strategy of a density dependent population at equilibrium $\bar{D}$.

Consider the case of a single species with two age classes governed by

$$Z_1(t+1) = f_1(E_1)Z_1(t) + f_2 Z_2(t)$$

$$Z_2(t+1) = \frac{p_1(E_1)Z_1(t)}{1+\alpha Z_1(t)}.$$

(9.52)

We wish to find a reproductive effort $\bar{E}_1$ which maximizes the net reproductive rate

$$R(\bar{E}_1) = f_1(\bar{E}_1) + \frac{p_1(\bar{E}_1)f_2}{1+\alpha Z_1}.$$

(9.53)

Note that reproductive effort during the last age class is total (i.e., $E_2 = 1$). Clearly the choice of $\bar{E}_1$ depends on the shape of the curve f_1 and p_1, which in turn are determined by the interaction between the species and the environment. The equilibrium $\bar{Z} = (\bar{Z}_1, \bar{Z}_2)$ is given by

$$\bar{Z}_1 = \frac{f_1(\bar{E}_1) + p_1(\bar{E}_1)f_2 - 1}{\alpha(1 - f_1(\bar{E}_1))}$$

$$\bar{Z}_2 = \frac{f_1(\bar{E}_1) + p_1(\bar{E}_1)f_2 - 1}{\alpha f_2}.$$

(9.54)

If the equilibrium is to be positive, we must have $f_1(\bar{E}_1) + p_1(\bar{E}_1)f_2 > 1$ and $1 > f_1(\bar{E}_1)$. At this equilibrium,

$$R(\bar{E}_1, \bar{Z}) = f_1(\bar{E}_1) + \frac{p_1(\bar{E}_1)f_2}{1+\alpha\bar{Z}_1} \equiv 1.$$

(9.55)

This equation just states the obvious fact that at ecological equilibrium, the net reproductive rate is equal to unity since total population size is neither increasing nor decreasing. For the phenotype $\bar{E}_1$ to be an ESS, any mutant phenotype $\hat{E}_1$ must be at a selective disadvantage when rare (Michod, 1979). This can be expressed as $R(\hat{E}_1, \bar{Z}) < R(\bar{E}_1, \bar{Z})$. In order to be more specific, assume that the fecundity $f_1(E_1)$ increases linearly with reproductive effort and that the probability of survival $p_1(E_1)$ decreases linearly with reproductive effort. That is, assume

$$f_1(E_1) = f_1 E_1$$

$$p_1(E_1) = p_1(1 - E_1).$$

Then

$$R(E_1, \bar{Z}) = f_1 E_1 + \frac{p_1 f_2(1 - E_1)}{1+\alpha\bar{Z}_1}.$$

(9.56)

We now show that under these assumptions $\bar{E}_1 = 0$ is an ESS. First notice that at $\bar{E}_1 = 0$, the equilibrium population size $\bar{Z}_1$ satisfies the relationship $1 + \alpha \bar{Z}_1 = p_1 f_2$. Thus, if $\hat{E}_1$ is any other reproductive strategy satisfying $0 < \hat{E}_1 \leq 1$, then

$$R(\hat{E}_1, \bar{Z}) = f_1 \hat{E}_1 + (1 - \hat{E}_1)$$
$$= (f_1 - 1)\hat{E}_1 + 1$$
$$< 1 \equiv R(0, \bar{Z}).$$

The last inequality follows since $f_1 < 1$ by assumption. This establishes that $\bar{E}_1 = 0$ is an evolutionary stable strategy. That is, a species whose growth is governed by Eq. (9.52) should expend no reproductive effort during its first age class, thus ensuring maximal survival to the second age class. During the second age class it will expend maximum reproductive effort.

We now derive this result in a different, but equivalent, manner. Assume that a species exists whose growth is governed by the equations

$$Z_1(t+1) = f_2 Z_2(t)$$
$$Z_2(t+1) = \frac{p_1 Z_1(t)}{1 + \alpha Z_1(t)}.$$
(9.57)

That is, the species expends no effort in reproduction during its first age class and the probability of survival to the second age class is density dependent. Suppose also that the population density of the species is at the ecological equilibrium

$$\bar{Z}_1 = (p_1 f_1 - 1)/\alpha$$
$$\bar{Z}_2 = (p_1 f_2 - 1)/\alpha f_2.$$
(9.58)

Now assume a mutant phenotype of the species appears which devotes a nonzero-fraction of its energy to reproductive effort E_1, $0 < E_1 < 1$, during the first age class. Also assume, as before, that fecundity increases linearly with reproductive effort and the probability of survival decreases linearly with reproductive effort. Let the population density of the new phenotype be denoted by $Y(t) = (Y_1(t), Y_2(t))$. Then the equations describing the population dynamics of the two phenotypes are

$$Z_1(t+1) = f_2 Z_2(t)$$
$$Z_2(t+1) = \frac{p_1 Z_1(t)}{1 + \alpha(Z_1(t) + Y_1(t))}$$
$$Y_1(t+1) = f_1 E_1 Y_1(t) + f_2 Y_2(t)$$
$$Y_2(t+1) = \frac{p_1(1 - E_1)Y_1(t)}{1 + \alpha(Z_1(t) + Y_1(t))}.$$
(9.59)

153

The question of whether or not the mutant phenotype can become established depends on the local stability properties of the equilibrium $(\bar{Z}_1, \bar{Z}_2, 0, 0)$. If this equilibrium point is locally stable, then the mutant phenotype cannot become established when it is rare. Linearizing Eq. (9.59) about the equilibrium $(\bar{Z}_1, \bar{Z}_2, 0, 0)$ and applying Eq. (9.58) yields the matrix

$$
\mathbf{A} = \begin{vmatrix}
0 & f_2 & 0 & 0 \\
\dfrac{1}{p_1 f_1^2} & 0 & \dfrac{1 - p_1 f_2}{p_1 f_1^2} & 0 \\
0 & 0 & f_1 E_1 & f_2 \\
0 & 0 & \dfrac{1 - E_1}{f_1} & 0
\end{vmatrix}.
$$

It is easily checked that $(-1)^4 \det (\mathbf{A} - \mathbf{I}) = (1 - 1/p_1 f_1)(1 - f_1) E_1$ if $f_2 = f_1$. This expression is positive since $p_1 f_1 > 1$ and $f_1 < 1$, which are necessary to ensure existence of a positive equilibrium. Thus, the equilibrium $(\bar{Z}_1, \bar{Z}_2, 0, 0)$ is stable, establishing that $E_1 = 0$ is an evolutionary stable strategy.

The above result is similar to one obtained by Cohen (1971) from analyzing growth and seed production in annual plants. A plant must divide its total energy between vegetative growth and seed production in such a way as to maximize the total seed yield. Cohen found that to maximize seed yield, a plant should have a sharp transition from 100 % vegetative growth to 100 % reproductive growth at some fixed switching time during the year. Vincent and Pulliam (1980) established the same result and showed that the optimal switching time is a function of both leaf mass and time remaining to the end of the growing season. The actual details of when the switch takes place, of course, depends on the details of the model used.

Cohen (1971) points out that several crop plants such as wheat, barley and maize do show a sharp transition between their vegetative and reproductive phases of growth. However, this is not universal. As Cohen notes "...*a very common pattern is an initial purely vegetative growth, followed by an extended period of mixed vegetative and reproductive growth, followed finally by a period of purely reproductive growth.*" An explanation of the discrepancy between predicted and actual growth patterns is that the model used to derive this result does not take into account some critical constraint on the way the plant can allocate its energy/resources. For example, it is plausible that some plants produce flowers and fruits only in the axils of normal leaves, so that vegetative growth and seed production must overlap.

9.10 Summary and Conclusions

Life history parameters (e.g., fecundity, mortality, interspecific competition) are age dependent. Analysis of the structure of a population should take into account the age dependent nature of these parameters. In simple terms, an age-structured population is a positive feedback system between the young and the adults of the population. The age structure in a population allows the positive feedback to be

154

delayed in time. After adults produce offspring it may be years before the young mature, enter the adult age class, and complete the positive feedback cycle by producing young themselves. Much of the diversity in natural ecosystems results from this delay in positive feedback, since the ecological niche of an age-structured population can have a temporal as well as a spatial component.

The classic Leslie matrix approach to modeling age-structured populations implies, except in the exceptional case of equilibrium, that the population decreases toward extinction or increases exponentially without bound. This model has been modified to incorporate a decrease in survival and fecundity when populations increase in density and an increase in survival and fecundity as population density declines. These modifications allow for age-dependent variations in the amount of positive feedback between the young and adults of the population to maintain stability.

Any organism has limited resources of time and energy at its disposal which must be divided between various life functions in such a way as to maximize fitness. The age structure of the population controls the amount of positive feedback resulting from any particular allocation of resources among the age classes. The theory of life history strategies is concerned with the identification of those age specific rates of reproduction and risk of death that maximize fitness. Analysis of positive feedback influences on life history patterns allow identification of the best strategies for maximizing fitness.

10. Spatially Heterogeneous Systems: Islands and Patchy Regions

Perhaps the single greatest discrepancy between traditional mathematical models of population dynamics and populations in the real world is the customary neglect of spatial heterogeneity in the models. Levin (1979) has remarked that, while the statistical description of spatial pattern in ecosystems is relatively well developed (s. especially Pielou, 1977), the description of the dynamics giving rise to such spatial patterns is not. All ecological systems exist on landscapes (or seascapes) of varying complexity, and the dynamics of populations and processes cannot be divorced from these spatial contexts, but mathematical models of spatially extended ecosystems are difficult and the numerical evaluation of the behavior of even only moderately realistic models is far beyond the capabilities of the fastest computers.

Nevertheless, attempts are being made to explore some of the implications of spatial extent of ecosystems through the use of simple models. Among the milestones of this burgeoning field of study are papers by Skellam (1951), Horn and MacArthur (1972), Levin (1974), Roff (1974), and a recent book by Okubo (1980).

The existence of spatial dimensions has two principal effects on populations. First, since the physical environment varies from point to point in space, rates of population growth, mortality, and interspecific interactions also vary, and, as a consequence, population density varies through space. Second, a portion of a population located at some point in space, A, affects a portion of the population at another point, B, only through movement of population members or transmission of signals through space. Therefore, the feedbacks of populations on themselves, directly or indirectly through interactions with other populations, are delayed and modified by the intervening space. In particular, the feedback networks of the competitive/mutualistic systems considered in Chap. 8 will become enormously complex because of their distribution through space.

The consequence of spatial distribution on communities of competing or mutualistic species will be the subject of this chapter. As will be seen, the effect of space in the models is often simply to add positive feedback loops. Therefore, the matrix methods introduced in earlier chapters can be used to assess the persistence and stability of such systems.

10.1 Classical Theory of Island Biogeography

The quantitative description of population dynamics involving heterogeneity is difficult. Progress has only been possible when systems that lend themselves to extreme simplification were studied. This is one reason for the success of the theory of island biogeography, as developed initially by Preston (1962) and MacArthur and Wilson (1963, 1967), which allows a great number of observations concerning island biota to be described by a small number of empirical relationships.

In simplest form, the theory of island biogeography is an attempt to describe two observed empirical relationships involving biota and physical space;

1) the relationship between island area and the number of species, and
2) the relationship between species number and the distance of an island from mainland sources of colonization.

The first relationship involves island area, A, and the steady-state number of species on islands, S. Given a number of islands of different sizes, but generally similar in other respects, such as distance from the mainland, the steady-state number of species (usually a group of closely related species is considered) on the islands tends to follow the equation

$$S = S_0 A^z \tag{10.1}$$

(Preston, 1962; MacArthur and Wilson, 1963, 1967). In Eq. (10.1) S_0 is a constant that depends on the species group being considered and z is a dimensionless parameter typically in the range 0.18—0.35 (Diamond and May, 1976).

The other principal relation is between the distance of islands from the mainland and the steady-state species number. Islands far from the mainland, but equal in other respects to islands closer to the mainland, usually have lower species abundances.

Simple empirical relationships describe the relevant data surprisingly well. In Fig. 10.1 the number of species is plotted as a function of area for islands that are otherwise similar. On this semi-log scale the line $ln(S) = ln(S_0) + zln(A)$ fits the data well for $S_0 = 18.9$ and $z = 0.18$. In Fig. 10.2 numbers of species on islands of similar size but a different distances, d, from the mainland are plotted (expressed as the ratio of species on given islands to the numbers on islands near the mainland), showing that

$$S/S_0 = 1 - \alpha d, \tag{10.2}$$

where S_0 (the number of species on an island close to the mainland) and α are constant parameter values, gives a good fit.

The idea that forms the basis of the theory of island biogeography is that the steady-state number of species on a given island is a result of a dynamic balance between extinction and colonization. If I is the colonization rate for a particular island and λS is the extinction rate (normally, as is done here, the extinction rate is assumed to be linear in species number, S, with λ being the extinction rate coefficient), then the rate of change of the number of species through time is

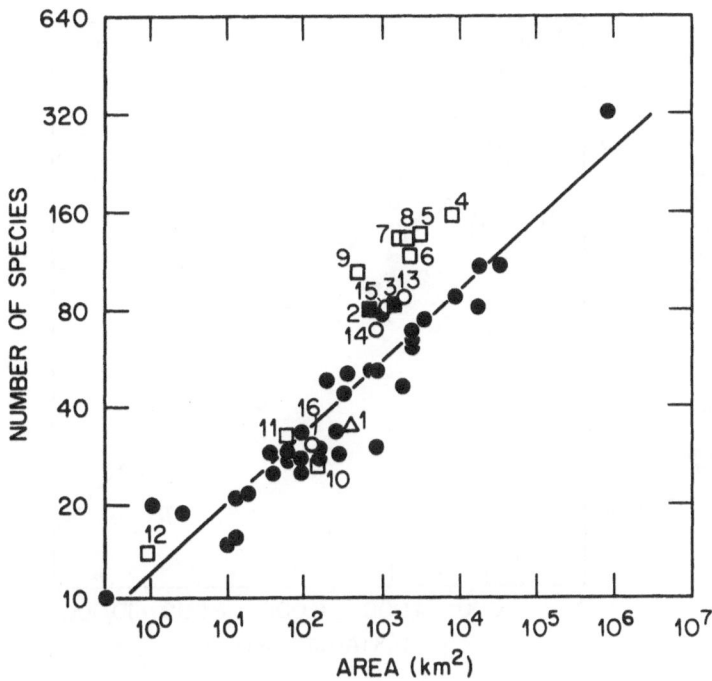

Fig. 10.1. Number of resident, non-marine, lowland bird species on New Guinea satellite islands as a function of island area. (Reproduced from Diamond, 1974, with permission of the author)

$$\frac{dS}{dt} = I - \lambda S, \qquad (10.3)$$

and the number of species in steady-state is,

$$\bar{S} = I/\lambda. \qquad (10.4)$$

From Eq. (10.4) the relationships between steady-state number and both island size and island distance from the mainland can be obtained, provided some of the underlying mechanics affecting I and λ are known.

Wilcox (1980) listed three main factors relating island area, A, to the extinction rate, λ. First there is the "*sample effect*", the well-known phenomenon that the larger the area that is sampled for species, the greater the number of species that will be found. Second, there are short-term insularization effects that occur immediately following the isolation of an island from the mainland. These account for the loss to sampling of species that were permanent residents of neighboring habitats but would occasionally visit the land area before it became detached from the mainland. Third, there are long-term insularization effects. These are effects that cause the gradual attrition of permanent species from the island. They may involve both slow elimination of species through competition or from stochastic fluctuations in

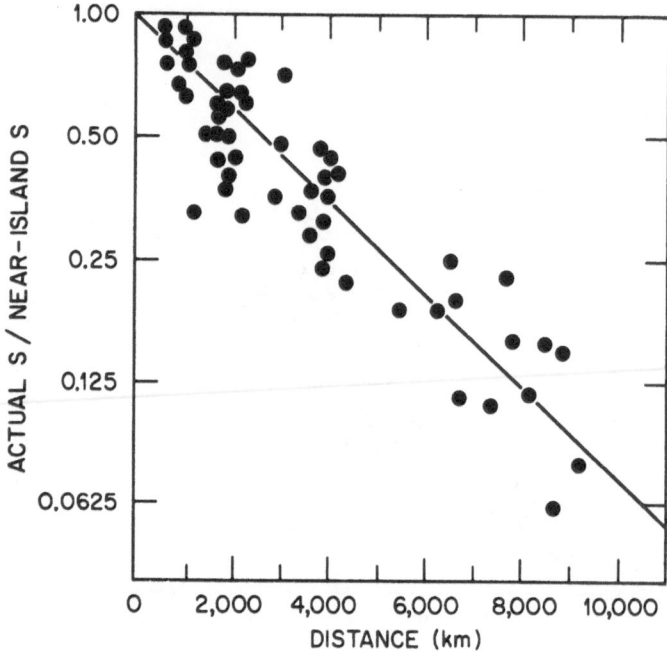

Fig. 10.2. Example of the linear relation between the number of species on islands of similar size and distance from colonization source. (Reproduced from Diamond, 1974, with permission of the author)

environmental conditions that eventually cause extinction of small populations. All three of these factors listed by Wilcox probably operate on islands, though in different proportions depending on particular circumstances.

The inverse relationships between species number and distance of islands from a source is anticipated from Eq. (10.4), since the colonization rate, I, depends inversely on distance. The colonization rate will also depend on the vagility of the species in question and on the degree of hostility of the environment surrounding the island (as discussed in the next section, "*island*" may have a more general meaning here than its normal one of land surrounded by water). Vagility also affects the area-species number relationship [Eq. (10.1)]. It has been found that z values are lower for species of higher vagility, which stems from the fact that sampling area size effects are more important than the other two factors in such cases (Wilcox, 1980).

Even deviations from the above simple relationships are explainable from theory. As Terborgh (1974a) has pointed out, the numbers of land bird species inhabiting islands that were part of a land bridge during the Pleistocene are often greater than the numbers on equivalent oceanic islands. Hence, these data do not lie on the species-area curve of Fig. 10.1. The reason for the discrepancy is that these land-bridge islands, by virtue of their recent connection with the mainland, initially had greater numbers of species than equivalent oceanic islands. As time passes, species should disappear until the land-bridge islands reach the expected equilibrium. Terborgh wrote an equation,

160

$$\frac{dS}{dt} = -kS^2, \tag{10.5}$$

to represent the rate of species extinction on supersaturated islands. He showed the parameter k to be appropriately $k \sim 10^{-4.9} A^{-0.5}$. Terborgh used this relationship to predict the number of bird extinctions on Barro Colorado Island, a Panamanian hill that became an island in 1914 when Lake Gatun was formed. His estimate of approximately 16.6 extinctions is consistent with observational estimates of 13 to 18 extinctions.

Elaborations on the basic theory of island biogeography have sought to elucidate some puzzling observations that are not explicable in terms of the mechanisms discussed so far. In particular, many species show a surprising patchiness in their occurrences. Diamond (1975, 1980) pointed out that many tropical bird and bat species may inhabit very disjunct habitat islands, being altogether absent from apparently suitable areas of habitat in between them. For example, the Tongan fruit bat *(Pteropus tonganus)* is widespread on tropical Pacific islands from Samoa to New Caledonia, and then occurs 1,050 km to the northwest on a single island (Rennell) and again 1,700 km further to the northwest on another single island (Karkar, near New Guinea). One possible explanation given by Diamond for the checkerboard pattern of this and other species is competitive exclusion, or the *"lockout effect"*. Mathematical description of the lockout effect is not possible from a theory based on Eq. (10.3) but can be attempted using the detailed models discussed later in the present chapter.

The theory of island biogeography had great initial success to the extent that it was accepted by much of the scientific community as a *"paradigm"* (Simberloff, 1976). In recent years, however, there have been second thoughts (even by some of its early practitioners, including Simberloff) on whether some of this success may have been more illusory than real. We shall not attempt to review the criticism, which is done in detail by F.S. Gilbert (1980).

Another problem with the traditional theory stems from its assumption of a background *"source"* of species, which makes the theory unsuitable for cases where there is no limitless source for colonization, but where a set of islands, or an *"island cluster"*, may be the only abode of at least some of the species. The essence of the island biogeography theory approach may illustrated by sketching the interactions represented by Eq. (10.3), as shown in Fig. 10.3. The mainland source provides a steady stream of colonists to the island, a positive influence. Since there is gradual extinction of species on isolated islands, a negative feedback loop of unit length operates from the island back on itself. The species on the island do not create a positive feedback to the source since the latter is assumed to be a constant, essentially infinite, reservoir of colonists. Hence, this model is an open loop system.

To alleviate the unreality (for some cases) inherent in the assumption of an infinite source, we have attempted an approach, described below, that does not make this assumption. The mathematics for such an approach is vastly more complicated, and only because the models fall into the category of positive linear systems can we make headway in the analysis of even simple systems. Attempts to predict numbers of species on islands or clusters of varying configurations is beyond

161

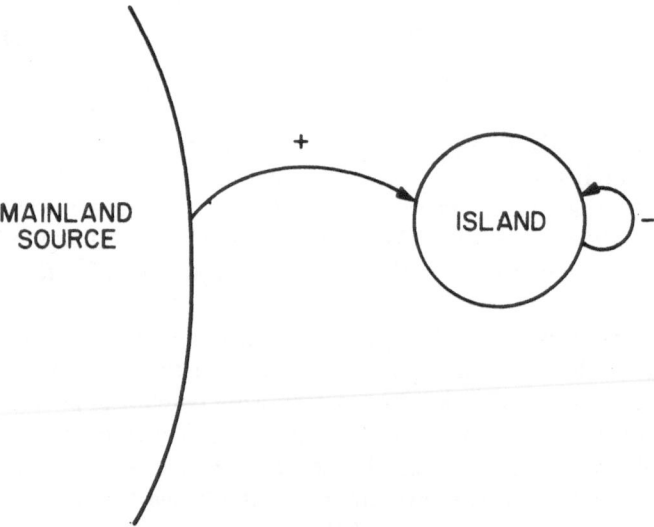

Fig. 10.3. Systems diagram of island biography theory model of the steady-state number of species on an island. Note that the model is an open loop system

the present capacity of our methods, but some progress is possible in determining whether or not a particular species or small group of species can persist stably on given clusters.

10.2 Island Clusters

Figure 10.3 underscores the simplifying assumptions of an infinite reservoir for a species source that traditional island biogeography theory makes. Many islands are far from any mainland sources of colonists, and they may normally receive colonists only from other, nearby, islands which themselves have small and vulnerable populations. The question can be raised as to whether or not traditional island biogeography can be applied in such cases. Studies of archipelagos (Hamilton et al., 1964; Johnson and Raven, 1973) indicate that, as the degree of isolation of an island from other islands increases, species abundance decreases. Now, however, the size of the nearest neighboring island or islands will itself be an important factor.

In later sections, where we look at nature reserves, it will be our assumption that it is unlikely for there to be infinite external sources continually replenishing the species on these reserves. In fact, a given reserve or spatial cluster of reserves may be the total habitat of certain species. Classical theory will not apply in such cases. The infinite reservoir of species in Fig. 10.3 must be replaced by one or more finite sources acting through positive feedback.

Consider the two-island case, and assume that S_1 and S_2 are the numbers of species on islands 1 and 2, respectively, at a given moment. The rate at which the number of species on each of the islands is changing will be equal to the difference between the rate at which new species are invading from the other island and their rate of extinction on the given island. The rate at which new species are invading, say, island 1 from island 2 is proportional to the number of species present on island 2 that are not simultaneously present on island 1. The analog is true for the number of species invading island 2. A set of equations based on these assumptions can be written for the two islands:

$$\frac{dS_1}{dt} = \{S_2 - J(S_1, S_2)\} I_{12} - \lambda_1 S_1, \tag{10.6a}$$

$$\frac{dS_2}{dt} = \{S_1 - J(S_1, S_2)\} I_{21} - \lambda_2 S_2, \tag{10.6b}$$

where I_{ij} is a rate coefficient of colonization of island j by island i, and $J(S_1, S_2)$ is the expected number of species present simultaneously on both islands. When island 1, say, contains at least all of the species that island 2 does, then $J(S_1, S_2) = S_2$. Then the term representing invasion of new species to island 1 goes to zero.

This approach may be useful for clusters of any number of islands if some rough idea of the form of $J(S_1, S_2)$ is known. Note that there can be no equilibrium of Eqs. (10.6a, b) if there are no new species being added to the system through time. This is because the normal tendency towards evenness of species between the two islands causes $S_1 - J(S_1, S_2)$ and $S_2 - J(S_1, S_2)$ to approach zero. Once the species on the two islands becomes identical, then there can be no further increase in S_1 or S_2 until there is first an extinction of at least one species, due to the $\lambda_i S_i$ term, which causes the species distributions to become uneven again. Hence, the total number of species, $S_T = S_1 + S_2 - J(S_1, S_2)$, is always decreasing, or at best staying the same. If, however, evolutionary time scales are considered, so that new species can be added through time, as the result of speciation, then positive terms, representing species creation, must be added to Eqs. (10.6). Only in this case can the number of species stabilize or increase through time. The stability conditions for positive linear matrices discussed in Chap. 2 should be useful in analyzing multi-island generalizations of the above model, though we shall not consider this model further here.

10.3 Insular Reserves

The theory of island biogeography has not been limited to oceanic or even necessarily "*water-surrounded*" islands, nor has it remained exclusively in the domain of academic interest. As Wilcox (1980) pointed out, "*One of the most profound developments in the application of ecology to biological conservation has been the recognition that virtually all natural habitats or reserves are destined to resemble islands, in that they will eventually become small isolated fragments of formerly much larger continuous habitat.*"

Almost every type of habitat, whether mountaintop, swamp, lake, river, etc., can be analogous, for certain species, to an oceanic island if it is surrounded by an environment of a type that is different enough to restrict the immigration and emigration of some of its species. The analogy is not total, because, for example, even when a habitat island such as a woodlot is surrounded by land of a different type, this surrounding land can harbor competing organisms, which surrounding water is less likely to do. Still, surrounding alien habitat of various types can be quite effective in isolating populations. Some bird species, for example, seem psychologically incapable of flying over such seemingly modest barriers as rivers, and can be effectively isolated to small regions.

Wilcox (1980) has suggested the term "*insular*", rather than "*island*", ecology be used to describe this more general class of phenomena involving isolated habitats. It may be expected, however, that some of the results of classical island biogeography theory may apply to terrestrial habitat islands. Vuilleumier (1970) showed that bird species living in habitat islands above the tree line in the northern Andes obey classical relationships between numbers of species, island size, and distance from large sources. Picton (1979) has found a strong relationship between the occurrence of large mammals and the size of their habitat areas in mountain areas of Montana.

Similar relationships have been found for fish in lakes (Barbour and Brown, 1974), birds in savanna patches (Schadde and Calaby, 1972; Schadde and Hitchcock, 1972), and many other cases.

Diamond (1975) used island biogeography to draw some conclusions concerning the design of nature reserves. Briefly stated, these are:

1) The area-species number relation implies that if 90 % of the area occupied by a habitat is converted by man into another habitat and the remaining 10 % is saved as an undivided reserve, one might expect to save roughly half of the species restricted to the preserved habitat type.
2) If one saves two reserves, the smaller reserve will retain fewer species if it is remote from the larger reserve than it would if it were near the larger reserve.
3) As the contrast between the preserved habitat type and the surrounding habitat type increases, the results of island biogeography become increasingly relevant.

The conclusions drawn from the theory of island biogeography to prescribe reserve sizes have not been free from criticism. Simberloff and Abele (1976) have criticized the idea, derived from island biogeography, that refuges should always consist of the largest possible area — as opposed to a group of smaller reserves with the same total area. A greater number of smaller reserves may actually increase the number of preserved species by providing separate refuges for different competitors, protecting some species from competitive exclusion. Also, a series of small reserves have less chance of being affected by a catastrophe such as a disease epibiotic, than does a single large reserve. Diamond (1976) pointed out that species have different characteristics, and that some, especially large vertebrates, may be more desirable to preserve than other species. Island biogeography theory, which treats all species equally, will not take account of these differences.

An important complicating factor involves the degree of dispersal that may occur between separate insular reserves. If the intervening territory is not too hostile, and if the inter-island distances are small, many species will be able to disperse from one habitat island to others. This has the advantage that islands may be recolonized following local population extinctions (Diamond, 1975). It may also have some of the disadvantages of one large reserve, the possibility of competitive exclusion and of spread of disease. Relative advantages of varying degrees of inter-island dispersal will be considered in the models analyzed later.

Another complicating factor in the design of nature reserves is the possible need for successional complexity within each reserve. Many animals depend, during their life cycles, on the resources provided by a variety of different vegetational successional stages. Reserves must be large enough, and subjected to enough disturbance, to provide a continuous variety of such stages (Kushlan, 1979; L.E. Gilbert, 1980).

Simple adaptations of island biogeography theory may not be adequate for the complexity of real systems of nature reserves. As Pickett and Thompson (1978) point out, "*for reserves the rich, extensive source areas do not, or will not, continue to exist (Meijer, 1973; Diamond, 1976; Terborgh, 1976). The reserves themselves must perform most if not all of this function as increasingly more land in all biomes is developed or disrupted. Because of this, the immigration rate, which is so important in maintaining equilibrium on true islands, is likely to decline significantly (Terborgh, 1974a; Willis, 1974; Diamond, 1976). Extinction will then become the dominant population process affecting equilibrium in reserves and species numbers will decline to a new level. There will still be immigrants, but such species will be widespread, fugitive types which do not need reserves for their survival (Willis, 1974; Terborgh, 1976; Whitcomb et al., 1976). Species requiring continuous habitat (Peterken, 1974) or specialized habitat types (Geist, 1971) for their survival or dispersal will not be able to disperse to sites where they have become extinct (Terborgh, 1974b).*"

Perhaps clusters of reserves can be designed in a way that will permit at least some inter-island dispersal of even the less vagile species. If this (perhaps optimistic) goal can be achieved then some of these species may be able to persist by continually recolonizing islands following extinctions. Methods for predicting the possibility of such persistence are given in later sections.

10.4 Modeling the Patchy System

Two general approaches have been developed for describing the interactions of a community of species distributed over a set of islands (or patches). In one approach each variable represents the fraction of islands inhabited by a particular species from a community of interacting species, while in the second approach each population number on each patch is allowed to be a mathematical variable. Cohen (1970) introduced the first method, and Levins and Culver (1971), Horn and MacArthur (1972), Slatkin (1974), Hastings (1977), and Zeigler (1977) have extended this approach. The density of a given species on a given island is ignored, it being assumed that some sort of carrying capacity level is reached on each island on a short time scale relative to inter-island migration processes. Extinction of given species on given patches can result from competition. The colonization or extinction on any island is stochastic, but when a large number of islands are considered, the fractions of islands on which colonization or extinction are occurring can be considered approximately deterministic, in these "patch occupancy" models.

Following for the moment Horn and MacArthur (1972), we consider two species, letting p refer to the first species and q to the second species. Also assume there are several, say n, different patch types, labelled $1, 2, \ldots, n$. On each patch type the competitive relationships between the two species can be different. The value p_i is the proportion of patches of type i occupied by species p and q_i is the proportion of patches of type i occupied by species q. Horn and MacArthur investigated competition and coexistence by using the criterion that if both species p and q will increase when each is rare and the other species is in equilibrium, then the two species can coexist.

To apply this criterion it is first necessary to develop equations for the rate of change of p_i and q_i for all values of i. It is assumed that random extinctions of a particular patch type i occur at a rate proportional to the number of patches occupied by that species, and that extinctions resulting from competition on patch type i occur at a rate proportional to the product of the proportion of patches of type i occupied by species p and the proportion occupied by species q. Patches of type i currently unoccupied by a given species are colonized or recolonized by members of the species dispersing from other islands at a rate proportional to the product of unoccupied patches of type i and all patches occupied by the given species. Expressed mathematically, the rates of change are

$$\frac{dp_i}{dt} = \sum_{j=1}^{n} m_{p,ij} p_j (1-p_i) - e_{p,i} p_i - c_{pq,i} p_i q_i \quad (i=1,2,\ldots,n), \qquad (10.7a)$$

$$\frac{dq_i}{dt} = \sum_{j=1}^{n} m_{q,ij} q_j (1-q_i) - e_{q,i} q_i - c_{qp,i} q_i p_i \quad (i=1,2,\ldots,n), \qquad (10.7b)$$

where $m_{p,ij}$ and $m_{q,ij}$ are coefficients of colonization of patch i from patch j for species p and q, respectively, $e_{p,i}$ and $e_{q,i}$ are coefficients of extinction on patch i, and $c_{pq,i}$ and $c_{qp,i}$ are coefficients of competition on patch i.

Now suppose species p is in equilibrium, with the equilibrium being represented by $\bar{\mathbf{p}} = (\bar{p}_1, \bar{p}_2, \ldots, \bar{p}_n)$, where $\bar{p}_i$ is a function of the $m_{p,ij}$'s and $e_{p,i}$'s. Species q is assumed to exist at very low levels. The set of Eq. (10.7b) then becomes approximately

$$\frac{dq_i}{dt} = \sum_{j=1}^{n} m_{q,ij} q_j - e_{q,i} q_i - c_{qp,i} p_i q_i \quad (i=1,2,\ldots,n). \qquad (10.8)$$

Species q can persist in the system of patches when at least one of the eigenvalues of this set of equations has a real part, λ_R, greater than zero. The matrix of the system is, for the special case $n=3$,

$$\mathbf{A} = \begin{vmatrix} m_{q,11} - e_{q,1} - c_{qp,1}\bar{p}_1 & m_{q,12} & m_{q,13} \\ m_{q,21} & m_{q,22} - e_{q,2} - c_{qp,2}\bar{p}_2 & m_{q,23} \\ m_{q,31} & m_{q,32} & m_{q,33} - e_{q,2} - c_{qp,3}\bar{p}_3 \end{vmatrix} \qquad (10.9)$$

166

which is a positive feedback matrix. There exists a $\lambda_R > 0$ if and only if for at least one $i, (-1)^i \det (\mathbf{A}_i) < 0$, where $\mathbf{A}_i$ is an $i \times i$ principal minor of the matrix $\mathbf{A}$. We shall not write out the expanded inequality since it is lengthy and offers no immediate insights. Horn and MacArthur considered the case $n = 2$. If

$$m_{q,12} m_{q,21} > (e_{q,2} + c_{qp,2}\bar{p}_2 - m_{q,22})(e_{q,1} + c_{qp,1}\bar{p}_1 - m_{q,11}) \qquad (10.10)$$

then species q can persist, or, if it is initially absent, invade. An analogous inequality implies that species p can invade a system in which species q is initially established in equilibrium.

Two situations especially favor coexistence. First, coexistence is promoted when one species has a competitive advantage on one type of patch (e. g., $c_{qp,1}\bar{p}_1 > c_{pq,1}\bar{p}_1$) and the other species has the advantage on the other type of patch ($c_{qp,2}\bar{p}_2 > c_{pq,2}\bar{p}_2$). Second, even if one species is at a competitive disadvantage on both patch types, it may be able to survive in the fashion of a fugitive species (Hutchinson, 1961) if its colonization rate is much higher than that of its competitor. For example, species q will have a sufficiently a high colonization rate to persist if $m_{q,12}$ and $m_{q,21}$ are large enough.

Armstrong (1976), who investigated experimentally and mathematically the coexistence of competitive species of fungi in a patchy laboratory environment, cautions that while these models *"show unequivocally that the fugitive species mechanism can indeed lead to coexistence, they are not sufficiently detailed to allow predictions concerning the types of situations in which the fugitive species mechanisms is likely to be important."* Armstrong suggest that *"further investigations of model systems are desirable in this connection..."*

More specific criticism was given by Levin (1974), who pointed out that patch-occupancy models may be unrealistic when migration rates can depend critically on inter-patch distances and sizes of individual colonies. In such cases, Levin offered a second approach, arguing that a detailed model of the m species distributed over the n patches is needed; in particular, a system of $m \times n$ equations of the form

$$\frac{dX_i^\mu}{dt} = f_i^\mu(\mathbf{X}^\mu) + \sum_{v \neq \mu}^n D^{v\mu}(X_i^v - X_i^\mu), \qquad (10.11)$$

where $i = 1, 2, \ldots, m$ and $\mu, v = 1, 2, \ldots, n$. The vector $\mathbf{X}$ represents all species in patch μ, so $f^{\mu i}(\mathbf{X}^\mu)$ represents the action of all species in that patch on species i. The function $D^{v\mu}(X_i^v - X_i^\mu)$ represents the relative migration effects between patches v and μ. The entire set of $m \times n$ equations must be considered simultaneously.

The advantage of Levin's (1974) model is that the detailed characteristics of real regions of islands can, in principle, be taken into account. Any actual patchy environment will have islands of various sizes, shapes, and inter-island distances. This implies that each patch will have unique mortality and colonization rates for the various species. The individual rates for each patch can be built into Levin's model, and the overall consequences for persistence and stability can be evaluated. We feel that the persistence of certain species may often depend critically on the sizes and inter-island distances of particular subjets of all the islands in a region, rather than on gross average characteristics.

167

As will be seen below, a model for a single species or group of competitive and mutualistic species interacting on a system of islands will have features of a positive feedback network. The results of positive feedback system theory can be applied to models of patchy environments, where they lead to simple criteria for the persistence of individual species in a region and the stability of the system at an equilibrium point. These persistence and stability criteria can be related to the basic parameters of the forest island system, extinction and colonization rates, which in turn are functions of island sizes and inter-island distances.

The nature of the present model is deterministic, as opposed to stochastic. It can easily be surmised that the deterministic approximation might not be valid for relatively rare species, present in small enough numbers to be strongly affected by random fluctuations. However, our main goal is to determine general tendencies, which should be deterministic on the average, even if detailed scenarios depend heavily on chance. This approach should be adequate for principal purposes. For example, assume that, in a particular patchy environment, some species in question are shown by means of a deterministic model to have a tendency to persist. Then it may only be necessary for conservation programs to assist nature by such means as restocking to offset occasional extinction caused by random fluctuations. If, on the other hand, the species have a tendency to approach extinction, the model suggests that either changes in the landscape pattern favoring persistence should be made or a program of continuous restocking undertaken. If for some reason, it is necessary to use a stochastic approach rather than a deterministic one, then Markov chain matrices equivalent to our differential equations (below) can be formulated.

Our procedure, using methods described by DeAngelis et al. (1979), is to first consider a single tree species in a region of forest patches. Criteria for both persistence and stability are found. Next, systems with a community of species, interacting competitively and mutualistically in a region of forest patches, are analyzed in the same way. Finally, we assess the prospects for applying our models to realistic situations.

10.5 A Single Species in a Patchy Region

Consider a spatial region with an arbitrary number of forest patches, N, with arbitrary sizes and spacing. These patches are usually assumed to be surrounded by non-forested land, although they could also be surrounded by forested land not suitable for the particular species in question. Since we are concerned with the collective effects of a large number of patches on guaranteeing persistence, we are particularly interested in tree species that, given enough time, would be eliminated by competition (and other factors) from any of the individual patches if there were no transfer of seeds from other patches. Such transfer occurs, however, so that recolonization can take place following local extinction. This being the case, we wish to determine two properties; 1) whether or not the species is persistent in a region, where persistence is defined for present purposes to be the tendency of a species to recover from a perturbation that reduces its biomass to nearly zero, and 2) given that a species is persistent and has a non-zero equilibrium point, whether or not the species is stable with respect to small perturbations away from this point.

Assume for simplicity that the growth of a single species in a particular forest patch can be described by the logistic equation,

$$\frac{dX_1}{dt} = r_1 X_1 - g_1 X_1^2,$$ (10.12)

where X_1 is the number of trees (seedling size and larger) of the species in the patch, r_1 is the maximum possible rate of increase, and r_1/g_1 is the patch carrying capacity (equivalent to K_1 in more conventional notation) when r_1 is positive, with the subscripts denoting "*patch* 1." The parameter r_1 represents the difference between the rate of recruitment of new trees of the species into the patch resulting from autochthonous seed sources and the density independent mortality of trees in the patch. This parameter is considered to be averaged over time periods long compared to the generation time of the species. In order for the species to disappear from an isolated patch over the long term, it is necessary and sufficient that $r_1 \leq 0$. If, for example, the species in question is a pioneer species that invades temporary gaps in the forest canopy, then the local r_1 is negative and its magnitude will reflect the time it takes for the canopy to close over the gap. Parameter g_1 measures the density-dependent effects causing increasing mortality and decreasing successful regeneration within the patch. Admittedly, this is an overly simple model, and we shall discuss generalizations later.

The next higher degree of landscape complexity is a region with two patches. A simple but reasonable assumption is that the flux of seeds from one patch to the next is proportional to the population on the source patch. We then have the equations

$$\frac{dX_1}{dt} = r_1 X_1 - g_1 X_1^2 + k_{12} X_2,$$ (10.13a)

$$\frac{dX_2}{dt} = r_2 X_2 - g_2 X_2^2 + k_{21} X_1,$$ (10.13b)

where X_1 and X_2 are the species population numbers in the two islands and k_{12} and k_{21} are the rates of transfer from one island to the next of seeds that eventually successfully germinate; i.e., the colonization rates.

The phase-plane diagram associated with Eqs. (10.13a) and (10.13b) will be similar to that for a two-species mutualistic system (Fig. 10.4). The origin, or $(0,0)$, is the point where the population is extinct. If $r_1 r_2 < k_{12} k_{21}$, where both r_1 and r_2 are negative, then there exists an equilibrium point, E, with positive values of X_1 and X_2. A similar diagram can be used to establish the fact that if either $r_1 > 0$ or $r_2 > 0$ there are always two equilibrium points; an unstable point $(0,0)$ and a stable one. The population is always persistent in the system. This case is not very interesting since seed dispersal between patches is not necessary to ensure persistence.

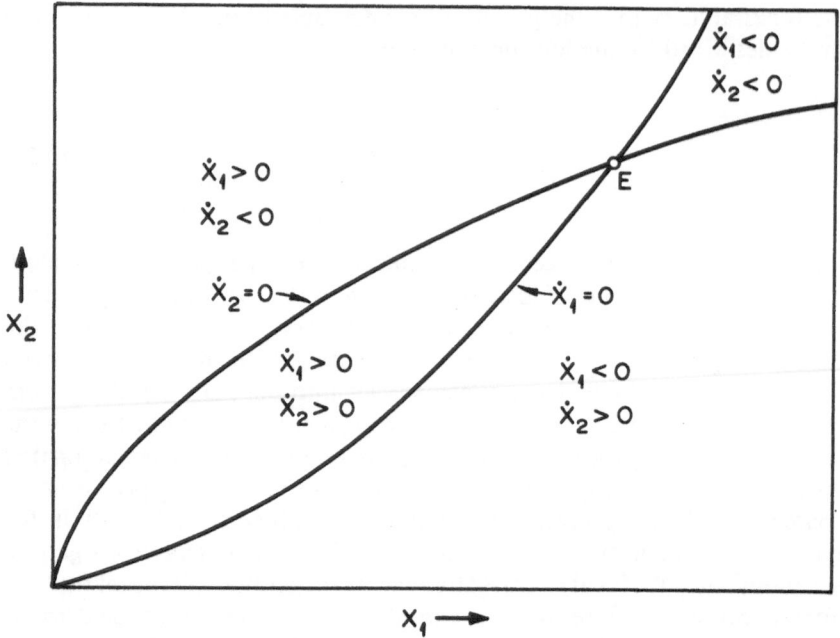

Fig. 10.4. Phase-plane diagram for a single species on two forest patches [Eqs. (10.13a, b)]. E is a stable equilibrium point and the origin is an unstable one. (Reproduced from DeAngelis et al., 1979, with permission of Academic Press, Inc.)

10.6 Time to Extinction on a Patch

It may be useful at this point to discuss the assumption that $r_1 \leq 0$ on individual patches (or islands; we will use the terms interchangeably). This is the assumption that on any given forest island in the absence of colonization there is a tendency towards extinction of the species in question on the time scale of interest. As MacArthur and Wilson (1967) point out, because the carrying capacity, K_1 (r_1/g_1 in our present notation), of the environment for a given species is finite, and since every individual organism has a chance of dying at any time, sooner or later it is statistically certain that any isolated population will eventually become extinct. Over time scales on which the species can become extinct, the concept of carrying capacity has little significance, so there is no paradox in the fact that when $r_1 < 0$, K_1 is also negative.

Hanson and Tuckwell (1978) have attempted to estimate the persistence times (or times to extinction) of populations subjected to large random fluctuations. Their basic model is the stochastic differential equation

$$dX(t) = rX(t)(1 - X(t)/K)dt - \varepsilon d\pi(\lambda, t), \tag{10.14}$$

where $d\pi(\lambda, t)$ represents population reductions arriving randomly as a Poisson

170

process. The constant ε controls the size of the disturbances, and λ is the Poisson parameter. Given an initial population size, $X_0 < K$, the authors were able to calculate mean times to extinction, T, as a function of the ratios X_0/ε and λ/r.

On islands times to extinction can be relatively short on an ecological time scale. We are concerned with cases in which a species can take advantage of temporarily favorable sites on an island to successfully colonize, but which will be eliminated by competition over the long term. The actual detailed dynamics of a species on an island are likely to be fairly complicated. Once a single organism of a species has become established, it can produce offspring that can fill up other available sites. The detailed dynamics will not be modeled, however. It will simply be assumed that the descendants of the initial propagule will persist with some mean survival time, T_1, on patch 1. In our deterministic model, this corresponds to a decrease in the expected number of organisms of a species on an island,

$$X(t) = N_0 \exp(-t/T_1) = N_0 \exp(r_1 t), \tag{10.15}$$

where N_0 is the initial number on the island before it is isolated from further colonization. Hence, $|r_1| = 1/T_1$ is the extinction rate associated with a particular species on an island. It may be possible to estimate this rate from data on real insular reserves, or perhaps through simulation models such as JABOWA (Botkin et al., 1972) and FORET (Shugart and West, 1977).

10.7 Persistence of a Species in a Two-patch Environment

The species is persistent if it always recovers from perturbations that take it close to zero; i.e., if $\dot{X}_1, \dot{X}_2 > 0$ in the vicinity of the origin, so that the origin is an unstable equilibrium point. Notice in Fig. 10.4 that if $r_1 r_2 < k_{12} k_{21}$, then eventually a trajectory in the X_1, X_2-phase plane will enter the region where $\dot{X}_1, \dot{X}_2 > 0$, so that the species persists. On the other hand, when the inequality is reversed, the population trajectory approaches the origin indefinitely, implying extinction. The origin is a stable equilibrium point in this case.

To reduce the confusion between the instability of the origin, which implies persistence, and the instability of E, the non-zero equilibrium point, we will no longer refer to the origin as being stable or unstable, but call it a repellor (instability) or an attractor (stability). When the origin is a repellor, the species in question is persistent.

In general, to determine whether the origin is a repellor or an attractor, it is necessary to calculate the eigenvalues of the system of equations linearized at $(0, 0)$. For the two-island case, this involves finding the eigenvalues of the matrix $\mathbf{A}(0)$,

$$\mathbf{A}(0) = \begin{vmatrix} r_1 & k_{12} \\ k_{21} & r_2 \end{vmatrix}. \tag{10.16}$$

The existence of a positive real part of at least one eigenvalue of this matrix implies $(0, 0)$ is a repellor and hence the species is persistent.

The two-patch system is presented largely for illustrative purposes. One might find it intuitively unreasonable that if a species cannot maintain itself over the long term on either of two isolated patches, it could do so when the patches are connected by a trickling of inter-patch seed dispersal. One might find it more plausible if looked at in the following way. An isolated population of some plant species in a habitat island that is cut off from outside sources will usually have no trouble providing more than enough seeds to maintain itself. However, suppose that occasionally, because of competition and other factors, there are no available spaces on the island for a seedling of a given species to become established. Then, despite the great fecundity of a species, it may vanish from the island. Later, however, temporary gaps for the species may open up by chance. At this point, even a very small seed flow from another island may be sufficient to reestablish the species on a given island. The idea of collective maintenance of the species seems more plausible for an environment with many patches (to be considered later), since a fairly steady supply of allochthonous seeds should then be available to each patch.

10.8 Stability of a Single-species, Two-patch System

The stability of the species at the equilibrium point, $E = (\bar{X}_1, \bar{X}_2)$, if it exists, can be determined by solving for the eigenvalues of the matrix of Eqs. (10.13a) and (10.13b) linearized about E,

$$\mathbf{A}(E) = \begin{vmatrix} r_1 - 2g_1 \bar{X}_1 & k_{21} \\ k_{12} & r_2 - 2g_2 \bar{X}_2 \end{vmatrix}. \tag{10.17}$$

If there is no positive real part of any eigenvalue, the system is stable. In the present case of two forest patches, one can see from the phase-plane diagram that E is always stable if it exists. This is not necessarily true for multi-patch systems, however. The matrices (10.16) and (10.17) reveal the positive feedback nature of the system.

10.9 Persistence of a Species in an N-patch Environment

If any value of r_i in an N-patch system is positive, the species persists on the ith patch and is, therefore, persistent in the entire system. Proof of this is not difficult and will be omitted here. The case is not very interesting, since it eliminates the need for collective effects of patch interaction for species persistence; so we shall assume that $r_i < 0$ for all values of i.

To determine if a species in an N-island region is persistent, we must solve for the eigenvalues of the $N \times N$ matrix

$$
\mathbf{A}(0) = \begin{vmatrix} r_1 & k_{12} & \cdots & k_{1,N-1} & k_{1N} \\ k_{21} & r_2 & \cdots & k_{2,N-1} & k_{2N} \\ \cdot & & & \cdot & \cdot \\ \cdot & & & \cdot & \cdot \\ \cdot & & & \cdot & \cdot \\ k_{N1} & \cdots & & k_{N,N-1} & r_N \end{vmatrix}, \qquad (10.18)
$$

where all the off-diagonal terms are greater than or equal to zero. Again, the existence of positive real parts of any eigenvalues of this matrix implies persistence. The special nature of the matrix $\mathbf{A}$ permits the use of simple criteria introduced earlier to determine the signs of the eigenvalues.

Our results concerning positive linear systems (Appendix B) give us a choice of conditions that guarantee that the eigenvalues of $\mathbf{A}$ all have negative real parts. The condition concerning the signs of the principal minors is a most convenient one, since determinants are easy to evaluate. This condition for all eigenvalues having negative real parts is written,

$$
(-1)^k \det \begin{vmatrix} a_{11} & a_{12} & \cdots & a_{1k} \\ a_{21} & a_{22} & \cdots & a_{2k} \\ \cdot & \cdot & & \cdot \\ \cdot & \cdot & & \cdot \\ \cdot & \cdot & & \cdot \\ a_{k1} & a_{k2} & & a_{kk} \end{vmatrix} > 0 \qquad (k = 1, 2, \ldots N), \qquad (10.19)
$$

where $a_{ii} = r_i$ and $a_{ij} = k_{ij}(i \neq j)$. If any one of these conditions (10.19) is not satisfied, then there exists at least one positive real part of an eigenvalue and the equilibrium point is a repellor.

The above results can easily be applied to our model of a tree species in a region of forest patches. Consider the imaginary system shown in Fig. 10.5 where all population growth rates, r_i, and colonization rates, k_{ij}, are known. Suppose the matrix $\mathbf{A}(0)$ is the community matrix evaluated at the origin, $\mathbf{X}_0 = (0, 0, \ldots, 0)$, where $\mathbf{X}_0$ is a vector of dimension N. Then to determine that the species is persistent, it is only necessary to find one subjet of p islands ($p \leq N$) for which

$$
(-1)^p \det |\mathbf{A}_p(0)| < 0,
$$

where $\mathbf{A}_p(0)$ is the $p \times p$ matrix associated with the p patches. This is illustrated in Fig. 10.5, where a small cluster of patches is assumed to violate condition (10.19), implying that the tree species will tend to persist in the region. The fact that persistence can be proven by looking at a small subset of the entire system distinguishes the single species forest patch model, as it also does multi-species mutualistic systems (Travis and Post, 1979), from multi-species competitive systems (e.g., May, 1973b; Strobeck, 1973), in which the whole system must always be considered. It also vastly simplifies the practical problems of designing models that guarantee persistence. The species need only be persistent in some small part of the region to persist in the whole region. Feedback effects from other parts of the region will always reinforce the repelling effect of the origin. Of course, as mentioned earlier, species for which any value of r_i is positive are persistent since they will persist on island i even if it is isolated.

Fig. 10.5. A region of forest patches. A small subset of patches in the center of the region is sufficient to maintain the species in this hypothetical case. The arrows denote major colonization links by means of seed dispersal. (Reproduced from DeAngelis et al., 1979, with permission of Academic Press, Inc.)

10.10 Multi-species, Multi-patch Systems with Competition and Mutualism

In nature, tree species usually occur in close proximity to one or more other tree species, with which they interact to some degree. For the sake of realism, we need to incorporate the effects of interspecific competition and mutualism among tree species into our model. This could be accomplished in an approximate way by implicitly including the effects of these interactions on the population growth rate, r_i, of the single species, and proceeding as we did in considering a single species. However, since the degree of interaction with other species depends on the population numbers of the other species, it is best to attempt a more general analysis that considers as variables all important tree species in the system that interact with the species of interest. Note that when one models competition $(-, -)$ and mutualism $(+, +)$, both commensalism $(+, 0)$ and amensalism $(-, 0)$ are automatically included as special cases.

There is the possibility of some confusion at this point. We have stated earlier that a single-species, N-patch model is formally analogous to an N-species mutualistic system. Now we are adding to our model the possibility of actual mutualism (in the traditional sense) between different species. We mean to distinguish between the two concepts. In the first case, the population of species i on some patch, P, interacts in a manner analogous to mutualism with the population of species i on some other patch, Q. In the second case, two species, i and j, on the same patch, P, may interact mutualistically or competitively.

Let us begin with the simple situation of two competing species on two forest patches. Let X_{1P}, X_{1Q}, X_{2P}, and X_{2Q} be the population numbers of species 1 and 2 on patches P and Q, respectively (Fig. 10.6). Note that all feedback loops are positive feedback loops. The equations for the dynamics of these two competing species can be written as

$$\frac{dX_{1P}}{dt} = r_{1P}X_{1P} - g_{1P}X_{1P}^2 + k_{1PQ}X_{1Q} - c_{12P}X_{1P}X_{2P}, \qquad (10.20a)$$

$$\frac{dX_{1Q}}{dt} = r_{1Q}X_{1Q} - g_{1Q}X_{1Q}^2 + k_{1QP}X_{1P} - c_{12Q}X_{1Q}X_{2Q}, \qquad (10.20b)$$

$$\frac{dX_{2P}}{dt} = r_{2P}X_{2P} - g_{2P}X_{2P}^2 + k_{2PQ}X_{2Q} - c_{21P}X_{2P}X_{1P}, \qquad (10.20c)$$

$$\frac{dX_{2Q}}{dt} = r_{2Q}X_{2Q} - g_{2Q}X_{2Q}^2 + k_{2QP}X_{2P} - c_{21Q}X_{2Q}X_{1Q}, \qquad (10.20d)$$

where k_{iPQ} is the colonization rate of species i on patch P from patch Q, and c_{ijP} is the competition coefficient of species j on species i on island P. This set of equations is similar to that considered by Levin (1974). The primary difference is that the transport term is not conservative here. Seeds from a source tree are usually so plentiful that those scattered to other islands do not constitute a significant loss from the source island.

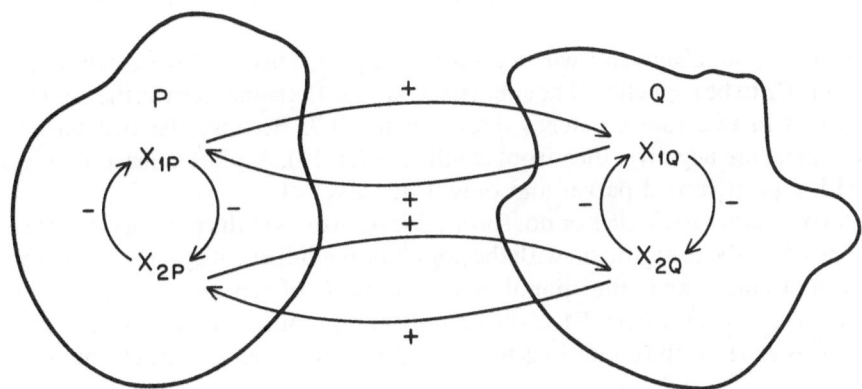

Fig. 10.6. Species 1 and 2 interact competitively within a patch $(-,-)$. Seed dispersal is represented as mutualistic interactions $(+,+)$. (Reproduced from DeAngelis et al., 1979, with permission of Academic Press, Inc.)

10.11 Persistence of a Species in a Two-species, Two-patch Environment

We assume that each of the two species in the two-patch system could persist in the absence of the other. It follows from our analysis in the earlier section concerning the persistence of one species on two forest patches that the above assumption is equivalent to the conditions

$$r_{1P}r_{1Q} < k_{1PQ}k_{1QP},$$

and $\qquad\qquad\qquad\qquad\qquad\qquad\qquad\qquad\qquad\qquad\qquad$ (10.21)

$$r_{2P}r_{2Q} < k_{2PQ}k_{2QP},$$

where all r's are assumed negative. Competition decreases the chances of each species persisting. Assume now that the population numbers of species 1 are close to zero on both patches and that the population numbers of species 2 on the two patches are given by the ordered pair $(X_{2P,E2}, X_{2Q,E2})$, where $E2$ is the equilibrium point of species 2 in the absence of species 1. To demonstrate the persistence of species 1, it is necessary to show that X_{1P} or X_{1Q} will tend to increase from zero in spite of the competitive presence of species 2. Notice that the linearized community matrix evaluated at the equilibrium point $\mathbf{X} = (0, 0, X_{2P,E2}, X_{2Q,E2})$ is decomposable; that is, it has the form

$$\mathbf{A}(E2) = \begin{vmatrix} \mathbf{A}_{11} & 0 \\ \mathbf{A}_{21} & \mathbf{A}_{22} \end{vmatrix}, \qquad\qquad (10.22)$$

where $\mathbf{A}_{11}$, $\mathbf{A}_{21}$ and $\mathbf{A}_{22}$ are square matrices. A consequence of the matrix taking the above form is that the two eigenvalues of $\mathbf{A}_{11}$ are also eigenvalues of the matrix $\mathbf{A}(E2)$. Therefore, persistence of species 1 is equivalent to the existence, for the submatrix

$$\mathbf{A}_{11} = \begin{vmatrix} r_{1P} - c_{12P}X_{2P,E2} & k_{1PQ} \\ k_{1QP} & r_{1Q} - c_{12Q}X_{2Q,E2} \end{vmatrix}, \qquad (10.23)$$

of at least one eigenvalue with a positive real part. This would cause the equilibrium point $E2$ to be a repellor. The submatrix $\mathbf{A}_{11}$ has the same form as the matrix for one species in two forest patches discussed in 10.7. Because the two terms on the diagonal are negative, then from conditions (10.19), $\mathbf{A}_{11}$ has at least one eigenvalue with a positive real part if and only if $\det(\mathbf{A}_{11}) < 0$.

To determine whether or not species 2 is persistent in the presence of species 1, we repeat the above argument with the population numbers of species 2 close to zero on both islands and the population numbers of species 1 being given by $(X_{1P,E1}, X_{1Q,E1})$, where $E1$ is the equilibrium point of species 1 in the absence of species 2. Thus, the conditions for the simultaneous persistence of both species are

$$(r_{1P} - c_{12P}X_{2P,E2})(r_{1Q} - c_{12Q}X_{2Q,E2}) < k_{1PQ}k_{1QP},$$

and $\qquad\qquad\qquad\qquad\qquad\qquad\qquad\qquad\qquad\qquad\qquad$ (10.24)

$$(r_{2P} - c_{21P}X_{1P,E1})(r_{2Q} - c_{21Q}X_{1Q,E1}) < k_{2PQ}k_{2QP},$$

quantifying the expected result that high colonization rates and low competition coefficients favor persistence (recall that the diagonal terms are negative).

In general, populations are not static in time and one should not necessarily expect the population of species 2, for example, to stay near $(X_{2P,E2}, X_{2Q,E2})$ at all times when species 1 is close to zero. For persistence to be ensured, the inequalities (10.24) should hold for all feasible values of (X_{2P}, X_{2Q}) or (X_{1P}, X_{1Q}), respectively. It is easy to see that if the first inequality of (10.24) is satisfied when (X_{2P}, X_{2Q}) equals some value (X_{2P}^*, X_{2Q}^*), it will also hold for all values such that $X_{2P} < X_{2P}^*$ and $X_{2Q} < X_{2Q}^*$. In other words, if species 1 can persist in the region when species 2 has the population numbers (X_{2P}^*, X_{2P}^*), it can also persist for smaller values of population of species 2. In most cases, we can safely take (X_{2P}^*, X_{2Q}^*) to be the equilibrium populations $(X_{2P,E2}, X_{2Q,E2})$.

It is interesting to compare conditions (10.24) for the persistence of two competing species with conditions (10.21) for the persistence of two species without competition. It can be seen that r_{1P}, r_{2P}, r_{1Q}, and r_{2Q} in (10.21) have been replaced by "effective" population growth rates that now include interspecific competition effects.

10.12 Persistence of a Species in an L-species, N-patch Environment

Determination of the persistence of species in an N-patch, L-species system is a generalization of the above approach. Consider the submatrix $\mathbf{A}_{ii}$,

$$
\mathbf{A}_{ii} = \begin{vmatrix} r_{iP,Eff} & k_{iPQ} & \cdots & k_{iPN} \\ & r_{iQ,Eff} & \cdots & k_{iQN} \\ \cdot & \cdot & \cdot & \\ \cdot & \cdot & \cdot & \\ \cdot & \cdot & \cdot & \\ k_{iNP} & k_{iNQ} & & r_{iN,Eff} \end{vmatrix},
$$
(10.25)

where

$$
r_{iP,Eff} = r_{iP} - \sum_{\substack{j=1 \\ j \neq i}}^{L} c_{ij} \bar{X}_{jP,E} < 0
$$
(10.26)

and where $X_{jP,E}$ is the equilibrium point in the absence of species i. The necessary and sufficient conditions for the persistence of species i are that at least one of the principal minors of $\mathbf{A}_{ii}$ violate conditions (10.19).

10.13 Stability of a Two-species, Two-patch Model

A second aspect of the forest island problem is the investigation of the stability about a feasible equilibrium point, E, of the system of L species, some interacting competitively and some mutualistically in an environment of N forest patches. Let us first consider two species in a two-patch region and look at local stability about a feasible equilibrium point, $E = (\bar{X}_{1P}, \bar{X}_{1Q}, \bar{X}_{2P}, \bar{X}_{2Q})$. The local stability about E is determined by the eigenvalues of the matrix obtained by linearizing Eq. (10.20) about the equilibrium point. The intra-patch species interactions are mutualistic (say, on patch P) when c_{ijP} and c_{jiP} are negative, and competitive when these terms are positive.

It is clear that when the interspecific interactions are mutualistic, the present matrix $A(E)$ has the same form as the matrix $A(E)$ of the section, "*Persistence of a species in an N-patch environment*"; that is, all off-diagonal elements are positive. Therefore, the mathematical techniques used in that section, in particular conditions (10.19), can be used in the present case. When the interactions are competitive, $A(E)$ is similar (in the mathematical sense; see Appendix F and Chap. 8) to a matrix with negative diagonal and positive off-diagonal elements, the similarity transformation being $A'(E) = SA(E)S^{-1}$, where

$$S = \begin{vmatrix} 1 & 0 & 0 & 0 \\ 0 & 1 & 0 & 0 \\ 0 & 0 & -1 & 0 \\ 0 & 0 & 0 & -1 \end{vmatrix}. \tag{10.27}$$

Since the eigenvalues of similar matrices are identical, the mathematics developed in earlier sections can again be used. The necessary and sufficient conditions for the local stability of two species competing within two patches are given by the inequalities (10.19).

10.14 Stability of an *L*-species, *N*-patch Model

We saw in an earlier chapter that linear positive feedback theory applied to systems in which a species only interacts with other species in such a way that an increase in species i affects species i positively through every causal chain. Every loop from a given species i (except loops of unit length) feeds back positively on species i. This concept applies to N-patch systems as well as spatially homogeneous systems. Consider Figs. 10.7a and b. In Fig. 10.7a three species interact competitively and two patches, while in Fig. 10.7b two species interact competitively on three patches. Each species is part of a positive feedback network.

The matrices in Fig. 10.7 show that the two systems are positive feedback matrices. That is, they can be transformed to positive matrices by means of similarity transformations (see Appendix F or Chap. 8; these matrices are often called Morishima matrices). A general matrix for an *L*-species, *N*-patch system has the form

$$\begin{vmatrix} B_{11} & B_{12} & \dots & B_{1L} \\ B_{21} & B_{22} & \dots & B_{2L} \\ \cdot & \cdot & & \\ \cdot & \cdot & & \\ \cdot & \cdot & & \\ B_{L1} & B_{L2} & \dots & B_{LL} \end{vmatrix}, \tag{10.28}$$

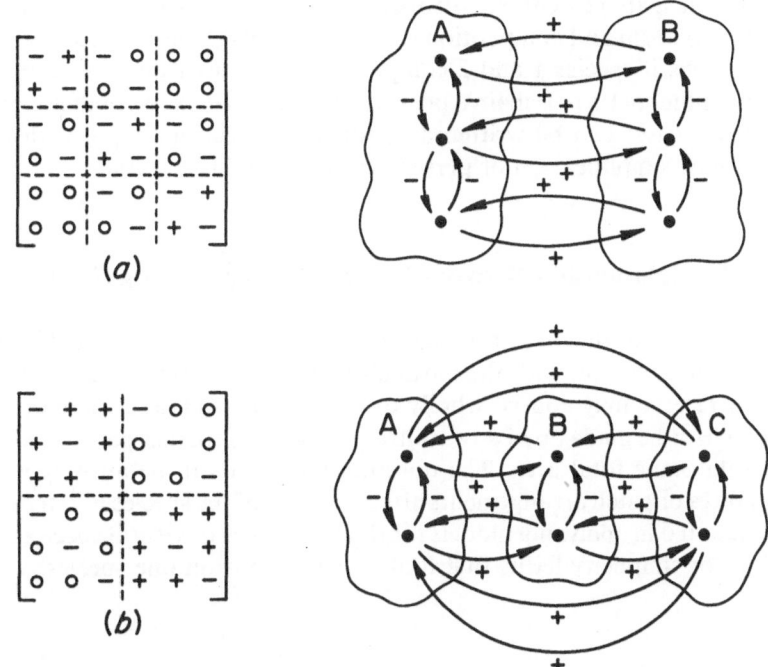

Fig. 10.7a and b. The matrix and graph representations of **a** three competing species on two forest patches and **b** two competing species on three forest patches. Note that if species 1 and 3 compete in **a** the system is not a pure positive feedback system. (Reproduced from DeAngelis et al., 1979, with permission of Academic Press, Inc.)

where each $\mathbf{B}_{ii}$ is similar to a positive matrix $N \times N$ and each $\mathbf{B}_{ij}$ is non-negative if $i+j$ is odd and non-positive if $i+j$ is even (or the matrix can be converted to a form for which these are true).

For example, suppose we consider a four-species community in a region of N patches where species 1 and 3, as well as the pair of species 2 and 4, compete, while species 3 and 4 and 1 and 2 are mutualistic;

$$\mathbf{B} = \begin{vmatrix} \mathbf{B}_{11} & \mathbf{B}_{12} & \mathbf{B}_{13} & 0 \\ \mathbf{B}_{21} & \mathbf{B}_{22} & 0 & \mathbf{B}_{24} \\ \mathbf{B}_{31} & 0 & \mathbf{B}_{33} & \mathbf{B}_{34} \\ 0 & \mathbf{B}_{42} & \mathbf{B}_{43} & \mathbf{B}_{44} \end{vmatrix}. \tag{10.29}$$

In this case, $\mathbf{B}_{12}, \mathbf{B}_{21}, \mathbf{B}_{34}, \mathbf{B}_{43} > 0$ and $\mathbf{B}_{13}, \mathbf{B}_{31}, \mathbf{B}_{24}, \mathbf{B}_{42} < 0$.

The criteria for stability for the above system are again (10.19). Remarkably, these criteria for a system of L species with "*limited competition*" (only certain patterns of competitive relationships are permitted; see Chap. 8) in an N-patch environment are simpler than the general criteria for a purely competitive system without spatial extent (see Strobeck, 1973). The drawback is that not all variations of mutualism and competition can be expressed as positive feedback matrices. For example, if species 1 and 2 compete, then species 1 and 3 cannot compete if the present model is to remain a positive feedback matrix. Nonetheless, a great number of real cases can be written as positive feedback matrices, with the consequent extreme simplification of persistence and stability criteria.

10.15 Relationship Between Reserve Design and Species Persistence

Let us assume that in a particular region a certain amount of land is to be set aside as a natural reserve, with the particular goal of preserving certain desirable species. The reserve may consist wholly of contiguous land, and therefore constitute one patch, or be made up of any arbitrary number of smaller patches, always adding up to the same total area. The question that needs to be answered is how various choices of patch arrangements affect survival of the species of interest. We are most interested in analyzing models for the persistence of several species simultaneously, but for simplicity let us first look at conditions on one species.

Single species

Consider a single species, say a seed-dispersed plant, and ignore possible feedback effects due to interactions with other species. If the reserve is a single patch, then the dynamics of the species can be approximated by Eq. (10.12). Define the size of the island by D. If D is larger than some critical size D_{crit} (Fig. 10.8), then $r > 0$, and the species will have a deterministic tendency to persist. In such a case, the patch can be said to have a carrying capacity, $K(D)$, defined by $r(D)/g(D)$, which is a function of patch size. If $r \leq 0$, the species inevitably tends towards extinction.

Hence, if a contiguous area of size greater than D_{crit} can be set aside as a reserve, it will guarantee the persistence of the species. However, it may not always be possible to acquire contiguous land. How will persistence be affected if the same total area, D_{total}, is divided into several patches, it being assumed that the species can colonize from one island to the next by seed dispersal?

The relevant mathematics to study this problem was given in the section *"Persistence of a species in an N-patch environment"*. To particularize this abstract theory to plausible real-world cases, however, requires additional considerations. Two factors are important in characterizing persistence of a species on a set of patches; 1) the values of r_i on each patch i, and 2) the inter-patch dispersal rates, k_{ij}. If $r_i > 0$ for any patch, the species persists on the set of patches, as was shown earlier. Therefore, this case is not very interesting. Let us assume, as earlier, that $r_i < 0$ for all patches. Inter-patch dispersal must overbalance the extinction rates on each patch for persistence to occur.

To be able to draw succinct conclusions, we must carefully limit our system. Assume that the shapes of the individual patches play no role in the extinction and dispersal phenomena; only patch sizes, D_i, and inter-patch distances, d_{ij}, are

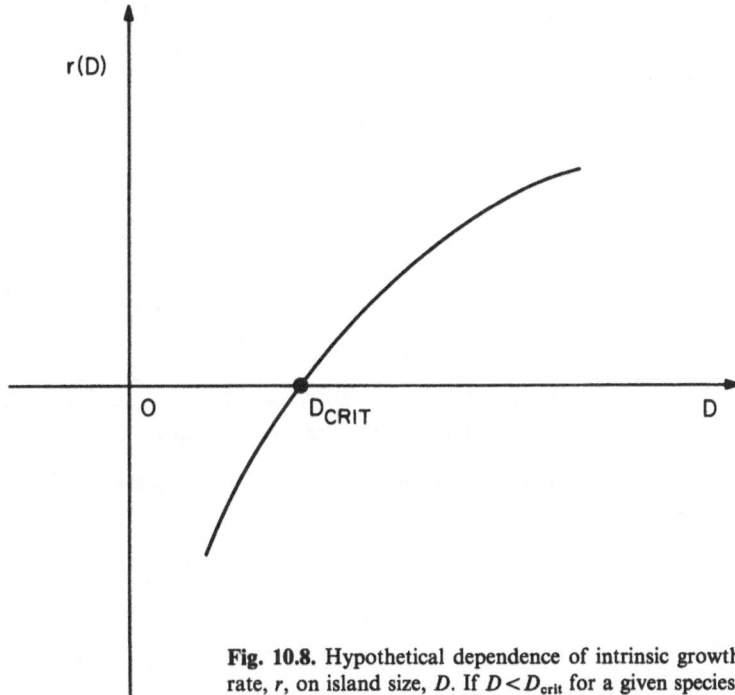

Fig. 10.8. Hypothetical dependence of intrinsic growth rate, r, on island size, D. If $D < D_{crit}$ for a given species, then the island is too small for the species to become established

important. In any realistic system, D_i will vary from patch to patch and d_{ij} will differ for different pairs of patch. It will be useful to prescribe fairly tight constraints on the variations for the sake of simplicity. However, we shall use Monte Carlo simulations to study many randomly selected systems obeying these constraints, and then draw conclusions from averages over these simulations.

First, assume a given simulation consists of N patches of average area, D_{ave}, such that

$$ND_{ave} = D_{total}. \tag{10.30}$$

The size of each patch is picked from a normal distribution about $D_{ave}(< D_{crit})$ with standard deviation $0.1 D_{ave}$. The spatial distributions of patches is determined as follows. Consider a rectangular region, say 100 km by 50 km. An initial patch is chosen in this region, using a uniform pseudo-random number generator to choose the longitudinal and latitudinal coordinates. Choose succeeding patches in the same manner, except that each patch must be no closer than some distance, d_{min}, from any other patch, and no further than d_{max}, from at least one other patch. The conditions on patch position eliminate anomalies that can occur if patches are chosen by chance to be unreasonably close to or far from each other. Figure 10.9 shows a typical pattern of patches generated by the computer.

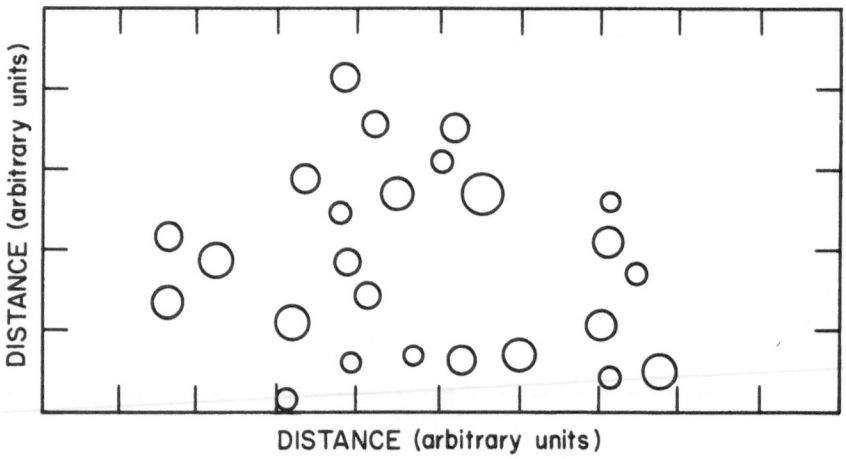

Fig. 10.9. Hypothetical cluster of islands generated by computer simulation for tests of species persistence

The extinction rate of the species on an individual patch will certainly depend on the size of the patch. While we have no information on what the relationship should be in reality, let us assume that

$$r = r_0(1 - \exp\left\{-\alpha(D - D_{\text{crit}})\right\}), \qquad (10.31)$$

where r_0 and α are constants. Hence, for $D < D_{\text{crit}}$, $r < 0$. In our simulations, we let $r_0 = 0.5$ (subject to small random variations), $D_{\text{crit}} = 5.0$, and $\alpha = 0.2$. The choice of α reflects our assumption as to how rapidly the extinction rate decreases with increasing patch size.

We also arbitrarily assumed that the dispersal rates, k_{ij}, depend on inter-patch distances, d_{ij}, as

$$k_{ij} = k'/d_{ij}^2, \qquad (10.32)$$

where k' is a constant.

We chose a series of average patch sizes, D_{ave}, and for each patch size we computed, using inequality (10.19), the average number of patches, N, necessary to sustain the species. This is plotted as the solid line in Fig. 10.10. Only for pairs of values (N, D_{ave}) above this line will the species persist. The dotted lines represent curves of equal total area, D_{total}. It is clear from the figure that the best strategy in the present case is to try to choose patch size as large as possible within the given practical constraints. If the parameter α had a smaller value (i.e., if the probability of extinction did not increase so rapidly with decreasing island size), the best strategy might be the reverse.

182

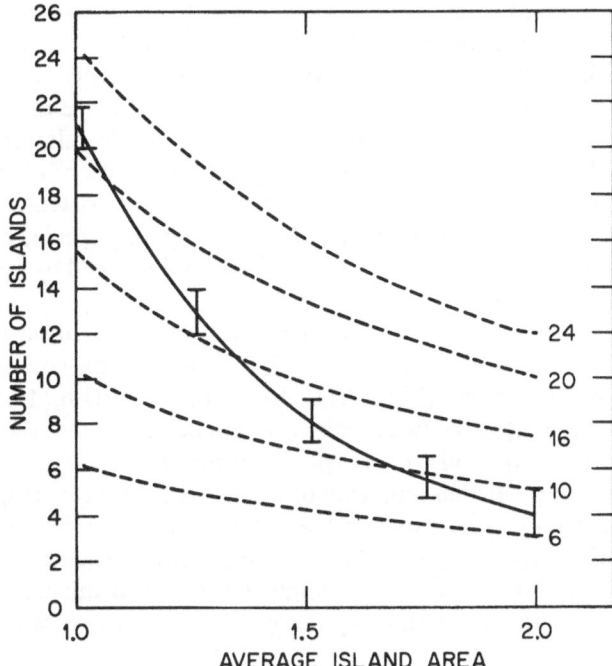

Fig. 10.10. Diagram showing relationship between number of patches in the hypothetical cluster, average patch size, and the probability of persistence of a given species. For all combinations of N and D_{ave} above the solid line, the species is, on the average, persistent. The dotted lines connect combinations having the same total area. The units of area are arbitrary because only the qualitative trend is desired

Multi-species competition

An even more interesting problem involves two or more competing species in a region of several patches. Classical competition theory indicates that coexistence of two species on a single patch is impossible if the species compete strongly enough. Consider the Lotka-Volterra model,

$$\frac{dX}{dt} = (r_1 - g_1 X - c_{12} Y) X, \qquad (10.33a)$$

$$\frac{dY}{dt} = (r_2 - g_2 Y - c_{21} X) Y. \qquad (10.33a)$$

If the product of the competition coefficients $c_{12}c_{21}$ exceeds g_1g_2, one of the species forced to extinction since the equilibrium point, if it exists, is unstable.

Simberloff and Wilson (1969, 1970), Root (1973), and Connell (1979), among others, have suggested that in patchy regions a wider variety of species might be maintained. Karlin and McGregor (1972), Levin (1974), Jackson and Buss (1975), and Gilpin (1975b) have provided a mathematical basis for these suggestions. Levin showed that when there are at least two patches, with inter-patch migration permitted, coexistence may occur even though it would not be mathematically possible on either patch separately. Jackson and Buss (1975) and Gilpin (1975b) proposed "*circular networks*" in which, for example, species 1 can eliminate species 2, species 2 will eliminate species 3, but species 3 will eliminate species 1. On a single island, one of the species will eliminate the other two. However, if several patches are connected by migration, mutual coexistence may be possible.

We can use our methodology to explore the relationship between the number of islands a system of reserves contains, the rates of dispersal of species populations among the reserves, and the possible coexistence of the competing species. Consider, for simplicity, only two competing species on a multi-patch system. The system dynamics will be approximately described by a generalization of Eqs. (10.20) to the multi-patch case.

In the simplest conceptualization of competition theory based on the Lotka-Volterra equations, there are four basic possible competitive relationships between the two species (Fig. 10.11). In the first case (Fig. 10.11a), where $r_1 c_{21} < r_2 g_1$ and $r_2 c_{12} < r_1 g_2$, the two species always coexist. In the second case (Fig. 10.11b), where $r_1 c_{21} > r_2 g_1$ and $r_2 c_{12} > r_1 g_2$, only one species will persist, that one being determined by the particular initial conditions. In the third case (Fig. 10.11c), where $r_1 c_{21} > r_2 g_1$ and $r_1 g_2 > r_2 c_{12}$, only species X can persist, while in the fourth case (Fig. 10.11d), where $r_1 c_{21} < r_2 g_2$ and $r_1 g_2 < r_2 c_{12}$, only species Y can persist.

A real set of patches might consist of a mosaic on which any of the above competitive relationships exist, because there will always be at least slight

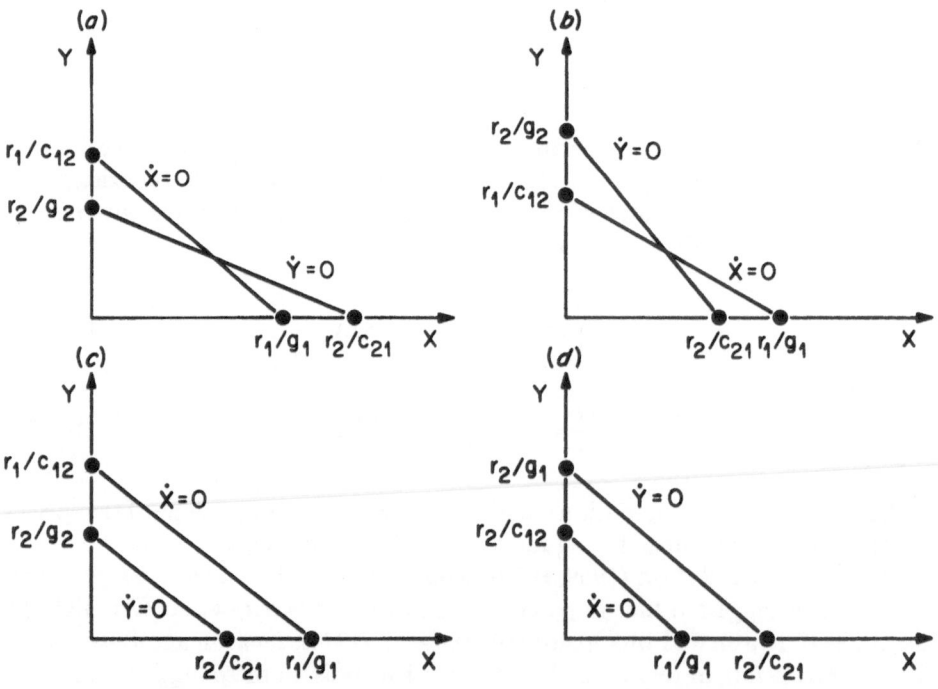

Fig. 10.11a–d. Phase-plane diagrams showing the possible ways the zero isoclines of two competing species can be oriented, assuming a Lotka-Volterra description of the dynamics. See text for detailed explanation

differences in habitat from patch to patch. An ideal arrangement of reserves would have enough variety in competitive advantages among the individual insular reserves to support all of the desired species. Unfortunately, the cost of gaining habitat variety often involves a sacrifice in the sizes of individual patches. Even if species X is a better competitor than species Y on patch 1 and species Y is the better competitor on patch 2, neither patch may by itself be large enough to support these species. Hence it may be necessary to locate the islands relative to one another such that enough inter-patch dispersal occurs that each species has at least a small population on the neighboring patch to provide recolonization potential should extinction occur on a given patch. By making possible a steady stream of colonists of each species from patch to patch, however, the arrangement of patches may raise the possibility of competitive exclusion of some species over the whole archipelago of patches.

A basic question then is what degree of opportunity of dispersal for given species should be designed into an archipelago of island reserves to make recolonizations by individual species likely but at the same time to keep the risk of competitive exclusion minimal. This will depend, of course, on the individual dispersal abilities of the species involved as well as their relative competitive abilities. The answer for any particular situation will not be obvious, but the mathematical model may enable us to draw some general conclusions.

Unlike in the previous example, we now assume that both of the species could survive on each of the individual patches (i.e., the r_i's are greater than zero).

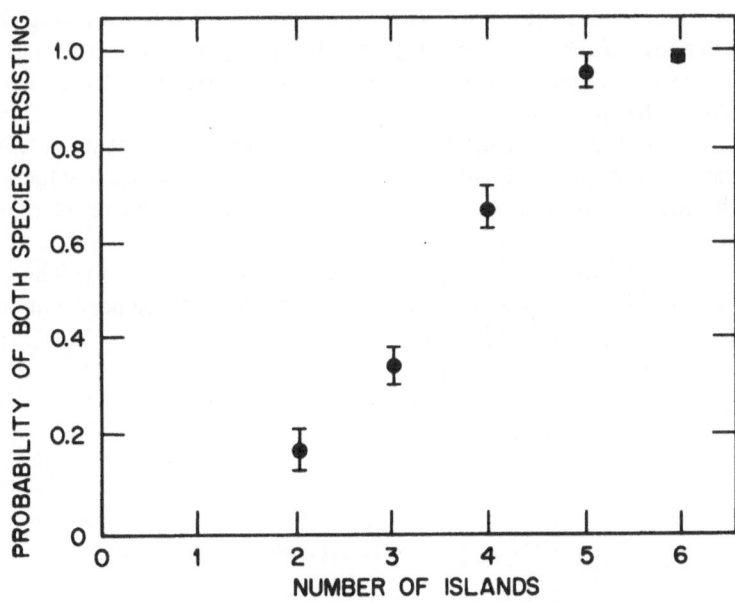

Fig. 10.12. Probability of the two competing species coexisting on the cluster of islands as a function of the number of islands in the hypothetical example. The bars represent the 95 % confidence interuals

However, we make the interspecific competition strong enough that the situation depicted in Fig. 10.11b pertains; that is, it is impossible for the two species to coexist stably on a patch. We ask the question: will the two species coexist if more than one island is present and the islands are connected by inter-island dispersal and, if so, how many islands are necessary for coexistence?

We sampled randomly constructed models using Monte Carlo methods. The values of r for both species were allowed to vary randomly over a narrow range about 0.5, while c_{12} and c_{21} vary between 1.0 and 1.5 and the dispersal coefficients k_{ij} between 0.001 and 0.05. In Fig. 10.12 the percentage of cases of coexistence of the two species is plotted as a function of the number of islands, showing an increase. Hence, even though the two species cannot coexist on any single island, they can coexist on a set of islands.

10.16 Summary and Conclusions

Classical island biogeography theory ignores feedback effects in species dispersal to islands, and so cannot address questions of species richness on clusters of islands or patches in which inter-patch dispersal is important. Two principal modeling approaches are suited for dealing with the dynamics of species in which possible habitats occur as insular clusters. In the first method the variables are percentages of clusters occupied. This approach has the disadvantage that details like individual island sizes and inter-island distances cannot be incorporated.

We use a second modeling approach in which the species characteristics, intrinsic rate of growth and carrying capacity, are specified separately for each patch in the cluster, and all inter-patch dispersal rates are specified. The first question asked was: ignoring interactions with other species (or incorporating them into the rate constants of the single species), can the species persist in the N-patch cluster? The theory of positive linear systems was used to provide relatively simple mathematical criteria for persistence.

The mathematical methodology of positive linear systems was extended to L species on N-patch clusters. This could be done for cases in which the species either all interact mutualistically or they obey certain patterns of competitive and mutualistic interaction.

Two hypothetical examples are worked out numerically. The examples indicate that, within limits, the methods of this chapter may be useful in predicting optimal designs of insular wildlife reserves.

11. Spatially Heterogeneous Ecosystems: Pattern Formation

11.1 Spontaneous Emergence of Spatial Patterns

No curious person can look out over the valley of a river or stream without wondering at the obstinate refusal of the water to flow in a straight line from higher to lower ground, even when the terrain is gentle enough to allow it the opportunity. Instead, the river meanders, snake-like down the valley. This phenomenon is so counterintuitive that the nineteenth century sociologist Lester Ward offered it as an example of nature's wastefulness of energy (see Worster, 1977, p. 175).

An explanation of the dynamics of river meanders was offered by Albert Einstein (see Stevens, 1974). Suppose that initially a stream is almost but not perfectly straight. Minor bends occur along its length. The water flowing around one such bend will be pushed by centrifugal force against the outer bank of the bend (gently, of course, since the bend is initially very slight). Since the surface waters of the stream are relatively unaffected by friction, they will tend to flow towards the outer bank. This would cause a buildup of water near the outer bank, but this buildup is prevented by a countercurrent in the deeper layers that flows towards the inner bank. The overall circulation pattern then consists of surface waters flowing towards the outer bank of the bend, then diving and flowing back along the bottom towards the inner bank.

As the current of water dives, it erodes the outer bank, later depositing some of the eroded material on the inner bank. This erosion and deposition increases the sharpness of the bend. What started as a very weak dynamic process caused by a gentle bend feeds back positively on itself, increasing the erosion and accelerating the formation of the bend. This process will occur all along the river, causing bends in opposite directions at fairly regular intervals. Hence a sinusoidal shape develops, eventually becoming so extreme that the stream occasionally short-circuits these loops by forming ox-bows. This dynamic action by the river causes it to expend energy uniformly along its length. Rather than being profligate, as Ward alleged, the river system follows a strict economy with regard to energy.

Much more complex than physical systems such as streams, but just as much subject to the effects of positive feedback, are the spatial distribution patterns of human populations. Humans are distributed across the earth in an extremely nonuniform manner. Some of the reasons for this inhomogeneity are obvious. Parts of the earth are highly favorable to human habitation, while others are quite hostile. This does not explain the particular patterns of human settlement that emerge, however. Even in regions of almost absolute uniformity, the human population does not spread evenly across the landscape, but establishes a pattern of a few large cities surrounded by numerous large and small towns, which are in turn surrounded by a great number of villages.

The German economist Christaller devised a conceptual model to explain this typical patterning (see Butler, 1980). Assume a uniform landscape (a flat plain, for example) over which farming people are initially evenly distributed and are more or less self-sufficient. Suppose now that the density of the population on this plain reaches a certain level above which it would be profitable for a tradesman to open a shop (a bakery, to use Butler's example) at some location. The threshold population density would have to be such that the number of people who are near enough the bakery to do their shopping there (rather than making their own bread, as they had in the past) is large enough for the baker to make a profit. When the population reaches that level, we can say that the region is "*unstable*" to development of nonuniformity in the form of a bakery.

As Butler (1980) put it, the location of the first bakery destroys the homogeneity of the plain. The baker will need assistants; other tradesmen will take advantage of the fact that people are attracted to the bakery, and will open up shops of other types nearby. A town will develop through this positive feedback.

Meanwhile other bakers will open up shops on other parts of the plain, but not too close to the original bakery, or to each other, since they will wish to avoid competition. In fact, just as in natural systems, the ideal patterning of bakeries, and consequently of market towns, will be such that they serve roughly hexagonal areas. Eventually, to serve the more complex needs of this pattern of small market towns, a few of these towns, again spaced rather evenly, will grow into cities.

The above examples are rather oversimplified, but they serve to show that all types of systems, ranging from physical to biological, can develop spatial nonuniformities resulting from initial slight irregularities that grow through processes of positive feedback.

In Chap. 10 we were concerned with one particular aspect of the spatial distribution of populations; the mutual reinforcement that populations of the same species, scattered over a cluster of "*islands*," give each other through migration and colonization. We were not concerned with the emergence of spatial patterns as such. The configurations of the clusters of islands or patches were assumed imposed by external circumstances and were assumed to be non-varying through time. We were interested in determining the survival of species in a region given a particular pattern of a cluster.

In the present chapter, we are concerned with the dynamics of the spatial patterns themselves. Such patterns can emerge spontaneously from the internal dynamics of ecological systems, even in homogeneous regions, similar to the way in which Benard convective cells, river meanders, and other dynamic structures may emerge in purely physical systems or in social systems such as that mentioned above. The study of the dynamics of spatial pattern formation in ecology, and the application of mathematical models, is a relatively new subject, but one that is growing rapidly. Okubo's (1980) book reviews much of the important work. Other excellent articles are those of Skellam (1973), Levin (1974, 1976, 1978a, b), and Whittaker and Levin (1977). Here we shall look at certain examples of pattern formation in which positive feedback plays an important part.

The above mention of the similarity of some ecological patterns with Benard cells has more than just metaphorical significance. It has been noticed that dense populations of free-swimming organisms such as ciliates and flagellates sometimes form polygonal cellular patterns characterized by upward flows of organisms in the center and downward flows of organisms at the edges (Loefer andMefferd, 1952; Winet and Jahn, 1972; Levandowsky et al., 1975; all cited by Okubo, 1980; Robbins, 1952, cited by Katchalsky et al., 1974). These patterns closely resemble the Benard cells described in Chap. 3. However, in the present case the mechanism that causes the instability does not seem to be temperature gradient. Instead, Platt (1961) has suggested that heavy organisms tend to swim towards the top of the water body. This creates a density inversion with the greater density (greater concentration of biomass with specific gravity greater than unity) at the surface, just as a temperature gradient in Benard's experiments resulted in a more dense fluid at the cooler top surface of fluid. If the biomass density gradient is great enough, a dynamic instability can develop from positive feedback, causing the observed convective patterns to appear. The energy driving the convective currents or organisms is their own locomotory energy rather than an external energy source, such as the energy supply that maintained the temperature gradient in Benard's experiments.

We shall next attempt to analyze mathematically some of the mechanics that can generate spatial patterns in ecological systems.

11.2 Diffusion Model

For an understanding of the formation of patterns such as those described above, it is necessary to model populations distributed in space. The dynamics of species populations in spatial domains has traditionally been described by means of partial differential equations (see Skellam, 1951, for one the early efforts). Let $X = X(s, t)$ be a species population at time t, distributed along the one-dimensional spatial axis, s (this can be extended to two or three dimensions if desired). We can write the equation

$$\frac{\partial X}{\partial t} = \frac{\partial}{\partial s} \left\{ D(s) \frac{\partial X}{\partial s} \right\} + r(s) X, \tag{11.1}$$

where $r(s)$ is the local growth rate of the population and $D(s)$ is a diffusion coefficient, which is also assumed to vary spatially in general.

Normally, Eq. (11.1) would be the starting point for mathematical and computer analysis. However, our goals here are primarily heuristic. We wish to show explicitly the positive feedback involved in spatial pattern development. To do this, it is convenient to switch from a continuous to discrete formulation; that is, we replace the continuous variable, s, by an evenly spaced sequence of points, s_i $(i = 0, \pm 1, \pm 2, \ldots)$.

In the discrete formulation the partial derivative $\partial X / \partial s$ can be written

$$\frac{\partial}{\partial s} [X(s) D(s)] \simeq \frac{D(s_{i+1}) X_{i+1} - D(s_i) X_i}{\Delta s}, \tag{11.2}$$

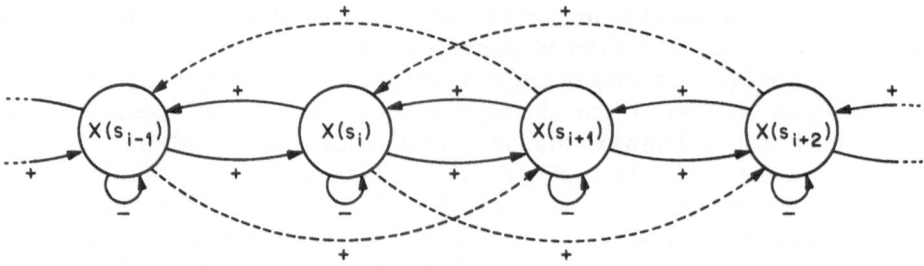

Fig. 11.1. System diagram of a species in one-dimensional space, represented by a line of discrete points [Eq. (11.2)]. The solid arrows indicate feedbacks that take place when only diffusion is operating. The dashed arrows indicate long range dispersal

where $X_i \equiv X(s_i)$, while $\partial/\partial s \{D(s)\partial X/\partial s\}$ can be written

$$\frac{\partial}{\partial s}\left\{D(s)\,\frac{\partial X}{\partial s}\right\} = \left\{D(s_{i+1})\left(\frac{\partial X}{\partial s}\right)_{i+1} - D(s_i)\left(\frac{\partial X}{\partial s}\right)_i\right\}\bigg/\varDelta s$$

$$= \left\{\frac{D(s_{i+1})X_{i+1}-D(s_{i+1})X_i}{(\varDelta s)^2} - \frac{D(s_i)X_i-D(s_i)X_{i-1}}{(\varDelta s)^2}\right\}, \quad (11.3)$$

where $\varDelta s = s_{i+1} - s_i$. Rearranging Eq. (11.3) slightly, we can write the discrete spatial version of Eq. (11.1) for the population X_i at point s_i as

$$\frac{\partial X_i}{\partial t} = \frac{D(s_{i+1})}{(\varDelta s)^2}\,X_{i+1} + \frac{D(s_i)}{(\varDelta s)^2}\,X_{i-1} - \frac{\{D(s_i)+D(s_{i+1})\}}{(\varDelta s)^2}\,X_i + r_iX_i. \quad (11.4)$$

There will be an equation of the form of Eq. (11.4) for every point in the sequence (except at the end points, where boundary conditions must be satisfied). The pattern is shown in Fig. 11.1, where the solid lines represent the feedbacks implicit in Eq. (11.4). All feedback loops with more than a single link (i.e., length greater than unity) are positive, while the loops of unit length (feedbacks of variables on themselves) are negative if

$$\frac{D(s_{i+1})+D(s_i)}{(\varDelta s)^2} > r_i \quad (11.5)$$

for all values of i.

Suppose the space occupied by the species is finite, bounded by the points 0 and L. An important question is whether or not a species in region $0 < s < L$, surrounded by unfavorable regions on both sides, can persist through time. This problem is relevant to patches of phytoplankton that have a positive growth rate in a particular region of a body of water, but which are diffusing into the surrounding inhospitable regions, a situation considered first by Kierstad and Slobodkin (1953). The population of phytoplankton can survive only if the region of positive growth rate is large enough to offset the losses due to diffusion. Levin (1979) studied this problem

190

in the special case where the diffusion coefficient, $D(s)$, and growth rate, $r(s)$, are constants. In this case, Levin showed that the population could persist only if $L \geq \pi (D/r)^{1/2}$. Levin's result is valid only for the one-dimensional model, but can be extended to two dimensions.

Next relax the assumption that $D(s)$ and $r(s)$ are constants with respect to s. When these parameters vary with s, analysis of the partial differential equation becomes difficult, but the difference equation approximation can be analyzed with relative ease. We need only study the properties of the matrix $\mathbf{A}$, where

$$
\mathbf{A} =
\begin{vmatrix}
a_{11} & a_{12} & 0 & 0 & \cdots & & 0 & 0 \\
a_{21} & a_{22} & a_{23} & 0 & \cdots & & 0 & 0 \\
0 & a_{32} & a_{33} & a_{34} & \cdots & & 0 & 0 \\
\cdot & \cdot & & & & & & \\
\cdot & \cdot & & & & & & \\
\cdot & \cdot & & & & & & \\
0 & 0 & 0 & 0 & \cdots a_{n-1,n-2} & a_{n-1,n-2} & a_{n-1,n} \\
0 & 0 & 0 & 0 & 0 & & a_{n,n-1} & a_{nn}
\end{vmatrix},
\tag{11.6}
$$

where $a_{ii} = r(s_i) - \{D(s_i) + D(s_{i+1})\}/(\Delta s)^2$, $a_{i,i-1} = D(s_i)/2(\Delta s)^2$, and $a_{i,i+1} = D(s_{i+1})/2(\Delta s)^2$. The one-dimensional grid interval length, Δs, must be chosen judiciously on a scale length small with respect to the variation in $D(s)$ and $r(s)$. Then no important spatial changes in these parameters between two adjacent grid points are likely to be missed.

Let $\mathbf{A}_i$ be the ith principal minor of $\mathbf{A}$. Recall from the previous chapters that if, for any value of i, $(-1)^i \det(\mathbf{A}_i) < 0$, then the system is unstable, and the species persists, since a small increment in the population greater than zero would tend to grow. It may not be necessary to evaluate a large matrix representing the whole region. As we discussed in Chap. 10 with respect to forest patches, a small subset of the entire region may be enough to maintain the species. For example, if r is especially large near points i and $i+1$, we might check the submatrix

$$
\mathbf{A}_2 =
\begin{vmatrix}
a_{ii} & a_{i,i+1} \\
a_{i+1,i} & a_{i+1,i+1}
\end{vmatrix}.
\tag{11.7}
$$

[Note that $\det(\mathbf{A}_2)$ is a principal minor of the matrix $\mathbf{A}'$ obtained by exchanging rows and columns such that i and $i+1$ replace 1 and 2, respectively. Hence $\mathbf{A}'$ is fully equivalent to matrix $\mathbf{A}$ and we can look at its principal minors.] If $(-1)^2 \det(\mathbf{A}_2) < 0$, then this part of the region is capable of sustaining the species, so the species will persist in the entire region. Of course, this discrete-spatial system is only an approximation of the continuous system, but it can be made arbitrarily close by choosing the grid interval smaller and smaller.

There is one sense in which Eq. (11.1) may be a poor representation of a population in space. This equation assumes that movement of the species is by gradual, point-to-adjacent-point diffusion. Actually, many species, even those in which dispersal takes place only in the seed stage, can travel considerable distances in brief instants of time. A more suitable model than Eq. (11.1) in many cases might be one that incorporated long scale movements, such as pictured by the dotted lines in Fig. 11.1, as well as the diffusional movements. Equation (11.4) could easily be extended to cover this situation, but Eq. (11.1) would have to be generalized to an integro-differential equation to incorporate more than just local movements, and would thus be exceedingly difficult to analyze.

We can conclude then that simple biological populations in space, involving reproduction and spatial diffusion, contain positive feedback loops. The strengths of the reproductive positive feedback loops relative to diffusional losses from the system determines whether or not the population will persist.

11.3 Pattern Formation Through Instability

We have not yet explicitly seen the role that diffusion can play in pattern formation. Turing (1952) was probably the first to show that in spatially homogeneous reacting chemical systems, the occurrence of diffusion could sometimes lead to an instability, giving rise to spatial inhomogeneities, and perhaps to a stable spatial pattern. Turing was interested primarily in the implications of this phenomenon in morphogenesis, but since models of chemical systems are analogous to some models of communities of biological species, there may also be ecological implications, as has been pointed out (Levin, 1974, 1977; Levin and Segel, 1976; Okubo, 1978; Mimura, 1979; Mimura et al., 1979).

Here we outline the concept of this diffusion instability, following Okubo's (1974, 1980) description, to show how it relates to positive feedback processes. Okubo considered two interacting species, X_1 and X_2, initially spread uniformly in a homogeneous space. The set of equations,

$$\frac{\partial X_1}{\partial t} = D_1 \frac{\partial^2 X_1}{\partial s^2} + F_1(X_1, X_2), \tag{11.8a}$$

$$\frac{\partial X_2}{\partial t} = D_2 \frac{\partial^2 X_2}{\partial s^2} + F_2(X_1, X_2), \tag{11.8b}$$

is assumed to describe the interacting species diffusing along one dimension, where D_1 and D_2 are (positive constant) diffusion rates. The precise nature of the interaction need not be specified at this point.

Suppose in the absence of diffusion the equilibrium point for species densities is $(\bar{X}_1, \bar{X}_2)$, where $F_1(\bar{X}_1, \bar{X}_2) = 0$, $F_2(\bar{X}_1, \bar{X}_2) = 0$, and suppose this equilibrium point is stable. Now we can assume some sort of perturbation, varying from point to point in space. To determine whether or not the system is stable to the perturbation, we can linearize the system near the equilibrium point, by setting $X_1(s, t) = \bar{X}_1 + x_1(s, t)$, $X_2(s, t) = \bar{X}_2 + x_2(s, t)$ in the equations and saving only first order terms in $x_1(s, t)$ and $x_2(s, t)$. The perturbation equations are then

$$\frac{\partial x_1(s, t)}{\partial t} = D_1 \frac{\partial^2 x_1(s, t)}{\partial s^2} + a_{11} x_1(s, t) + a_{12} x_2(s, t), \tag{11.9a}$$

$$\frac{\partial x_2(s, t)}{\partial t} = D_2 \frac{\partial^2 x_2(s, t)}{\partial s^2} + a_{21}x_1(s, t) + a_{22}x_2(s, t), \qquad (11.9b)$$

where $a_{ij} = dF_i/dX_j|_{(\bar{X}_1, \bar{X}_2)}$.

Any spatial perturbation can be represented mathematically as a series of sine and cosine terms, or, equivalently, as a series of terms of the form exp (iks), where i is the square root of -1 and where, in a finite region, k can take on any integral value. We can thus study the stability of the system to spatial perturbations by examining, for any value of k, assumed perturbations of the form

$$x_1 = \exp(iks) \quad \text{and} \quad x_2 = \exp(iks). \qquad (11.10)$$

Substituting these into Eq. (11.9a, b), we obtain the stability matrix,

$$\mathbf{A}' = \begin{vmatrix} a_{11} - D_1 k^2 & a_{12} \\ a_{21} & a_{22} - D_2 k^2 \end{vmatrix}. \qquad (11.11)$$

Suppose that $a_{21}a_{12} > 0$; that is, the species involved are competitive or mutualistic. Suppose further that in the absence of diffusion $(D_1 = D_2 = 0)$, the system is stable. The assumption of stability in the absence of diffusion implies, from our stability criteria relevant to positive linear systems, that the principal minors must obey certain inequalities; i.e.,

$$a_{11} < 0, \qquad (11.12a)$$

$$a_{11}a_{22} - a_{21}a_{12} > 0. \qquad (11.12b)$$

We now ask if there are any integral values of k for which the system is unstable when diffusion is included; that is, we ask what integral values of k satisfy either of the conditions that would imply system instability; i.e.,

$$(a_{11} - D_1 k^2) > 0, \qquad (11.13a)$$

$$(a_{11} - D_1 k^2)(a_{22} - D_2 k^2) < a_{12}a_{21}. \qquad (11.13b)$$

Since from Inequalities (11.12a, b) it follows that $a_{11}, a_{22} < 0$, it is apparent that, if Inequalities (11.12a, b) are satisfied, Inequalities (11.13a, b) cannot be satisfied. Hence, mutualistic and competitive two-species systems, unlike predator-prey systems (Okubo, 1980), will not exhibit diffusive instabilities. It is possible to generalize this result to positive feedback systems of any number of species.

193

11.4 Congregation of Colonial Organisms

One way in which patterns can emerge in spatially distributed populations is through aggregation. In the case of insects, this aggregation is often facilitated by pheromone attractants. Spatial models have been made for certain types of aggregation, such as that by colonial organisms like slime molds (Keller and Segel, 1970) and social insects (Deneubourg, 1976). Since the two models are similar, we shall consider only the model of the building of a termite nest due to Deneubourg and described by Nicolis and Prigogine (1977) and Prigogine and Stengers (1984).

It has been proposed that the early stages of termite nest building can be divided into two phases (Grasse, 1959). At first, building material is deposited more or less randomly. When a deposit in one place becomes great enough relative to its surroundings, however, termites will preferentially deposit more material near that place, leading to the gradual erection of a wall or pillar.

Deneubourg (1977) hypothesized that chemotaxis, or directional motion induced by variations in chemical concentration, accounted for the aggregation, as is also the case for slime molds, and showed, by means of a simple model, how this could operate. Let C be the concentration of the insects depositing material and P be the concentration of the material itself. The chemical substance responsible for the chemotaxis is mixed with the building material. Its concentration is represented by H. All three variables may change in both space and time. The chemical attractant diffuses through space and attracts termites towards regions of higher H. Termites lay down more building material in these regions, further increasing P and, thus, indirectly, increasing H in these regions. The equations for the entire system can be written as

$$\frac{\partial P}{\partial t} = k_1 C - k_2 P, \tag{11.14a}$$

$$\frac{\partial H}{\partial t} = k_2 P - k_4 H + D_H \nabla^2 H, \tag{11.14b}$$

$$\frac{\partial C}{\partial t} = F^e - k_1 C + D \nabla^2 C + \gamma \nabla \cdot (C \nabla H), \tag{11.14c}$$

where $D_H \nabla^2 H$ and $D \nabla^2 C$ are terms representing two-dimensional diffusion of, respectively, attractants and insects (since there is some degree of randomness in the insect movement patterns), $\gamma \nabla \cdot (C \nabla H)$ represents the chemotactic force on the insects ($\gamma < 0$), F^e is the flux of insects bringing new building material from outside the region being modeled, and the k_i's are constants, with k_1 representing the rate of material deposition by insects, k_2 a coefficient for transition from building material to attractant, and k_4 a coefficient of the denaturing of the attractant.

It can be shown (see Nicolis and Prigogine, 1977) that there are two possible steady state solutions of this set of equations. One of these is the spatially homogeneous state, $C_0 = F^e/k_1$, $P_0 = F^e/k_2$, $H_0 = F^e/k_4$; that is, the state that exists before termite nest construction begins. Only if this state is unstable can

194

aggregations develop. We shall not analyze this model in detail since it is done elsewhere, but we attempt to point out the positive feedback characteristic that gives rise to the spatial inhomogeneity. First, linearize about the equilibrium state by letting $C = C_0 + c$, $H = H_0 + h$, and $P = P_0 + p$, and then, for the sake of convenience, replace the continuous spatial formulation by a one-dimensional spatially discrete formulation, so that the term $\gamma C_0 \nabla^2 h$ is replaced by $(\gamma C_0 / \Delta^2)\{h(s_{i+1}) - 2h(s_i) + h(s_{i-1})\}/2$, where Δ is the grid spacing distance between cells and s_i is the discrete spatial variable. Let us, for the moment, assume that diffusional effects are realtively small and can be ignored; $\nabla^2 H = 0$, $\nabla^2 C = 0$. The linearized model then takes the mathematical form,

$$\frac{dp(s_i)}{dt} = k_1 c(s_i) - k_2 p(s_i), \qquad (11.15a)$$

$$\frac{dh(s_i)}{dt} = k_2 p(s_i) - k_4 h(s_i), \qquad (11.15b)$$

$$\frac{dc(s_i)}{dt} = -k_1 c(s_i) + (\gamma C_0 / \Delta^2)\{h(s_{i+1}) - 2h(s_i) + h(s_{i-1})\}. \qquad (11.15c)$$

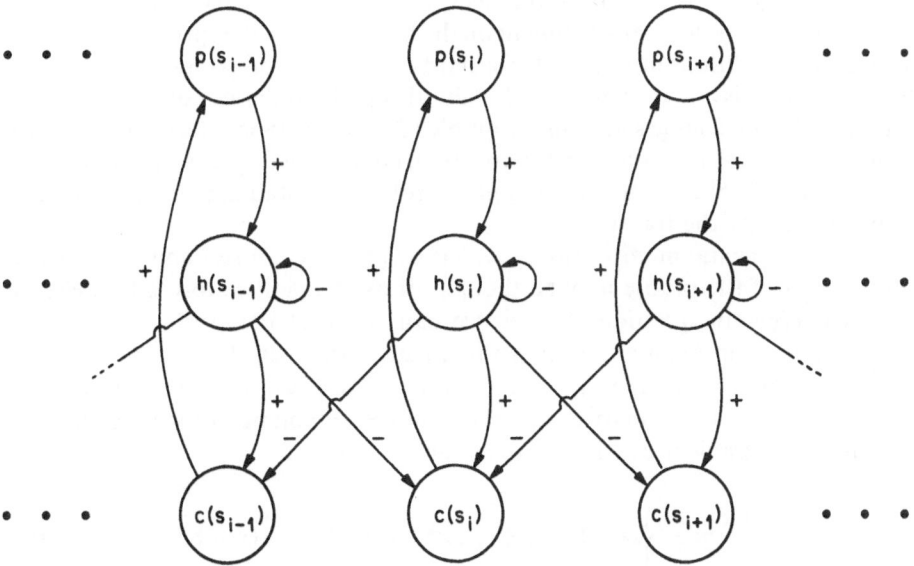

Fig. 11.2. The feedback loops associated with the social insect colony growth described by Eq. (11.15a–c). Note that all feedback loops of length greater than 1 are positive, including such loops as $h(s_i) \to c(s_{i+1}) \to p(s_{i+1}) \to h(s_{i+1}) \to c(s_i) \to p(s_i) \to h(s_i)$

It is easy to see from the feedback diagram (Fig. 11.2) that the set of Eq. (11.15a–c) is a system analogous to a set of competing positive feedback subsystems (do not forget that $\gamma < 0$). If the positive feedback of this network is strong enough to overcome the denaturing loss term, $-k_4 h(s_i)$, then a small perturbation anywhere in the system will continue to grow. For example, if either p or h or c is increased in given spatial cell, say cell i, all three variables in that cell will increase in size at the expense of neighboring cells. The omitted diffusion terms are negative feedback in character and will tend to suppress the spatial structuring unless the positive feedback is strong enough.

11.5 Boundary Formation by Competition

When more than one species is present, complex spatial patterning can develop as a result of the coupling of spatial dispersion with predator-prey, competitive, and mutualistic interactions. Spatial variation in environmental conditions (e.g. temperature, moisture) normally leads to gradual replacement of individual species by others along the gradient. Most transition zones, or ecotones, between floristic types are rather broad and diffuse, but many are remarkably narrow (e.g, Oosting, 1955). A frequent explanation for the steepness of such ecotones is that the environmental gradients are also steep and that interspecific competition enhances the effect (Emlen, 1972, p. 319). For example, the sharp altitudinal boundary between hardwood and boreal forests at about 792 m in the Green Mountains of Vermont (and at other elevations elsewhere) is attributed by Siccama (1974) in part to a climatic discontinuity near this elevation. The boundary is reinforced by modifications of the local environment on the boreal side by the spruce and fir trees, which make conditions inhospitable for hardwoods. The forest-prairie interface in North America is a case where strikingly abrupt boundaries sometimes occur, although climatic changes are imperceptible (Transeau, 1935). The reason for the abruptness of these boundaries has never been completely explained, although it is usually assumed that fire and grazing play a role in maintaining pure prairie up to the boundaries of the forest.

The environmental modifications created by the boreal forest and the susceptibility to fires of the prairie may be thought of as competitive forces between the ecosystem types in question. Hence, we can say that competition acts as an organizing force in maintaining sharp boundaries between these types.

The fact that sharp boundaries can result from competition has been established theoretically from mathematical models. A model of competing species along an environmental gradient consists of the coupled equations

$$\frac{\partial X}{\partial t} = \frac{\partial}{\partial s}\left\{ D_1(s)\,\frac{\partial X}{\partial s}\right\} + r_1(s)X - g_1(s)X^2 - c_1(s)XY, \qquad (11.16a)$$

$$\frac{\partial Y}{\partial t} = \frac{\partial}{\partial s}\left\{ D_2(s)\,\frac{\partial Y}{\partial s}\right\} + r_2(s)Y - g_2(s)Y^2 - c_2(s)XY, \qquad (11.16b)$$

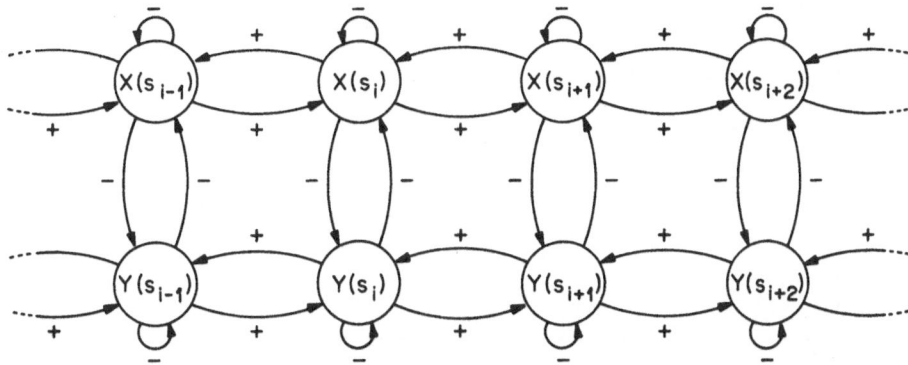

Fig. 11.3. System diagram of two competing species in a one-dimensional space, represented by a line of discrete points

where X and Y are the densities of the two competing species and $c_1(s)$ and $c_2(s)$ are competition coefficients. The positive feedback aspects of this system can be shown by transforming it to a spatially discrete form analogous to Eq. (11.4), as represented graphically in Fig. 11.3.

Suppose that the environmental conditions along the gradient cause the coefficients $r_i(s)$, $g_i(s)$, and $c_i(s)$ to change in such a way that, for small values of s, species X is favored over species Y, while for large values of s, species Y is favored over species X. In particular, for small values of s

$$r_1(s)/c_1(s) > r_2(s)/g_2(s), \tag{11.17a}$$

$$r_1(s)/g_1(s) > r_2(s)/c_2(s), \tag{11.17b}$$

and for large values of s these inequalities are reversed. If in the region intermediate to these extremes, Inequality (11.17a) holds, but Inequality (11.17b) is reversed, then there is a smooth transition from dominance by one species to dominance by the other species along the gradient (Fig. 11.4a). Competition acts to sharpen the ecotone beyond what it would be without competition, but it is still rather smooth.

On the other hand, if Inequality (11.17b) holds in the intermediate region, but Inequality (11.17a) is reversed, then the positive feedback between X_i and Y_i is stronger than the self-limitation effects, $-g_1(s) X_i^2$ and $-g_2(s) Y_i^2$, and the process of competitive exclusion operates in this area (Fig. 11.4b). If there were no diffusion of populations, only one species could occupy a given point in space along the gradient, the species occupying each point being determined by initial conditions. An unlimited number of stable spatial species distributions could exist. If $D_1(s)$ and $D_2(s)$ are non-zero, neither species can be entirely excluded from anyplace along the environmental gradient, but positive feedback effects will produce a sharp, almost discontinuous ecotone (Zeeman, 1974; Yamamura, 1976).

The discontinuities predicted by this model are evident in many cases in nature; for example, shrub-grass boundaries. Grasses and shrubs tend to be competitively

197

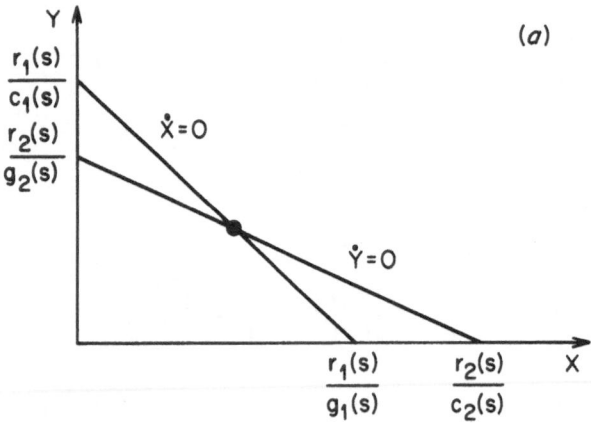

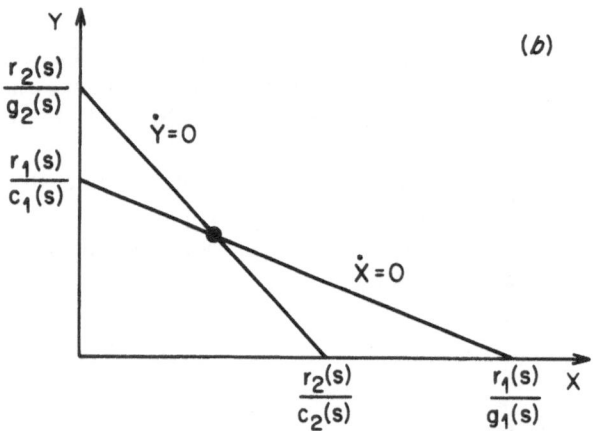

Fig. 11.4a and b. Zero isoclines for two competing species along an environmental gradient. If the isoclines in the region of species overlap have the configuration shown in **a**, the transition from dominance of one species to dominance of the other will be smooth. If they have the configuration shown in **b**, the transition will be sharp

exclusive because, among other reasons, (1) grasses reduce moisture in top soil layers, preventing shrub seedling growth, and (2) shrubs shade out grass seedlings (Ricklefs, 1976).

While interspecific competition can lead to such spatial discontinuities as discussed above, intraspecific competition can lead to other types of spatial regularities. As is well known, intraspecific competition is expressed as territoriality in many species. This territoriality can easily be observed in the regular spacing of some desert plants that exclude others from a certain radius of their stem by root competition for water. Some animals, especially birds, are behaviorally adapted to defend territories, at least during the breeding season. Territoriality is generally explained in terms of the enhanced fitness conferred on the holders of territories, who guarantee for themselves a stable food supply (e.g., Wilson, 1975). Often the territories take on almost precise geometrical shapes, such as the visible hexagonal territories of male mouthbreeder fish on the sandy bottoms of aquatic waterbodies (Barlow, 1974), though usually they are approximately circular. In any case, territories ensure a fairly even spacing of organisms.

One could feasibly model the emergence of spatial territorial patterns in almost the same way that boundaries between competing species were modeled above. It would only be necessary to reinterpret X and Y as measures of ownership of space.

Territoriality on the part of one species can alter the entire ecological community. For example, Mech (1977) has studied the territoriality of wolves. Wolf packs in northern Minnesota inhabit territories of 125 to 310 km^2, surrounded by "*buffer zones*" of about 2 km, which they keep away from to avoid possibly dangerous encounters with wolves of other packs. These buffer zones provide refuges for a chief prey item of the wolves, white-tailed deer. The relatively high concentration of deer in the buffer zones is especially marked when the deer population undergoes a decline. (One might also expect some spatial patterns in the vegetation since browsing by deer will be greatest in the buffer zones.) Mech speculated that the existence of the buffer zone refuges could soften the severity of predator-prey cycles by allowing sufficient deer to survive any population decline caused by wolf predation. As the deer start to build back up in population level, they again invade the center areas of the wolf territories. Predator-prey cycles in time and space might be discerned over the long term.

Wiens (1976) illustrates how the even spacing of competing harvester ant colonies (*Pogonomyrmex*) imparts the same regular patterning to many other aspects of the local ecosystem. By clearing the vegetation in a disc around their mound, the ants reduce transpiration from that disc and, therefore, increase the soil water below. The increased soil water, the aeration of the soil due to the ant's digging, and the nutrients imported from their foraging help increase the growth of plants around the periphery of the disc. Rabbits may add to the concentration of nutrients by using the mounds as latrines. The spatial pattern can be self-perpetuating for long time periods, some ant colonies lasting as long as forty years. Majer (1976) has also ascertained, by experimental manipulation, that ant colony mosaics consisting of diverse ant species in Ghana cocoa farms are maintained by interspecific competition, possibly involving the use of pheromones. This patterning of colonies affects other ecosystem aspects since the ant species vary in their utilization of prey species. As in the case of wolf territories studied by Mech (1977), there are "no-man's lands" between ant colonies that act as buffer zones and incidentally provide refuges for some prey, thus preventing the exhaustion of resources.

11.6 Summary and Conclusions

Spatial patterns can emerge from the dynamic interactions among the components of almost any conceivable system, ranging from physico-chemical systems to human social systems. Such emergence of pattern can occur in an initially uniform spatial environment when a perturbation that disturbs the uniformity is reinforced by positive feedback.

Spatial diffusion tends to act as a negative feedback in a spatially distributed system. An isolated population occupying a given locale and undergoing diffusional emigration must have a higher intrinsic growth rate to survive than it would need if there were no diffusion. Diffusion can trigger instabilities in predator-prey systems, but we show that it will not destabilize two-species mutualistic or competitive systems.

Two types of formation of spatial pattern by organisms, nest building by colonial insects and the formation by competing vegetational types of sharp gradients, are examined in some detail through analysis of models.

12. Disease and Pest Outbreaks

Until recent times, the periodic irruptions of infectious diseases have been significant factors in the lives of humans and in the history of civilized society. Today many of the serious infectious diseases of humankind have been controlled. However, the increased control of human infectious disease has not lessened the importance of studying and understanding infectious diseases, because agricultural crops as well as forest resources are continually damaged by outbreaks of pests and disease. Strategies for their elimination or control are urgently needed.

The impetus for many studies of the population dynamics of infectious diseases, parasites and pests is a desire to understand the outbreak of epidemics. The sudden appearance, spread and subsequent disappearance of a disease may be viewed as an instability due to the positive feedback nature of infected individuals spreading disease to susceptible individuals. The positive feedback that leads to the spread of a disease can arise from many population mechanisms. These include (1) enhanced susceptibility to infection due to pathogen or pest induced physiological weakening or lowered immunological defense, (2) mutualistic interactions with other pest or disease organisms, (3) continued invasions of pests or infected individuals from other locations, and (4) reservoirs of infections in alternate host or vector populations.

This chapter will consider each of these types of positive feedback in the disease process. The next section considers the case where pest outbreaks are furthered by physiological weakening of the host. The Section 12.2 examines the effect of mutualistic interactions of disease and pest organisms. The Section 12.3 presents and analyzes a model of the progress of a directly communicated disease in a host population. This model is then used in the next sections to examine the role of interactions between separate population centers on threshold conditions for the persistence of disease in host populations. The final section uses a suitable modification of the general model of section four to examine the dynamics of diseases borne by vectors that have their own population processes.

12.1 Physiological Effects in the Host Species

Competition for food and other essential resources may be important in an individual host's survival, particularly at high host densities. Parasitic infection often reduces the competitive fitness of a host and, therefore, may be responsible for further physiological weakening, usually through malnutrition, and increased mortality of weak hosts. Viewed at a population level, the transition from a mild infestation of pests or disease to a sizable irruption may be due to individuals that are weakened, so that their capacity to resist further spread of invaders is lowered.

An example of the effects of bark beetles in tree populations is given by Varley et al. (1974). Immature beetles first feed on the twigs and foliage of the trees. Females then excavate galleries in the cambium layers of the trunk and larger branches of trees that have been sufficiently weakened by these attacks, where they deposit eggs. Unlike healthy trees, that can flood the galleries with sap or resin, the weakened trees succumb. The next generation of beetles, hatched from the deposited eggs, will then migrate to nearby trees and feed on them until they are similarly weakened enough for another cycle of reproduction.

Behavior patterns of some insects have evolved to take advantage of the power of numbers in weakening host trees. Varley et al. (1974), citing McNew (1971), noted that the female of the Western pine beetle emits a pheromone that attracts other pine beetles of both sexes, apparently so they can concentrate their attacks on one tree at a time. The tree's defenses are overwhelmed and it subsequently becomes the source of a local outbreak of beetles.

Ludwig et al. (1978) incorporated physiological weakening of host spruce trees as an important feedback mechanism in their model of spruce budworm outbreaks. The mechanism is built in by considering two variables to describe a forest stand, the total surface area of branches, S, and the *"energy reserve"*, E, of the stand. The surface area of branches is roughly proportional to the rate of photosynthesis in the stand, and so this surface area is proportional to the rate of replenishment of the energy reserve, at least up to a point. Carbohydrates from the energy reserve go into making new branches. Hence the two variables interact in analogy to a pair of mutualistic species. Ludwig et al. (1978) modeled the dynamics of their interaction by the equations

$$\frac{dS}{dt} = r_s S \left(1 - \frac{SK_e}{K_s E} \right), \tag{12.1}$$

$$\frac{dE}{dt} = r_e E \left(1 - \frac{E}{K_e} \right) - p \frac{B}{S}. \tag{12.2}$$

In these equations r_s and r_e are the maximum rates of increase of the variables and K_e is the carrying capacity of energy reserves, p is the rate of energy consumption by budworm, and K_s is the maximum value for S. The carrying capacity for live branches is $K_s E/K_e$ which is thus assumed dependent on the current reserve of energy in the stand. The symbol B is the density of spruce budworms, also assumed to be a variable, but effectively a parameter in the present context. Note that the negative effect of budworm stress on the energy reserve is proportional to B/S, the number of budworms per branch. The zero isoclines of the equations for branch area and energy reserve can be plotted on a phase diagram (Fig. 12.1). For low budworm densities the analogy with a two species mutualistic system becomes obvious. There are three equilibrium points, 0, C, and D. The points 0 and D are always stable, while C is always unstable. For a range of initial values of E and S

lying above the separatrix (not drawn) in the plot, the variables reinforce each other to drive the system to equilibrium D. If the initial values of E and S lie below the separatrix, the mutual interaction will drive both towards extinction.

We need not consider in detail the rest of the model by Ludwig et al. (1978). It is sufficient to say that, if conditions are favorable (i. e., if S is large enough) the spruce budworm will escape its predators and undergo a *"catastrophic"* increase. The effect of this increase will be to alter the isoclines in Fig. 12.1 to the new configuration shown in Fig. 12.2. The equilibrium point D no longer exists, so the positive

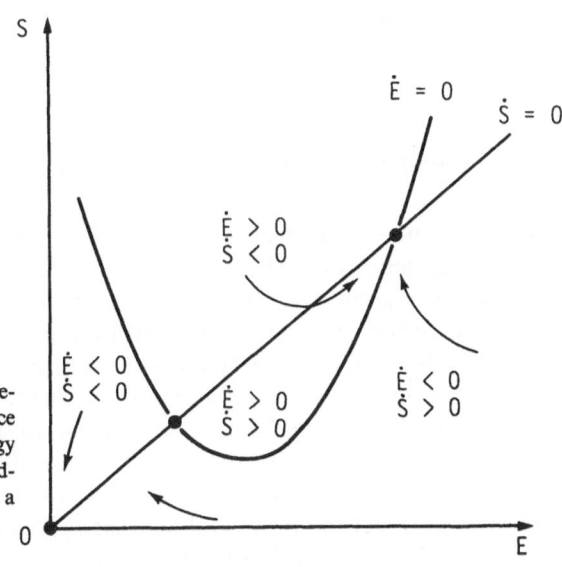

Fig. 12.1. Isoclines for equations representing the interaction between surface area of branches (S) and the total energy reserve (E) of a forest stand when budworm densities are low. In this case a stable equilibrium (D) is possible

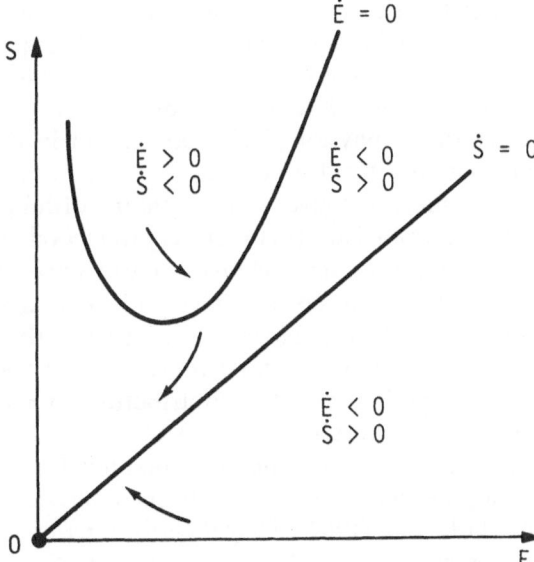

Fig. 12.2. Isoclines of the equations for surface area (S) and stand energy reserve (E) when the budworm levels are high. Here there is no stable equilibrium other than extinction

feedback between the branch surface area and energy reserve drives both toward zero. The energy reserve has been so reduced by the budworms that it cannot supply sufficient energy to maintain life processes in the branches, with the result that the branches are unable to photosynthesize at a rate sufficient to replenish the energy reserves. Normally, as the complete model would describe, the stand does not diminish all the way to extinction, but recovers at some point after the local budworm population has starved.

Other host-parasite systems may also exhibit mutual reinforcement due to debilitation of the host. Anderson (1980), speaking about host-parasite systems in general, pointed out that parasitic infections can lead to malnutrition of the host. As its food level decreases, the host's ability to mount an effective immunological response may further decrease, leading to a greater parasite pathogenicity.

12.2 Mutualistic Interactions of more than one Pathogenic Agent

The causes underlying an outbreak of a disease or pest are often complex, and may involve interactions among two or more agents. For example, Alexander (1971) reports: "The stimulation of pathogenesis is quite noteworthy with *Entamoeba histolytica*; when introduced into germfree guinea pigs, the protozoan does not cause amoebic infection or lesions, while the disease develops when *E. histolytica* is inoculated either into germfree animals together with several bacteria species or into nongermfree guinea pigs (Phillips, 1968), and, in some cases, 'harmless' organisms, through their cooperative efforts, produce disease, or two weakly virulent pathogens do more harm than either does alone. Among the animal pathogens, such synergistic actions may involve either a virus and a bacterium or two dissimilar bacteria, and it is now widely recognized that certain animal or human diseases are more prevalent than usual during epidemics caused by other pathogens, as with bacterial pneumonia and influenza (Arndt and Ritts, 1961; Loosli, 1968). Synergism is likewise evident among plant pathogens, be they viruses or fungi (DeVay, 1956; Kassanis, 1963). The extent to which these effects result from an influence of one microorganism on another or, alternatively, arise because of some structural or physiological modification in the host, predisposing it to a new infection, is difficult to assess."

Graham (1967) gives some examples of fungal-insect mutualisms. Some of the most successful bark beetles in very wet wood are those that transmit blue-staining fungi. These fungi may act on the tree to make it more suitable for the beetle. Yeasts are also known to be associated with bark and ambrosia beetles. When they are introduced into tree phloem by bark beetles, they provide the beetles with digestible food. Although there is no definite proof, the yeasts may also produce substances that attract beetles, thus contributing to mass attacks by which the beetles overwhelm host resistance.

An example of a complex case that involves both synergistic interactions among pathogens and also feedback effects of physiological weakening of plants, is given by Clark et al. (1967). They describe a pathological condition that involves early degeneration and death of cacao trees in West Africa. According to Clark et al.

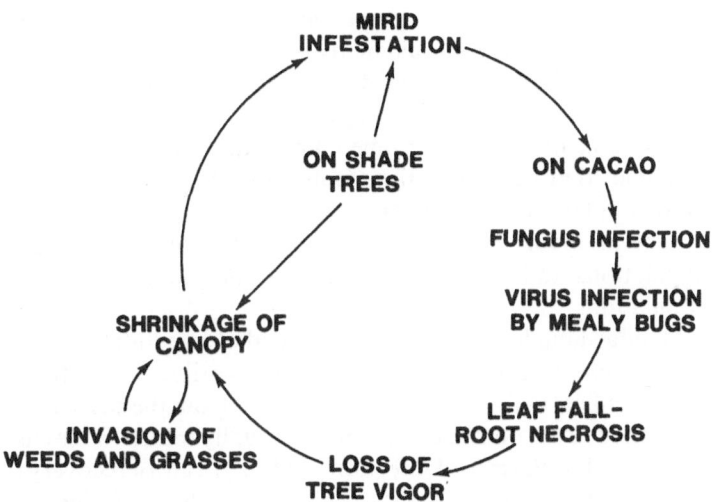

Fig. 12.3. Interaction pattern of organisms that contribute to loss of plant vigor in cocoa plantations

(1967): *"Two agents were held primarily responsible for the crisis: a virus causing the disease known as 'swollen shoot' virus disease (SSVD); and mirid bugs (capids) causing defoliation and loss of tree condition, often aggravated by a fungus infection. Although both agents could act independently, it was found that their effects usually combined to produce compound and far-reaching sequels, and, therefore, that they were best regarded as co-acting elements in a single complex process which determined the premature degeneration of cacao plants (e.g., Kay et al., 1961)."*

As a result of the attacks of these agents on the cacao trees, leaf fall and root necrosis occur, resulting in loss of tree vigor and shrinkage of the canopy. This contributes to a positive feedback cycle causing further damage, since mirids feed on the vegetative parts of exposed or inadequately shaded cacao trees (Fig. 12.3). Planting cacao trees in dense plantations also encouraged a positive feedback cycle that did not occur so easily under natural forest conditions. The spread of SSVD depends on the mealybug, which must transmit the virus from one tree to another in the short time they remain virulent. Dense stands of cacao trees increase the ease of dissemination.

In summary, Clark et al. (1967) describe the problem of the West African cacao as resulting from synergism among (1) human use of cacao, (2) polyphagous mirid, (3) fungus, (4) plant viruses, (5) polyphagous mealybugs, and (6) susceptibility of cacao under cropping conditions.

12.3 Models of a Directly Communicated Disease or Parasite

The spread of disease in a population depends on the relative rate of contact that susceptible individuals within a local population have with infected individuals either within the local population or from other populations that are reservoirs of the disease. These reservoirs may be other populations of hosts or populations of vector organisms. The contact with other populations containing infected individuals will cause a mutual increase in the contact rate of susceptible and infected individuals, thus generating positive feedback between these groups. All populations of hosts or vectors, whether they be separate populations of host or populations of vectors or some combination of both, have their own population dynamics that determine the amount of influence the positive feedback between local populations will have on the progress of the disease. In this section we develop a general model describing the dynamics of the disease process in isolated populations. In later sections we extend this model to include other host or vector populations.

Standard models describing the spread of infectious diseases typically divide the host population into three categories; susceptibles (X), infecteds (Y), and immunes (Z). Members of the host population move from the susceptible class to the infected class, then to the immune class and eventually return to the susceptible class at rates determined by the rate of infection [$f(X, Y)$], rate of recovery from infection (v) and the rate of loss of immunity (γ), respectively. New individuals are introduced through reproduction as susceptibles at a per capita rate determined by a function $g(X, Y, Z)$, which decreases toward zero as the total population increases toward some maximum size. These flows of individuals from one class to another are depicted in Fig. 12.4. The recovery of infected individuals and loss of immunity together make possible a positive feedback loop, which is broken if either the disease is always fatal or immunity is permanent. Another positive feedback loop occurs due to reproduction of susceptible individuals by infected and recovered individuals (provided the newborn individuals are not already infected). Both of these positive feedback loops are countered by a negative feedback loop between susceptible and infected individuals as well as by losses from each class due to death. These death-

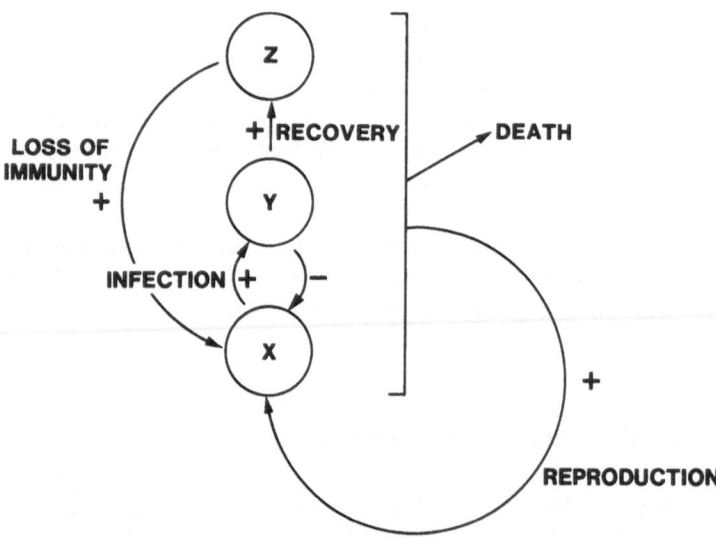

Fig. 12.4. Pattern of flow of individuals between susceptible (X), infected (Y) and recovered or immune (Z) classes in a communicable disease cycle. Here we assume all newborn individuals are susceptible

206

related losses include the infection free death rate of the host population (b, assumed to be the same for each class) and the increased death rate of infected individuals (α). Representing all of these gains and losses from each host population class as positive and negative terms in differential equations, we have

$$\frac{dX}{dt} = g(X, Y, Z)(X + Y + Z) - bX - f(X, Y) + \gamma Z, \qquad (12.3)$$

$$\frac{dY}{dt} = f(X, Y) - (\alpha + b + v) Y, \qquad (12.4)$$

$$\frac{dZ}{dt} = vY - (b + \gamma) Z. \qquad (12.5)$$

The usual procedure in analyzing the dynamics of these equations is first to assume that the total population size remains constant (N). This simplifies the analysis since the three-equation system may then be reduced to a two-equation system by eliminating (12.5) and substituting $N - (X + Y)$ for Z into (12.3). The reduction allows one to employ powerful mathematical tools to explore the global dynamics of the system of equations (see Hethcote, 1976; Copasso and Serio, 1978).

Such a simplification is inadequate for the study of disease in natural animal populations, and human populations when infectious disease is a major cause of mortality. Anderson (1979) and Anderson and May (1979) forcefully argue that a broader understanding of the population biology of infectious diseases is possible if the host population is taken as a dynamic variable rather than a constant. They present a detailed analysis of several systems of equations similar to (12.3)–(12.5).

In order to examine the dynamics of the disease process described above we will analyze the stability properties of Eqs. (12.3)–(12.5). To accomplish this we will introduce particular forms of the functions $g(X, Y, Z)$ and $f(X, Y)$. Assume the per capita birth rate to be logistic with a maximum fecundity, a, when the population is small and decreasing to zero as the population approaches a maximum population size ($1/c$):

$$g(X, Y, Z) = \begin{cases} a[1 - c(X + Y + Z)] & \text{if } X + Y + Z < 1/c, \\ 0 & \text{if } X + Y + Z \geq 1/c. \end{cases} \qquad (12.6)$$

The mathematical form that $f(X, Y)$ takes depends on the natural history of the particular disease in question. For example, some diseases are spread by airborne droplets while others are spread by contaminated fecal matter. The rate $f(X, Y)$ would certainly be different in the two cases. As a general rule, however, we can say that $f(X, Y)$ will roughly increase with increasing X and Y. Let us assume that

$$f(X, Y) = \beta XY, \qquad (12.7)$$

where β is a transmission coefficient.

When the disease is not present in the population, the infected and immune classes are zero and the equilibrium size of the susceptible class ($\bar{X}$) is found by setting the nonzero portions of (12.3) equal to zero and solving for $\bar{X}$:

$$\bar{X} = \frac{a-b}{ac}. \tag{12.8}$$

We assume that $a > b$ so that this equilibrium will be positive. In order to determine whether or not a disease can become established in a population of susceptible individuals, it is sufficient to demonstrate that the infected class can grow from an initially small number of infected individuals. We accomplish this by determining the conditions under which the equilibrium $(\bar{X}, 0, 0)$ is locally unstable. The stability matrix of (12.3)–(12.7) evaluated at this equilibrium point is

$$S = \begin{bmatrix} a - 2ac\bar{X} - b & a - 2ac\bar{X} - \beta\bar{X} & a - 2ac\bar{X} + \gamma \\ 0 & \beta\bar{X} - (\alpha + b + v) & 0 \\ 0 & v & -(b + \gamma) \end{bmatrix}. \tag{12.9}$$

The eigenvalues of the stability matrix S are $\lambda_1 = a - 2ac\bar{X} - b$, $\lambda_2 = \beta\bar{X} - (\alpha + b + v)$, and $\lambda_3 = -(b + v)$. Using (12.8), we see that $\lambda_1 = -(a - b)$ is always negative, as well as λ_3. If the infected class is to increase, then λ_2 must be positive or

$$\bar{X} = \frac{a-b}{ac} > \frac{\alpha + b + v}{\beta}. \tag{12.10}$$

This implies that there exists a threshold population size that must be realized before a disease may become epidemic in a population. The threshold depends on how fast infected individuals are removed from the infected class compared to the transmission rate of the disease. If the transmission rate is comparatively large, then the threshold population size is quite small. The reverse is true when the transmission of the disease is quite slow.

When (12.10) is satisfied, the infected and recovered classes will increase toward a new equilibrium $(\hat{X}, \hat{Y}, \hat{Z})$ where there will be infected individuals constantly in the population. This can be determined by setting the left hand side of Eqs. (12.3)–(12.5) equal to zero and solving for $\hat{X}$, $\hat{Y}$, and $\hat{Z}$. When the disease is present the equilibrium size of the susceptible population is

$$\hat{X} = \frac{\alpha + b + v}{\beta}. \tag{12.11}$$

The recovered or immune class is proportional to the number of infected individuals; that is

$$\hat{Z} = \frac{v}{b + \gamma} \hat{Y}, \tag{12.12}$$

and the equilibrium size of the infected class is obtained by solving the following equation for $\hat{Y}$:

$$A\hat{Y}^2 + B\hat{Y} + C = 0, \qquad (12.13)$$

where

$$A = ac\left(1 + \frac{v}{b+\gamma}\right)^2,$$

$$B = 2a\left(1 + \frac{v}{b+\gamma}\right)\left(\frac{\alpha+b+v}{\beta}\right) - ac\left(1 + \frac{v}{b+\gamma}\right) + \frac{\lambda v}{b+\gamma} - (\alpha+b+v),$$

$$C = -\left[(a-b)\left(\frac{\alpha+b+v}{\beta}\right) - ac\left(\frac{\alpha+b+v}{\beta}\right)^2\right].$$

Using the quadratic formula, we see that (12.13) has a real solution for $\hat{Y}$ if

$$B^2 - 4AC > 0. \qquad (12.14)$$

If B is negative then (12.14) is enough to ensure that $\hat{Y}$ has a positive value. If B is positive then we need the additional condition

$$C < 0. \qquad (12.15)$$

This additional condition, however, is the threshold condition for a disease to increase when rare [see Eq. (12.10)] and is therefore already satisfied. Thus, we can conclude that, if the disease-free equilibrium population is larger than a certain threshold size, then it will be unstable to the introduction of the disease. The number of infected individuals will increase to a new positive equilibrium and the disease will remain endemic in the population. If the equilibrium population size is below the threshold the infection will gradually disappear and the population will approach its infection-free equilibrium.

12.4 Effects of Spatial Heterogeneity on Disease Outbreak Threshold Conditions

Mathematical models have been used to show that an isolated host population must have a critical population size for certain contagious diseases to be maintained in the population over a long period of time. Bartlett (1957, 1960) suggested that an isolated human community must have at least 250,000 to 300,000 inhabitants for measles to be maintained. Maintenance of a disease, however, depends not only on population size, but also on the spatial arrangement of settlement. In a densely compacted population, most members of the population will soon be exposed to the infection, which may then be maintained in the population. When the population is spatially heterogeneous, on the other hand, the contagion may not become

209

established because the transmission rate between population centers is too low. A useful analogy may be provided by isolated trees as opposed to those in clusters. During brush fires isolated trees tend to survive because only grass is burning around them. Trees in groups prove mutually destructive, their burning foliage raising temperatures in a self-reinforcing way (Edlin, 1976).

We look at a deterministic model of a host population in which spatial heterogeneity is taken into account by dividing the population into several local groups that interact slightly. The dynamics within each group are described by the equations presented in the previous section. The degree of mixing between groups can be specified, relaxing the assumption of uniform mixing of all individuals. The form of the equations is such that results of positive feedback theory can be used to predict persistence of diseases as a function of the degree of interaction between subpopulations.

The incorporation of spatial heterogeneity into epidemic models can have two diametrically opposite effects. Models that incorporate continuous spatial distribution of individuals that interact strongly with neighbors and weakly with more distant individuals demonstrate a damping effect of geographic dispersion due to an effective lowering of the rate of infection (Bailey, 1975; Noble, 1974). On the other hand, the fade-out of a disease in a subpopulation below the threshold size may be countered by reintroduction from other subpopulations (Black, 1966). Thus, interaction between subpopulations can effectively raise the infection rate so that the disease may persist in the total population even when it would fade out of each separate subpopulation. It is on this latter possibility that we focus our attention.

Let us assume that the total population exists in a region of m population centers. The members of each center make short visits to at least some of the other centers. To model this realistically, we could introduce complex model equations, but we shall attempt to simplify matters as much as possible so that our basic point can be clearly made. Actually, only certain fractions of each population will visit certain other population centers. However, we shall make the simplifying assumption that all members of each center spend the same amount of time visiting other centers, although the time spent visiting depends on the center. While visiting the other centers, infected visitors will have the chance of transmitting the disease to susceptibles in the visited center, while susceptible visitors have the chance of acquiring the disease from infected members of the visited population center.

We point out here that we are not modeling the migration of individuals from one population to another since we assume that visitors return to their home subpopulation. We are, in a sense, modeling the migration of the disease itself, rather than the migration of hosts.

Equations (12.3)–(12.5) are each generalized to a set of m equations,

$$\frac{dX_i}{dt} = g_i(X_i, Y_i, Z_i)(X_i + Y_i + Z_i) - b_i X_i - \sum_{j=1}^{m} f(X_i, Y_i, T_{ij}) + \gamma_i Z_i$$

$$(i = 1, \ldots, m), \tag{12.16}$$

$$\frac{dY_i}{dt} = \sum_{j=1}^{m} f(X_i, Y_j, T_{ij}) - (\alpha_i + b_i + v_i) Y_i \quad (i = 1, \ldots, m), \tag{12.17}$$

210

$$\frac{dZ_i}{dt} = v_i Y_i - (b_i + \gamma_i) Z_i \quad (i = 1, \ldots, m). \tag{12.18}$$

The parameters b_i, α_i, v_i, γ_i and the function g_i are the same as before, only now they are specific to particular subpopulations. The functions $f(X_i, Y_j, T_{ij})$ represent the rate of contagion in subpopulation i resulting from contact with infected individuals of subpopulation center j. The parameter T_{ij} represents the time of contact as a fraction of the unit of time used in the model. This is incorporated to express the idea that the rate of contagion between two subpopulation centers should increase with the fraction of time members from each center are in contact.

We now want to determine conditions under which interactions between subpopulation centers can cause a disease to become established when each center, in isolation, is incapable of supporting the disease. This is determined by examining local stability of the equilibrium point $\bar{N} = (\bar{X}_1, \bar{X}_2, \ldots, \bar{X}_m, 0, 0, \ldots, 0, 0, \ldots, 0)$. The perturbed equations obtained from (12.16)–(12.18) in the neighborhood of N are

$$\begin{bmatrix} \dfrac{d\mathbf{X}}{dt} \\[2ex] \dfrac{d\mathbf{Y}}{dt} \\[2ex] \dfrac{d\mathbf{Z}}{dt} \end{bmatrix} = \begin{bmatrix} \mathbf{A}_{11}(\bar{\mathbf{N}}) & \mathbf{A}_{12}(\bar{\mathbf{N}}) & \mathbf{A}_{13}(\bar{\mathbf{N}}) \\[2ex] \mathbf{A}_{21}(\bar{\mathbf{N}}) & \mathbf{A}_{22}(\bar{\mathbf{N}}) & \mathbf{A}_{23}(\bar{\mathbf{N}}) \\[2ex] \mathbf{A}_{31}(\bar{\mathbf{N}}) & \mathbf{A}_{32}(\bar{\mathbf{N}}) & \mathbf{A}_{33}(\bar{\mathbf{N}}) \end{bmatrix} \cdot \begin{bmatrix} \mathbf{X} \\[2ex] \mathbf{Y} \\[2ex] \mathbf{Z} \end{bmatrix}, \tag{12.19}$$

where $d\mathbf{X}/dt$, $d\mathbf{Y}/dt$, $d\mathbf{Z}/dt$, $\mathbf{X}$, $\mathbf{Y}$, and $\mathbf{Z}$ are $(m \times 1)$ vectors and $\mathbf{A}_{ij}$, $(i, j = 1, 2, 3)$ are $(m \times m)$ matrices of the form (see Equation p. 212).

A disease can become established if the vector $\mathbf{Y} = (Y_1, Y_2, \ldots, Y_m)$ can increase when each of the Y_i's are arbitrarily close to zero, $\mathbf{Z} = (0, 0, \ldots, 0)$ and the vector $\mathbf{X} = (X_1, X_2, \ldots, X_m)$ is equal to $\bar{\mathbf{X}} = (\bar{X}_1, \bar{X}_2, \ldots, \bar{X}_m)$. This is equivalent to requiring that the linear system (12.19) be unstable at the equilibrium point $\bar{\mathbf{N}}$.

The diagonal submatrix $\mathbf{A}_{21}(\bar{\mathbf{N}})$ contains elements that express the effect of infected individuals on susceptible individuals at the equilibrium point. At this point, however, there are no infected individuals so this submatrix is zero ($\mathbf{A}_{21}(\bar{\mathbf{N}}) = 0$). Since $\mathbf{A}_{23}(\bar{\mathbf{N}})$ is also zero, the total matrix $\mathbf{A}(\bar{\mathbf{N}})$ is therefore decomposable into 3 pieces. The set of eigenvalues of $\mathbf{A}(\bar{\mathbf{N}})$ is the union of the subsets of eigenvalues of $\mathbf{A}_{11}(\bar{\mathbf{N}})$, $\mathbf{A}_{22}(\bar{\mathbf{N}})$ and $\mathbf{A}_{33}(\bar{\mathbf{N}})$. Since we assume by our choice of $g_i(X, Y, Z)$ that each of the subpopulations is stable in the absence of the disease, the submatrix $\mathbf{A}_{11}(\bar{\mathbf{N}})$ consists of negative diagonal elements. By inspection, the submatrices $\mathbf{A}_{11}(\bar{\mathbf{N}})$ and $\mathbf{A}_{33}(\bar{\mathbf{N}})$ contribute eigenvalues with negative real parts. If the equilibrium point $\bar{\mathbf{N}}$ is unstable, then the submatrix $\mathbf{A}_{22}(\bar{\mathbf{N}})$ must have at least one eigenvalue with a positive real part. We have established the following theorem:

Theorem. A disease can become established in a spatially heterogeneous population if and only if the submatrix $\mathbf{A}_{22}(\bar{\mathbf{N}})$ of Eq. (12.19) has at least one eigenvalue with a positive real part.

$$A_{11}(N)=\text{diag}\left[\, g_i+(X_i+Y_i+Z_i)\frac{\partial g_i}{\partial X_i}\Big|_N \quad -b_i-\sum_{j=1}^{m}\frac{\partial f(X_i,Y_j,T_{ij})}{\partial X_i}\Big|_N \,\right]$$

$$A_{12}(N)=
\begin{bmatrix}
g_1+(\bar X_1+\bar Y_1+\bar Z_1)\dfrac{\partial g_1}{\partial Y_1}\Big|_N & \cdots & -\dfrac{\partial f(X_1,Y_m,T_{1m})}{\partial Y_m}\Big|_N \\[2ex]
\vdots & & \vdots \\[1ex]
-\dfrac{\partial f(X_m,Y_1,T_{m1})}{\partial Y_1}\Big|_N & \cdots & g_m+(X_m+Y_m+Z_m)\dfrac{\partial g_m}{\partial Y_m}\Big|_N \;\; -\dfrac{\partial f(X_m,Y_m,T_{mm})}{\partial Y_m}\Big|_N
\end{bmatrix}$$

$$A_{13}(N)=\text{diag}\left[\, g_i+(\bar X_i+\bar Y_i+\bar Z_i)\frac{\partial g_i}{\partial Z_i}\Big|_N +\gamma_i \,\right]$$

$$A_{21}(N)=\text{diag}\left[\, \sum_{j=1}^{m}\frac{\partial f(X_i,Y_j,T_{ij})}{\partial X_i}\Big|_N \,\right]$$

$$A_{22}(N)=
\begin{bmatrix}
-(\alpha_i+b_i+v_i)+\dfrac{\partial f(X_1,Y_1,T_{11})}{\partial Y_1}\Big|_N & \dfrac{\partial f(X_1,Y_2,T_{12})}{\partial Y_2}\Big|_N & \cdots & \dfrac{\partial f(X_1,Y_m,T_{1m})}{\partial Y_m}\Big|_N \\[2ex]
\vdots & & & \vdots \\[1ex]
\dfrac{\partial f(X_m,Y_1,T_{m1})}{\partial Y_1}\Big|_N & \dfrac{\partial f(X_m,Y_2,T_{m2})}{\partial Y_2}\Big|_N & \cdots & -(\alpha_m+b_m+v_m)+\dfrac{\partial f(X_m,Y_m,T_{mm})}{\partial Y_m}\Big|_N
\end{bmatrix}$$

$$A_{23}(N)=0$$

$$A_{31}(N)=0$$

$$A_{32}(N)=\text{diag}\,[v_i]$$

$$A_{33}(N)=\text{diag}\,[-(b_i+\gamma_i)].$$

The effect that the number of infected individuals has on the rate of infection will always be positive, that is

$$\left.\frac{\partial f(X_i, Y_j, T_{ij})}{\partial Y_j}\right|_{\bar{N}} > 0. \tag{12.20}$$

This results in $A_{22}(\bar{N})$ having the form of a positive feedback matrix that permits the use of simple criteria to determine stability. For example, if for every set of positive constants $u_1, u_2, \ldots, u_m$ there exists a subpopulation i such that

$$\sum_{\substack{j=1 \\ j \neq i}}^{m} u_j \left.\frac{\partial f(X_i, Y_j, T_{ij})}{\partial Y_j}\right|_{\bar{N}} > u_i(\alpha_i + b_i + v_i), \tag{12.21}$$

then the disease can become established in the total population (see Appendix B). A rough interpretation of this inequality is that for at least one subpopulation the rate at which susceptible individuals contract the disease must exceed the rate at which infected individuals are removed from the subpopulation through recovery (v_i) or death ($\alpha_i + b_i$). This intuitive condition, applicable to any number of interacting subpopulations, parallels the threshold condition for the single isolated population.

Similarly, we can derive an inequality which can be interpreted as stating that for at least one subpopulation i, the rate at which infected individuals spread the disease among various subpopulations must be greater than the rate at which the infected of subpopulation i are removed. This condition is not as obvious as the previous one, because it does not equate the creation and removal of infecteds in a single subpopulation.

12.5 Design of Immunization Programs

May and Anderson (1984) considered the problem of determining an optimal immunization program in a spatially heterogenious population. They assumed that the population was divided into m subgroups, with one transmission within groups and another, lower transmission rate, between groups. An optimal immunization program is one that eradicates the disease by immunizing the smallest overall number in each cohort of newborns. Let $f(X_i, Y_i, T_{ij}) = B_{ij} X_i Y_j$, $g_i(X_i, Y_i, Z_i) = \mu$, $\gamma_i = \alpha_i = 0$, $b_i = \mu$ in Eqs. (12.16) and (12.17), and define the force of the infection as

$$\lambda_i = \sum_{j=1}^{m} B_{ij} Y_j. \tag{12.22}$$

Suppose a constant fraction p_i of newborns in subpopulation i are immunized at birth. Then Eqs. (12.16)–(12.18) become

$$\frac{dX_i}{dt} = \mu(1 - p_i) N_i - (\lambda_i + \mu) X_i, \tag{12.23}$$

213

$$\frac{dY_i}{dt} = \lambda_i X_i - (v_i + \mu) Y_i, \tag{12.24}$$

$$\frac{dZ_i}{dt} = v_i Y_i + \mu p_i N_i - \mu Z_i. \tag{12.25}$$

The population analyzed by May and Anderson (1984) has separate transmission parameters for intragroup contacts, $B_{ii} = \beta$, and intergroup contacts $B_{ij} = \varepsilon\beta$, $0 < \varepsilon < 1$. Let $f_i = N_i/N \left(N = \sum_{i=1}^{m} N_i \right)$ be the fraction of the total population in the i-th subgroup. Then the overall fraction of the total population immunized is

$$P = \sum_{i=1}^{m} f_i p_i. \tag{12.26}$$

Also define

$$q_i = 1 - p_i$$

$$s_i = f_i q_i$$

$$\varrho_i = \beta N(v_i + \mu).$$

The authors proved that the optimal immunization policy is defined by

$$q_i f_i = \frac{1}{\varrho_i(1 + \varepsilon + M\varepsilon)} ;$$

that is, the fraction in each subpopulation which must be immunized is a constant.

May and Anderson (1984) concluded by showing that the total fraction of the population immunized under this criterion is less than that obtained for a uniformly mixed population with the same average transmission parameter.

12.6 Shape of the Contagion Rate Function

We shall now explore the relationship between the functional form of the rate of contagion and whether or not a disease can become established in a population.

Modifying Eq. (12.7) to account for the fraction of time two populations are in contact, we propose that the rate of contagion have the form

$$f(X_i, Y_i, T_{ij}) = \beta(T_{ij}) X_i Y_j, \tag{12.27}$$

where T_{ij} represents the function of time populations i and j are in contact $\left(\sum_{i=1}^{m} T_{ij} = 1 \right)$ and $\beta(T_{ij})$ reflects the infectability. If a population i does not

214

contact any other population, then $T_{ii}=1$, $\beta(T_{ii})=\beta(1)=\beta_0$ and $T_{ij}=0$, $\beta(T_{ij})=\beta(0)=0$. Suppose that infectability is linearly proportional to the fraction of time two populations are in contact; that is

$$\beta(T_{ij})=\beta_0 T_{ij}. \tag{12.28}$$

To determine whether the disease can spread in the total population, we examine the matrix

$$
\mathbf{A}_{22}(\mathbf{\bar N})=
\begin{bmatrix}
\beta_0 T_{11}\bar X_1 -(\alpha_1+b_1+v_1) & \beta_0 T_{12}\bar X_1 & \cdots & \beta_0 T_{1m}\bar X_1 \\
\beta_0 T_{21}\bar X_2 & \beta_0 T_{22}\bar X_2 -(\alpha_2+b_2+v_2) & \cdots & \beta_0 T_{2m}\bar X_2 \\
\vdots & \vdots & & \vdots \\
\beta_0 T_{m1}\bar X_m & \beta_0 T_{m2}\bar X_m & \cdots & \beta_0 T_{mm}\bar X_m -(\alpha_m+b_m+v_m)
\end{bmatrix}
\tag{12.29}
$$

for eigenvalues with positive real parts. Employing inequality (12.21) and letting $u_1=1/\bar X_1$, $u_2=1/\bar X_2,\ldots,u_m=1/\bar X_m$, we may determine when this is the case. If there exists at least one subpopulation, i, such that

$$\sum_{j=1}^{m} \beta_0 T_{ji} > (\alpha_i+b_i+v_i)/\bar X_i, \tag{12.30}$$

then $\mathbf{A}_{22}(\mathbf{\bar N})$ is unstable to the introduction of the disease. Because $\sum_{j=1}^{m} T_{ji}=1$, inequality (12.30) may be rewritten as

$$\bar X_i > \frac{\alpha_i+b_i+v_i}{\beta_0}. \tag{12.31}$$

This is exactly the condition established by Eq. (12.10) for determining when a disease can become established in an isolated population. Thus, in the case when infectability is linearly proportional to the fraction of time subpopulation centers are in contact, the threshold condition for the maintenance of a disease in an entire population is identical to the threshold condition for the maintenance of the disease in each of the isolated subpopulation centers. In other words, the disease cannot become endemic in the entire population unless it is endemic in some isolated subpopulation.

This important property of disease in heterogeneous populations is not a consequence of the assumption that the rate of contagion is of "mass action" type as given by (12.7). In fact, it is easily established that the same principle holds when the rate of contagion has the form

$$f(X_i, Y_i, T_{ij})=\beta_0 T_{ij} X_i h(Y_i), \tag{12.32}$$

215

where $h > 0$, $h(0) = 0$ and $h'(Y_i) < 0$. Copasso and Serio (1978) have suggested such an interaction term to explain a cholera epidemic spread in Bari.

If the basic infectability β_0 is the same in each subpopulation we see that the dynamics are the same as if each subpopulation was separate. This is intuitively correct, because it does not matter whether a susceptible contracts the disease at home or while traveling if the probability of becoming infected is the same everywhere.

For visiting to enhance the probability of an outbreak, it is necessary that the visiting increase the risk of contracting the disease. This could be the case if there existed one or more subpopulations in which the disease was endemic; that is, above threshold size. Hethcote (1976) shows, for a model of a disease without immunity, that migration (as opposed to visiting) could cause a disease to remain endemic in a subpopulation below threshold if the subpopulation interacted with another that was sufficiently far above its threshold size. Another possibility lies in the intuitive concept that visiting individuals actually experience higher contact rates due to traveling. This could come about for example if the infection rate is a concave function of the length of time an individual spends in a given community rather than a linear function (Post et al., 1983).

12.7 Comparison with other Spatially Heterogeneous Models

Several models, similar to the one introduced here, have been considered by others. Rushton and Mautner (1955) present solutions to a special case of a model of a simple epidemic in many communities. Watson (1972) developed a stochastic model to evaluate the severity of an outbreak in a population divided into m subpopulations. A deterministic analog of Watson's model is similar to a special case of the model presented in this chapter with linear rates of contagion and without reproduction. Using computer simulations, Watson determined that when the total population size became large, the probability of a major outbreak involving most subpopulations increased to certainty. The population size required, however, was large enough to ensure that most subpopulations were larger than the threshold size required to support an epidemic in isolation. This is in concordance with the analytical results presented in the previous section.

Lajmanovich and Yorke (1976) analyzed equations similar to (12.16)–(12.18) with linear infection rate functions, analogous to (12.7). They showed that, depending on parameter values, either the disease will disappear from the population or the disease will be endemic with the infective and susceptible levels approaching unique constant values, regardless of the initial conditions. That is, the authors established uniqueness and global stability conditions for the equilibrium. This result holds for epidemic models where the population size in each center may be assumed to be constant and the rate of infection is proportional to the product of the susceptible and infected populations. Specifically, their result holds for both forms of the infection rate function we have discussed [Eqs. (12.7), (12.32)]. However, their model is an S–I–S (susceptible→infected→susceptible) model, which does not have a recovered class, so their results do not apply directly to our model.

The theorem of Section 5, on the other hand, may be applied to more general epidemic models than those presented here, and by Hethcote (1978) and Lajmanovich and Yorke (1976). For example, Copasso and Serio (1978) introduced an interaction term in which the dependence upon the number of infectives incorporates a nonlinear bounded function that may express saturation or psychological effects. In this sense, the theorem in this chapter has wider application than the theorem of Lajmanovich and Yorke (1976).

The introduction of spatial heterogeneity into epidemic models does not by itself alter the general qualitative behavior found in single population models with homogeneous mixing. Bailey (1975) pointed out that deterministic models incorporating spatial heterogeneity display the same type of asymptotic behavior as deterministic models with homogeneous mixing. Our model is not an exception. Because of this difficulty, most investigators have resorted to stochastic models or time delays to describe the periodic behavior exhibited by such diseases as measles and influenza (Bartlett, 1960). However, there can be no doubt that non-homogeneous mixing affects the quantitative behavior of epidemics. The current work demonstrates that spatial heterogeneity can have a pronounced effect on threshold values. This effect depends on the nature of the interaction between infected and susceptible populations. As a further example, Rushton and Mautner (1955), Watson (1972), and others have discussed the effect of nonhomogeneous mixing between populations on disease dynamics during the fade-out process.

12.8 Host-vector Models

The dynamics of diseases are intimately related to their facilities for transmission from infected hosts to susceptibles. Many important human diseases, such as malaria, sleeping sickness, plague, yellow fever, encephalitis and typhus are spread by arthropods, technically known as "*vectors*". Aphids are known to transmit 164 of the 620 known plant viruses (Eastop, 1977). Various beetles spread the fungus *Endotheca parasitica*, the chestnut blight organism, from tree to tree. Vector species are subject to their own population dynamics, which may be so critical to the spread of the disease that human attempts to eradicate the disease often center on controlling the vector. Any attempt to accurately model a vector-spread disease, human or otherwise, must take into account the dynamics of the vector population.

There is a formal similarity between models for a directly communicated disease with two interacting host populations and parasitic diseases that involve two populations. We can use the results of the previous section to establish a threshold condition for endemism of the disease or parasite. Because host populations cannot infect themselves directly, $\beta_{11} = \beta_{22} = 0$. The threshold which must be met for the disease to become endemic is

$$\beta(T_{12})\beta(T_{21}) > \frac{(\alpha_1 + b_1 + v_1)(\alpha_2 + b_2 + v_2)}{\bar{X}_1 \bar{X}_2}. \tag{12.33}$$

The details of a particular parasitic disease may be quite complicated, involving several forms of the parasite in the host or the vector, as well as the possible

217

multiplication of parasites outside of the host-vector populations. Various complications can be added to the equations to account for realistic features of a disease that are considered important without altering the applicability of the theorem of the previous section. For example, the general analysis is applicable to models incorporating age-specific interactions or interactions between multiple host and vector populations.

Aron and May (1982) have considered a simple model specific to malaria. They assumes constant total population N for humans and N' for mosquitos, with Y and Y' being the number of malaria infectives and $N-Y$ and $N'-Y'$ being the numbers of susceptibles, respectively. Infective individuals upon recovering are immediately susceptible. Aron and May (1982) make the reasonable assumption that mosquitos tend to have a constant biting rate, a bites per day, independent of the density of human hosts. The rates of change of Y and Y' are then

$$\frac{dY}{dt} = abY'[(N-Y)/N] - rY, \tag{12.34}$$

$$\frac{dY'}{dt} = \frac{aY}{N}(N'-Y) - \mu Y', \tag{12.35}$$

where b is the probability of a susceptible acquiring malaria from a bite and r and μ are removal rates. They transformed these equations into equations for fractions of infectives $y = Y/N$ and $y' = Y'/N'$:

$$\frac{dY}{dt} = \left(\frac{abN'}{N}\right)(1-y)y' - ry, \tag{12.36}$$

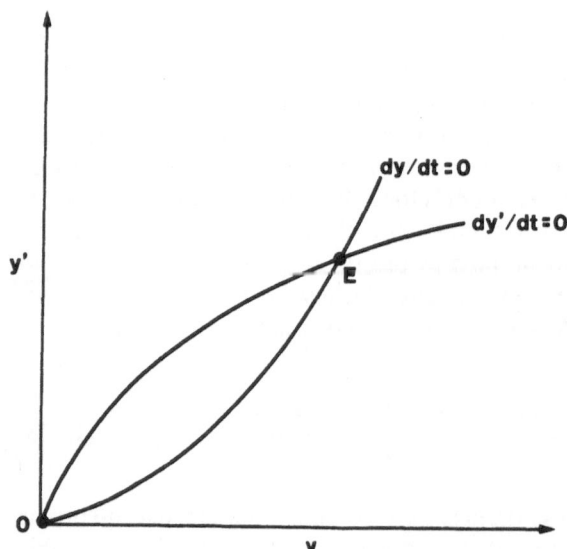

Fig. 12.5. Isoclines of the transformed equations of Aron and May's (1982) model of the spread of malaria. y refers to the fraction of infected humans and y' refers to the fraction of infected mosquitoes. The isoclines here show equation parameters that result in an endemic level of malaria in both population, that is the equilibrium, E, is stable

$$\frac{dy'}{dt} = ay(1 - y') - \mu y'.$$
(12.37)

The zero isoclines of this set of equations are plotted in Fig. 12.5. Note their similarity to isoclines of mutualistic interactions.

12.9 Summary and Conclusions

The sudden appearance, spread and subsequent diappearance of a disease or parasitic organism may be related to positive feedback at several levels ranging from physiological response mechanisms to migration of parasite vectors. Parasitic infections reduce the competitive fitness of individual hosts and may be responsible for further physiological weakening. In many cases this is further enhanced by the presence of other disease organisms acting in concert.

The spread of disease in a population depends on the rate of contact between susceptible individuals and disease organisms or infected individuals. The mixing between these groups is not always uniform but depends on the spatial distribution of individuals and the patterns of contact between them. A simple model may be constructed for small homogeneous groups in which mixing among susceptible and infected individuals may be portrayed as random. This model has the property that below a critical population density of susceptibles, the contact rate between susceptible and infected individuals is too low to sustain the disease. In heterogeneous groups, positive feedback between separate subpopulations can effectively raise the contact rate, provided the infection rate function has a certain form, so that the disease may persist in the collection of subpopulation, even when it would fade out of each individual subpopulation.

13. The Ecosystem and Succession

13.1 The Ecosystem

Up to this point, the types of positive feedback considered have involved single species or communities of species. The ecosystem, however, includes not only interactions among species, but also interactions between species and their physical and chemical environment, therefore involving such factors as energy flow, nutrient cycles, soil conditions, and microclimate.

An ecosystem has been defined as a *"unit of biological organization in a given area (that is, a 'community') interacting with the physical environment..."* (Odum, 1969). The term "ecosystem" was coined by Tansley (1935) in the hope of bringing rigor to ecology and providing and objective entity to study, like the cell of the microbiologist or the organism of the traditional biologist. The degree to which the ecosystem is such a unit of biological organization depends to some extent on the number and nature of its internal feedbacks. We will examine some of these feedbacks after first looking at the phenomenon of ecological succession, one of the most important properties of ecosystems.

Ecological succession may be defined as the process by which an ecosystem changes temporally from one state to another. We will first discuss some of the main generalities concerning succession from the point of view of positive feedback processes and next consider various models of ecological succession. We will then examine the question of whether or not the ecosystem can be viewed as a unit of biological organization.

13.2 Succession as a Positive Feedback Process

In Chaps. 3–5 we showed that many physico-chemical and biological systems could be viewed as self-organizing. In these systems, an external source of energy is available to the system; for example, the externally imposed temperature gradient in the case of Bénard's experiments. A given system, initially at a certain level of dynamic complexity can, at some unpredictable time, jump spontaneously to a level of greater complexity, if the supply of external energy is great enough. This change comes about when a slight perturbation to the system triggers a positive feedback instability, allowing the external energy to drive the system away from its original state towards a new dynamic state. In Bénard's experiments a small perturbation of the fluid might be sufficient, while in a case of chemical autocatalysis a slight trace of autocatalyzing substance may be all that is needed.

Ecological succession is removed by many orders of magnitude in complexity from simple physico-chemical systems, and differs in many respects from biotic evolution. Nevertheless, it might well be viewed as a type of self-organization. An ecosystem shares with the physico-chemical systems considered earlier the basic thermodynamic property of being an open, non-equilibrium system. It receives short wave length energy from which it extracts work before it allows the "degraded" energy of longer wavelength to radiate into space. There is a continual input of water from the atmosphere or other external sources and nutrients from the air and rocks, as well as a balancing output of materials from the ecosystem. In the context of ecological succession, the appearance of a new invading species can be looked on as a new fluctuation to which the system may be unstable. The progressive replacement of simpler communities by more complex ones seems analogous to the outbreaks of instability leading to more highly structured dissipative systems. While such an analogy is loose at best, its summarization of the apparent regularities in ecological succession may provide a key to reducing the complexity of the ecosystem level of organization to a comprehensible system. Thinking in this direction was started by, among others, Margalef (1963), Connell and Orias (1964), Odum (1969), and Gutierrez and Fey (1975, 1980), who have applied cybernetic ideas to empirical observations of succession.

Gutierrez and Fey (1980) have propounded the self-organizing view of ecological succession in great detail and have accorded positive feedback a major role in the process. They view positive feedbacks as dominant in the early stages of succession, being outweighed by negative feedbacks as the climax stage is approached.

These authors trace cybernetic concepts (whether explicitly formulated as such or not) of succession back to Cowles (1911), who conceived of succession as primarily an autogenic process (resulting from internal dynamics) rather than an allogenic process (driven by external geochemical forces). Clements (1916), however, was probably the first to fully appreciate the aspects of mutual reinforcement that occur in succession. In his view, *"habitat and population act and react upon each other, alternating as cause and effect until a state of equilibrium is reached."* As Gutierrez and Fey (1980) stress, Clements used the term *"react"* to mean the way in which the plant or community of plants affects and modifies the environment.

An illustration of the influence of biota in modifying the physical environment and the positive feedback of the environment on the biota is sand dune succession on the southern end of Lake Michigan (see Olson, 1958). At the beginning of biotic succession, the wind-formed dunes are highly mobile and, thus, inhospitable to most plants. Early in the course of a successional sere, Marram grass invades the sand and sends down roots. Sand accumulates about the stems, which grow higher in response. Eventually a dune is stabilized in this manner by the pioneer plants. Dying plants add humus to the sand, increasing the nutrient and moisture storage and making it more hospitable to later colonists. The eventual climax on these sand dunes should be a mesophytic forest of beeches, maples, and hemlocks.

Reciprocal influences may even extend to weather patterns. Forests in dry, mountainous areas of the world can create their own microclimates. Moisture-laden clouds passing over bare mountains often do not drop rain because of the heat reflected from the bare rocks. Forest cover diminishes this albedo, so that local rain occurs, improving conditions for forest growth in these areas. Hence amounts of local rainfall and the magnitude of forest cover are mutually reinforcing (Edlin, 1976). In desert areas, shrubs and trees can create microenvironmental modifications that enhance the survival of a ground flora community (Noy-Meir, 1973). Solar radiation, temperature, and wind are all reduced, while organic and inorganic nutrients, as well as sand stabilization (in *"sandy"* deserts), are increased.

A further elaboration of the cybernetic view of ecosystem succession, recounted by Gutierrez and Fey (1980), was Lindeman's (1942) realization that the process arises from the interaction of mutual causalities between the living community and the physical environment. This, together with a few other basic propositions, form what Gutierrez and Fey call the *"classical hypothesis"* of succession. They compare this with a *"contemporary hypothesis"*, largely due to theorists such as E. P. Odum, R. H. Whittaker, and K. E. F. Watt. The contemporary hypothesis stresses the greater role of the community as opposed to the physical environment in generating succession, but differs relatively little from the classical hypothesis. For example, Whittaker (1970), describing primary succession on a rock surface exposed by a landslide, relates how "... *succession proceeded by back-and-forth interplay between organisms and environment: as one dominant species modified the soil and microclimate in ways that made possible the entry of a second species, which became dominant and modified environment in ways that suppressed the first and made possible the entry of a third dominant, which in turn altered the environment.*"

This more or less holistic, cybernetic view of ecological succession embodied in the classical and contemporary hypotheses has been challenged by the counterargument that close examination shows there to be no consistent community or ecosystem level patterns in succession and that the observed phenomena are explainable in terms of different growth and mortality rates, mature plant sizes, abilities to withstand stress, and perhaps dispersal abilities of the component plant species (Drury and Nisbet, 1973). Drury and Nisbet point out many exceptions to predictions of the contemporary hypothesis that support their species-level view.

Despite criticisms of the contemporary hypothesis by Drury and Nisbet (1973) and others, efforts have been made to couch the contemporary hypothesis in mathematical language so that its full explanatory powers can be exhibited. Gutierrez and Fey (1980) have made an attempt to do so by means of a computer simulation model that we review briefly below.

The model described in detail by Gutierrez and Fey (1980) discusses only secondary succession. The fundamental dynamic hypothesis is exhibited by the diagram in Fig. 13.1. There are four basic feedback loops in the model:

1) a nutrient recycling loop (positive feedback),
2) a biomass growth loop (positive feedback),
3) a community diversification loop (positive feedback), and
4) a loop involving primary production in relation to available soil nutrients (negative feedback).

The first loop expresses the fact that as biomass increases, soil also increases, increasing the total amount of available nutrients. The second loop is merely Blackman's *"compound interest law"* of plant growth (Chap. 6). The third loop, community diversification, includes three different types of diversity; species

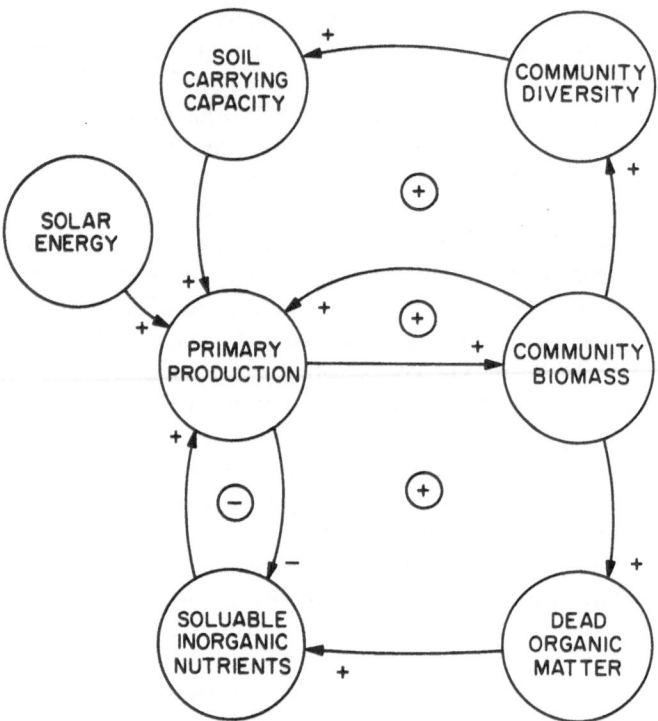

Fig. 13.1. A general systems model of secondary succession. (Reproduced from Gutierrez and Fey, 1980, with permission of MIT Press)

diversity, biochemical diversity, and stratification and spatial heterogeneity, which, the authors assume to be mutually reinforcing. Odum (1969) has commented: *"Whether or not species diversity continues to increase during succession will depend on whether the increase in potential niches resulting from increased biomass, stratification, and other consequences of biological organization exceeds the counter effects of increasing size and competition."* Hence, increasing niche specialization will lead to increased soil carrying capacity, at least up to a certain limit. Ultimately, as succession proceeds, the gain in these positive loops will decrease and the negative feedback effect of primary production in depleting nutrient levels will increase.

Each of the feedback loops in Fig. 13.1 is composed of complex subloops, which are described by Gutierrez and Fey (1980). The model is used by those authors to simulate secondary autogenic grassland succession. As the authors point out, the *"simulated patterns are consistent (or, at least, not inconsistent) with each of the trends known to be associated with ecological succession."*

The work of Gutierrez and Fey (1980) is probably the most sweeping effort thus far to analyze ecological succession in cybernetic terms, but much interesting work has also been done by others. For example, Connell and Orias (1964) confronted the problem of explaining changes in species diversity through evolution, but they used the same fundamental ideas of positive and negative feedback loops that Gutierrez and Fey used (Fig. 13.2). Their premise is that one cannot specify a priori the number of "*potential*" niches a particular physical environment will have because

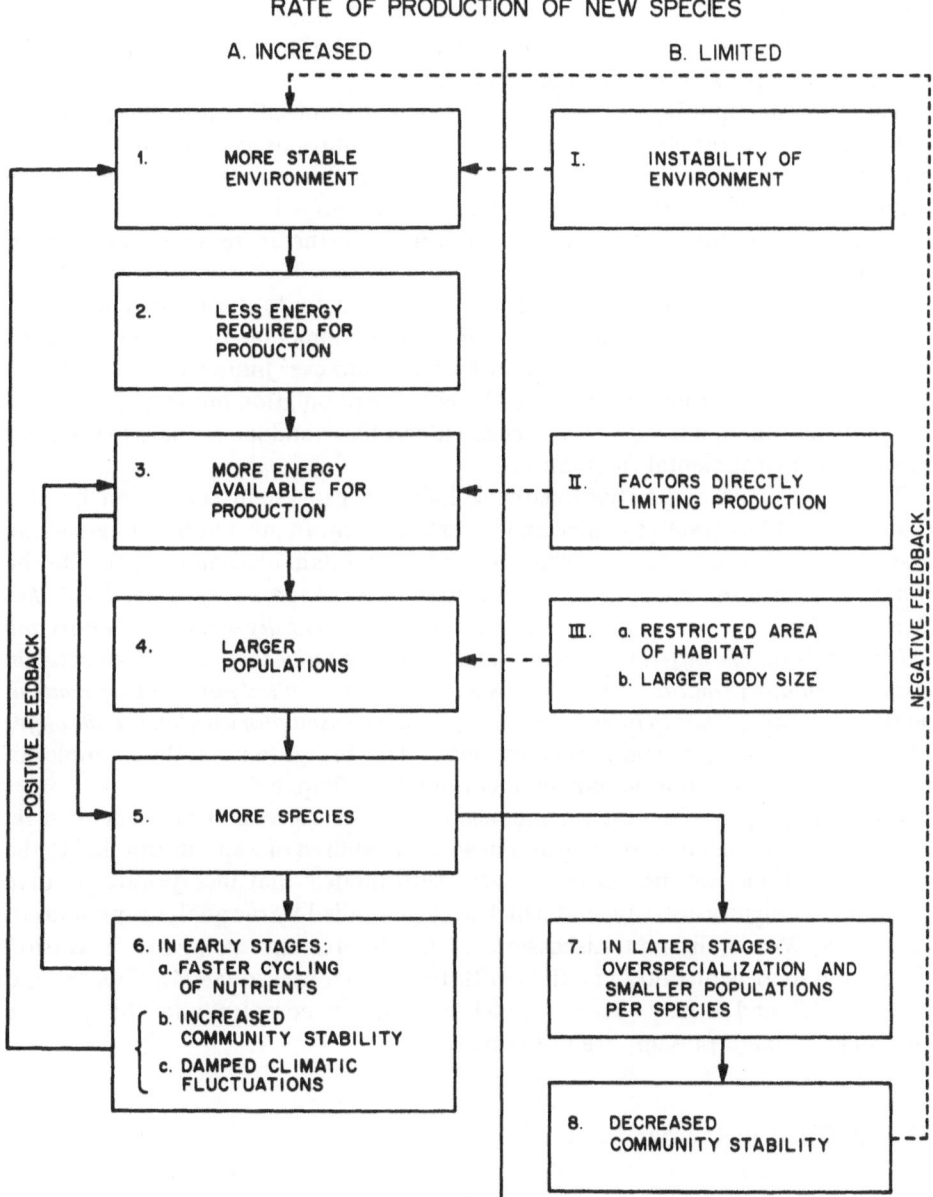

Fig. 13.2. A model used to explain changes in species diversity through either successional or evolutionary time scales. (Reproduced from Connell and Orias, 1964, with permission of The University of Chicago Press)

niche space is continuously created by the changing biota. [Also see Whittaker (1969) for a similar argument.]

The causal sequence in the positive feedback loop is as follows. Assume there is enough environmental stability to permit the ecosystem in its early stages of evolution to survive. The ecosystem stabilizes the environment further, allowing more energy to be channeled into production, which results in greater population sizes. Since an invading species that utilizes an indigenous one is more likely to be successful if the indigenous species has a large population than a small one, more species can be packed into the community.

The increasing number of species includes heterotrophs, first mainly herbivores, but later carnivores also, causing more rapid nutrient cycling and possibly (although current ecological theory is divided on this point) greater ecosystem stability. Microclimate fluctuations are also more damped by the greater biomass. A more stable environment provides conditions for further increases in biomass and diversity.

This positive feedback loop operates only for a finite duration. Ultimately, biomass is limited and the degree to which environmental fluctuations can be buffered is also limited. The division of biomass into ever more species eventually approaches an upper limit asymptotically because population numbers per species become small enough to make the species vulnerable to sudden random extinctions caused by environmental fluctuations.

The importance of positive feedback loops in ecosystem succession has led Gutierrez and Fey (1980) to make an interesting point. In most human engineering systems, negative feedback controls are used to maintain or achieve goals. On the other hand "*Natural ecological control seems to take the form of positive loops that drive populations and related variables toward environmentally determined limits and hold them there. The difference is important because negative loop control can result in a slow, oscillatory, and/or energy inefficient response. Limited positive loop control, on the other hand, tends to be rapid, efficient, and non-oscillatory within broad ranges of parameter values and relationship formats.*" This brings to mind the examples of insect flight and the ram jet engine mentioned in Chap. 5.

Like other types of ecological models, models of succession are very diverse, reflecting differences both in style and in what is required of a specific model. In the next few sections, we discuss some succession models that incorporate positive feedback in a significant way, and which may be studied by the analytic methods of this book. We shall omit discussion of the interesting ecological succession simulation models of Botkin et al. (1972), Ek and Monserud (1974), Shugart and West (1977) and others. These models are highly complex and the positive feedbacks, though present, are not conspicuous.

13.3 A Clementsian Model

Let us begin by considering a model for primary succession in the classical, Clementsian sense. A deterministic model is most suitable for this process. Let X_1, $X_2, \ldots$ quantify amounts of a sequence of vegetation types as they appear during succession, and let Y be a quantitative measure of environmental quality (e.g., a combination of soil conditions, microclimate, etc.). The reciprocal influences between pioneer vegetation and environment may be represented by the coupled equations

$$dY/dt = g(Y, X_1),$$ (13.1a)

$$dX_1/dt = f_1(Y, X_1).$$ (13.1b)

We assume that in the initial stages of succession (small Y and X_1) these variables are mutually reinforcing. Both Y and X_1 may also have internal feedbacks. For example, there may be a negative feedback on Y representing soil erosion, and so forth. For small values of Y and X_1 these equations may be linearized near zero:

$$dy/dt = a_{11}y + a_{12}x_1$$
$$dx_1/dt = a_{21}y + a_{22}x_1,$$ (13.2)

where $a_{11} = \partial g / \partial Y$, etc. Define by $\mathbf{A}_1$ the matrix

$$\mathbf{A}_1 = \begin{vmatrix} a_{11} & a_{12} \\ a_{21} & a_{22} \end{vmatrix} = \begin{vmatrix} - & + \\ + & - \end{vmatrix}.$$

We have assumed that a_{11} and a_{22} are both less than zero, meaning that environmental conditions cannot improve in the absence of conditioning by biomass and that biomass of the pioneer species cannot survive in the absence of at least weakly favorable environmental conditions. When the principal minor, $\det(\mathbf{A}_1)$, satisfies the condition $(-1)^2 \det(\mathbf{A}_1) < 0$ (see Chap. 2 or Appendix B), the pioneer vegetation and its environment (e.g., Marram grass and stabilizing sand dunes) mutually reinforce each other's changes.

Eventually, as the vegetation reaches a density high enough that density-dependent limitation sets in, this reciprocal reinforcement will decline and a stable equilibrium may be reached. Meanwhile, a new vegetation type, X_2, will have invaded. It will initially be mutualistic with the improved environment and competitive with vegetation type X_1. Hence, the new set of equations is

$$dY/dt = g(Y, X_1, X_2),$$ (13.3a)

$$dX_1/dt = f_1(Y, X_1, X_2),$$ (13.3b)

$$dX_2/dt = f_2(Y, X_1, X_2).$$ (13.3c)

Let us assume that by this stage of succession, the feedback between Y to X_1 is no longer very strong. If we can ignore this feedback (i.e., set $a_{21} = a_{12} = 0$), then close to the Y, X_1 equilibrium, the stability matrix will be

$$\mathbf{A}_2 = \begin{vmatrix} a_{11} & a_{12} & a_{13} \\ a_{21} & a_{22} & a_{23} \\ a_{31} & a_{32} & a_{33} \end{vmatrix} = \begin{vmatrix} - & 0 & + \\ 0 & - & - \\ + & - & - \end{vmatrix},$$ (13.4)

where a_{23} and a_{32} represent possible competition between vegetation types X_1 and X_2. Matrix $\mathbf{A}_2$ can easily be shown to be a positive feedback matrix by performing a similarity transform, $\mathbf{A}_2' = \mathbf{S}^{-1}\mathbf{A}_2\mathbf{S}$, where

$$\mathbf{S} = \mathbf{S}^{-1} = \begin{vmatrix} 1 & 0 & 0 \\ 0 & -1 & 0 \\ 0 & 0 & -1 \end{vmatrix}.$$

If $(-1)^3 \det(\mathbf{A}_2) < 0$, then a new vegetation type can successfully establish itself.

In a similar way, one can consider the invasion and establishment of succeeding species in this chain of obligatory succession. The model is obviously simplistic, and probably not appropriate to the great majority of successional processes, but it may be a useful starting point for more complex ideas.

13.4 Markov Chain Models

The modern non-deterministic view of succession has been reflected in a trend towards stochastic models. In these models the strict sequential replacement of one community type by another is abandoned in favor of an approach that assigns probabilities of transition from one type of another. Succession, simulated in this way, may have general trends, but transitions in the opposite direction are also possible, as is the case in real ecosystems.

The simplest model of this type is the Markov chain model, consisting of a matrix of transition probabilities for a specified time period. Such models have been used by Waggoner and Stephens (1970) and Horn (1975a, b), among others. We have written earlier (Chap. 9) about Leslie matrix models as examples of discrete-time positive feedback models. The mathematical results for those systems are relevant also to Markov chain models of succession.

Jeffers (1978) proposed a Markov chain model for the successive changes in a raised mire or bog. Jeffers denoted the four possible states of the system as

1) bog dominated by *Sphagnum*.
2) *Calluna-Cladonia* association with birch and pine seedlings,
3) pine and birch woodland, and
4) *Molinia-Pteridium* dominated association on which grazing by large herbivores is occurring, where succession is from wetter to drier facies.

The matrix of transition probabilities (for twenty year time intervals) is

	1. Bog	2. *Calluna*	3. Woodland	4. Grazed
1. Bog	$p_{11}=0.65$	$p_{21}=0.30$	$p_{31}=0.0$	$p_{41}=0.0$
2. *Calluna*	$p_{21}=0.29$	$p_{22}=0.33$	$p_{32}=0.28$	$p_{42}=0.40$
3. Woodland	$p_{31}=0.06$	$p_{23}=0.30$	$p_{33}=0.69$	$p_{43}=0.20$
4. Grazed	$p_{41}=0.0$	$p_{24}=0.07$	$p_{34}=0.03$	$p_{44}=0.40$

As the sums of all the columns of the Markov chain matrix are always 1.0, the largest real eigenvalue is always 1.0. Thus, the system is always stable in the sense that a set of probabilities of the system being in particular community types exists, and this set of probabilities does not change through time. Corresponding to the eigenvalue $\lambda_0 = 1.0$, is the eigenvector, $\mathbf{X}_0$, the elements of which are the mean expected proportions of the various ecosystem types in the steady state. The simple constant-coefficient (i.e., stationary) Markov chain model has the advantage of simplicity, but questions have been raised as to how true a representation of reality it may be. Horn (1974), with respect to forest succession models, and Usher (1979), with respect to models of termite succession and predator-prey dynamics, have suggested that the assumption of constant parameters might not be valid.

Consider Usher's (1979) model for Ghanian termite succession on baitwood blocks, for example. Usher recognized seven successional states:

1) no termite activity,
2) *Ancistrotermes*,
3) *Macrotermes*,
4) *Microtermes*,
5) *Pseudocanthotermes*,
6) other termites, and
7) unknown termites.

Usher found that probabilities of transition from one state to the next had very little correlation with the donor stage, but a strong positive correlation with the recipient stage, especially for transition to *Pseudocanthotermes*; that is, transition to the recipient stage proceeded faster as the relative size of the recipient class increased. That is exactly what we expect from a deviation-amplifying system, which we view succession to be in its early stages. Usher hypothesized that stationary Markov chains are probably not good approximations of ecological succession and that transition probabilities, p_{ij}, should, in general, be considered functions of the size of the recipient class. Therefore, he suggested that the transitional donor-dependent Markov chain model is not appropriate and changes should be made in the model.

Precisely this sort of modification has been incorporated into another Markov chain model of community succession (Abugov 1982). There are three stages in Abugov's model: (1) cleared space, X_1, (2) a fugitive species, X_2, and (3) a competitive species, X_3. The state of an area at some time t is given by the vector $(X_1(t), X_2(t), X_3(t))$, representing the fraction of a region of land occupied by the three stages $(X_1 + X_2 + X_3 = 1)$. This state is related to the stage at the following time interval, $t+1$, by the Markov chain equation

$$
\begin{vmatrix} X_1(t+1) \\ X_2(t+1) \\ X_3(t+1) \end{vmatrix} = \begin{vmatrix} p_{11} & p_{12} & p_{13} \\ p_{21} & p_{22} & p_{23} \\ p_{31} & p_{32} & p_{33} \end{vmatrix} \begin{vmatrix} X_1(t) \\ X_2(t) \\ X_3(t) \end{vmatrix}.
\qquad (13.5)
$$

In Abugov's model, unlike the constant-coefficient Markov chain model, some of the transition probabilities, p_{ij}, depend on the X_i's:

229

$$p_{11} = 1 - p_{21} - p_{31} \qquad p_{12} = P \qquad\qquad p_{13} = P$$

$$p_{21} = \frac{K_{21}(1-P)X_2}{X_1 + \beta} \quad p_{22} = \frac{K_{21}(1-P)X_2}{\beta} \quad p_{23} = 0 \qquad (13.6)$$

$$p_{31} = \frac{K_{31}(1-P)X_3}{X_1 + \beta} \quad p_{32} = \frac{K_{31}(1-P)X_3}{\beta} \quad p_{33} = 1 - P,$$

where $\beta = K_{21}X_2 + K_{31}X_3$. The constant P is a disturbance rate that sets the system back to earlier stages, and K_{21} and K_{31} are constants describing relative colonization and competitive abilities of the fugitive and competitive dominant species.

It is possible in this model for succession to be arrested at stage 2 or even stage 1 under certain conditions of parameter values and initial conditions. If, for example, the fraction of land covered by stage 2, the fugitive species, becomes too great, it can totally dominate and exclude stage 3 by positive feedback monopolization. Suppose that the initial values of X_1 and X_2 are large and close to equilibrium, and that $X_3 \ll 1$. If

$$Q \equiv \frac{K_{31}(1-P)X_1}{X_1 + \beta} + \frac{K_{31}(1-P)X_2}{\beta} + 1 - P < 1, \qquad (13.7)$$

then

$$X_3(t+1) = QX_3(t) < X_3(t). \qquad (13.8)$$

Inequality (13.8) implies that X_3 will continually decrease in the future because the subsequent changes in X_1 and X_2 will be small and will not change the sign of Q in any future iterations.

The above analysis can be generalized to larger systems if desired. For example, consider an m-state successional model, the dynamics of which will be described by an $m \times m$ matrix $\mathbf{P}$. One may want to know if the system can recover from a perturbation that reduces fractions of land covered by all but the lowest state to small remnants; that is, will the equilibrium point $\mathbf{X} = (1, 0, 0, \ldots, 0)$ be a repellor? To determine this, it would be necessary to linearize about $(1, 0, 0, \ldots, 0)$ the $(m-1) \times (m-1)$ submatrix, $\mathbf{P}_s$ formed by omitting the first row and first column. It can be shown that succession can proceed, at least past the lowest stage, if

$$(-1)^k \det (\mathbf{P}_{s,k} - 1) < 1 \qquad (13.9)$$

for any k, where $\mathbf{P}_{s,k}$ is the kth principal minor of $\mathbf{P}_s$.

This modeling exercise underscores the difference between a stationary Markov process, in which the transfer coefficients, p_{ij}, are constant, and a non-stationary process, in which the current state of the system influences the transition probabilities. The latter type of model implicitly assumes that the course of succession is not generated solely by a set of constant properties of competing vegetation types, but that the properties change as the biotic environment changes.

230

13.5 A Model of a Fire-dependent System

Succession does not always lead to an approximately stable steady state, but can involve cycles in which a state that is *"earlier"* than the theoretical climax is perpetuated by recurrent fires or other disturbances. Apparently, fire plays a major role in maintaining certain species and associations. Pielou (1979) cites the observation by Seddon (1974) that in parts of southeastern Australia the dominant trees are sclerophyllous eucalypts, but the undergrowth consists of lush mesophytic vegetation. Pielou points out that one of the possible explanations for why the canopy layer does not match the ground layer in a community undergoing succession to a temperate rain-forest type is that frequent fires favor the eucalypts (see also Spurr, 1964). Fires increase soil leaching and sclerophylls are better adapted than mesophylls to the poorer soils. Fires, infertile soil, and sclerophyllous forest form a mutual causal cycle that prevents invasion by the rain forest. There apparently exists the same type of symbiosis of fire with moist grasslands and with longleaf pine (*Pinus palustris* Mill.) stands in the southeastern United States (Gourou, 1947).

Mutch (1970) advanced the hypothesis that fire-dependent (abbreviated *"fire"* from now on) communities burn more readily than non-fire-dependent (*"non-fire"*) communities because natural selection has favored development of characteristics that make the former more flammable. Mutch based this hypothesis on laboratory combustion tests of the litter of various species types.

The competition between fire and non-fire communities in a region can be visualized as a feedback model (Fig. 13.3). The fire system consists of a seedling state, X_1, a mature state, X_2, and a burned state, X_3, leading cyclically from one to the next. Assume the non-fire seedlings, X_4, and the mature trees, X_5, would take

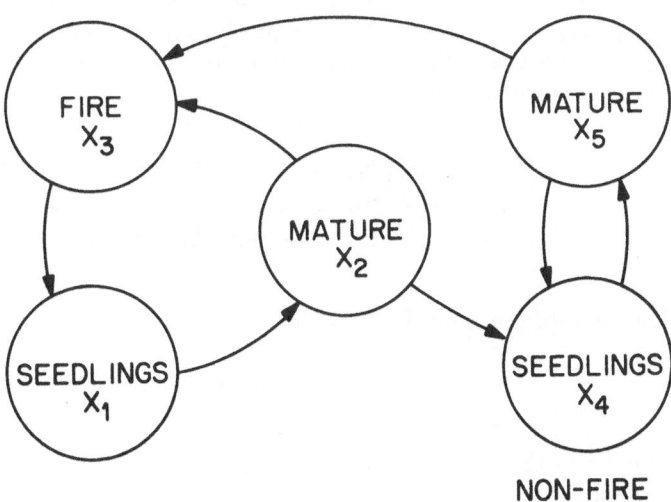

Fig. 13.3. A systems model for two competing communities, one of which is a fire-dependent community and one of which is not. The arrows represent possible paths of replacement

231

over from mature fire species in the absence of fire. However, with fire included, there is a transfer from non-fire species back to the fire community. Let us write the discrete-time matrix equation,

$$
\begin{vmatrix} X_1(t+1) \\ X_2(t+1) \\ X_3(t+1) \\ X_4(t+1) \\ X_5(t+1) \end{vmatrix} = \begin{vmatrix} p_{11} & 0 & p_{13} & 0 & 0 \\ p_{21} & p_{22} & 0 & 0 & 0 \\ 0 & p_{32} & p_{33} & p_{34} & p_{35} \\ 0 & p_{42} & 0 & p_{44} & p_{45} \\ 0 & 0 & 0 & p_{54} & p_{55} \end{vmatrix} \begin{vmatrix} X_1(t) \\ X_2(t) \\ X_3(t) \\ X_4(t) \\ X_5(t) \end{vmatrix}, \qquad (13.10)
$$

where each p_{ij} is non-negative and where $X_i(t)$ is the fraction of the spatial region in state i.

If Eq. (13.10) is considered to be a Markov chain model, with constant transfer coefficients, p_{ij}, we can conclude that all five states have some probability of occurrence. Were it not for the possibility of fire, represented by the transfer coefficients p_{34} and p_{35}, eventually only the non-fire community would exist. Mathematically, the subset of states 4 and 5 would constitute an absorbing subset.

The model is more complex if the transfer coefficients, p_{ij}, are not constants, but depend on what fractions, $X_1, \ldots, X_5$, of each state currently exist in the spatial region. In particular, assume that the fire transfer rates, p_{34} and p_{35}, are small but increase with increasing amounts of mature fire community trees, X_2; that is

$$
p_{34} = p'_{34} X_2
$$

$$
p_{35} = p'_{35} X_2 .
$$

The dependence can be assumed approximately linear, at least when X_2 is small.

Consider an established non-fire community. We wish to determine whether or not a fire community can invade. Assume the initial state of the system is then $(0, 0, 0, \bar{X}_4, \bar{X}_5)$, where $\bar{X}_4$ and $\bar{X}_5$ are equilibrium values of X_4 and X_5. Expanding the equations for X_1, X_2, and X_3 about this equilibrium point, one can obtain the linearized matrix equation,

$$
\begin{vmatrix} x_1(t+1) \\ x_2(t+1) \\ x_3(t+1) \end{vmatrix} = \begin{vmatrix} p_{11} & 0 & p_{13} \\ p_{21} & p_{22} & 0 \\ 0 & p_{32}+p'_{34}\bar{X}_4+p'_{35}\bar{X}_5 & p_{33} \end{vmatrix} \begin{vmatrix} x_1(t) \\ x_2(t) \\ x_3(t) \end{vmatrix}. \qquad (13.11)
$$

The equilibrium point is unstable if

$$
(-1)^3 \det (\mathbf{P}-\mathbf{I}) < 0,
$$

where $\mathbf{P}$ is the 3×3 matrix in Eq. (13.11).

That is, if

$$
p_{13}p_{21}(p_{32}+p'_{34}\bar{X}_4+p'_{35}\bar{X}_5) > -(p_{11}-1)(p_{22}-1)(p_{33}-1).
$$

The larger p'_{34} and p'_{35} are, the more likely Eq. (13.11) is to be unstable, and the more likely a fire-adapted community will encroach on the non-fire community. As in the preceding example, the model used is a deviation from the stationary Markov process model of ecosystem succession.

An analogous self-reinforcing process called paludification occurs on alluvial areas of Alaska (Miles, 1979, p. 50), which are covered with quaking bogs and spruce forest. Mosses, especially *Sphagnum* spp., cover the ground. As moss cover develops, it insulates the soil, leading to permafrost. The permafrost retards drainage, promoting further *Sphagnum* growth. This process kills the trees. Eventually, however, the successional process may reverse itself.

13.6 Positive Feedback Loops in Ecosystems

Ecosystems close to equilibrium are just as likely as early successional stages of ecosystems to contain positive feedback loops, though in the former case the positive loops will normally be balanced by negative feedback loops that prevent the system from leaving its steady state equilibrium. As Alexander (1971) states, "*The very stability of climax communities argues for the existence of interdependent relationships. In a heterogeneous community containing a multitude of potential nutrients it is probably true that completely independent species sacrifice something to be self-reliant.*" Whitehead (cited by Worster, 1977) perceived the importance of positive feedback loops in a forest: "*Each tree may lose something of its individual perfection of growth, but they mutually assist each other in preserving the conditions for survival. The soil is preserved and shaded; and the microbes necessary for its fertility are neither scorched, nor frozen, nor washed away. A forest is a triumph of the organization of mutually dependent species.*" Ecologists agree that intricate relationships exist, involving many multilink pathways among individuals and species.

Wilson (1980) presents a realistic, though hypothetical, example of long feedback chains (Fig. 13.4). The figure has been drawn in a slightly different style from Wilson's original figure and some feedbacks have been omitted to emphasize long positive feedbacks loops. The plant, X_1, is affected positively by a microbe, X_2, that changes phosphorus from a nonusable to a usable form. The plant is attacked by a herbivorous insect, X_4, against which it retaliates by producing a toxin, X_3. This toxin also inhibits the microbe. Two things tend to oppose this negative influence and make the soil more hospitable for the microbe, X_2. The plant produces an inhibitor, X_6, which is released into the soil and that suppresses a competitor, X_7, of X_5, allowing X_5, another microbe, to obtain more resources, X_8. The microbe, X_5, is a detoxifier of X_3. The plant releases an inhibitor, X_9, that suppresses a competitor of X_2, X_{10}, for resources, X_{11}, thereby indirectly benefiting X_2.

Note that loops $1\rightarrow6\rightarrow7\rightarrow8\rightarrow5\rightarrow3\rightarrow2\rightarrow1$, $1\rightarrow3\rightarrow4\rightarrow1$, and $1\rightarrow9\rightarrow10\rightarrow11$ $\rightarrow2\rightarrow1$ are positive feedback loops, whereas loops $1\rightarrow4\rightarrow1$ and $1\rightarrow3\rightarrow2\rightarrow1$, and $1\rightarrow6\rightarrow7\rightarrow8\rightarrow5\rightarrow3\rightarrow4\rightarrow1$ are negative feedback loops. There are also several other feedback loops of length 2 that have been omitted. It will be conceptually simpler to condense the model shown in Fig. 13.4 to a more manageable model, by replacing

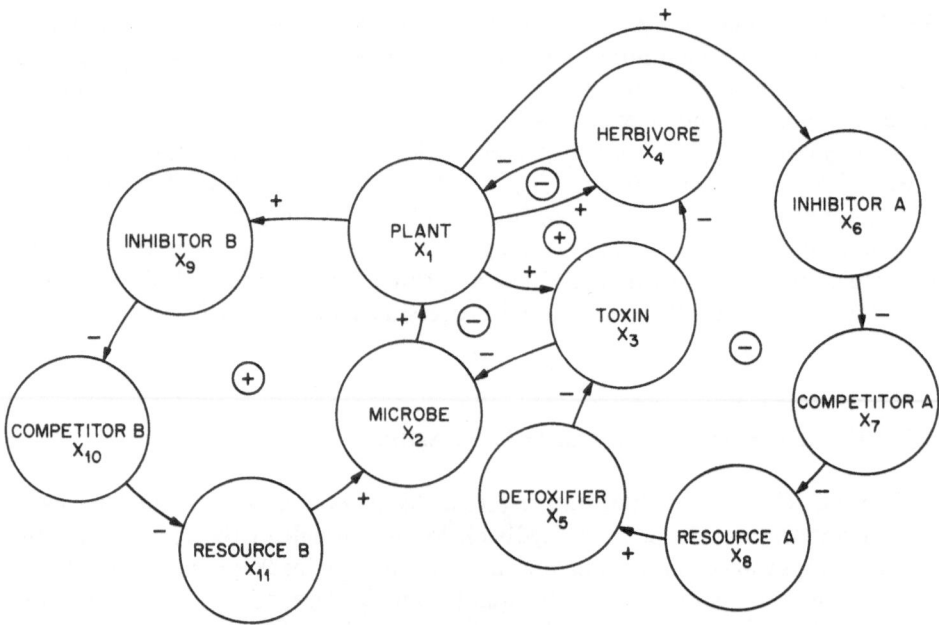

Fig. 13.4. Some hypothetical feedback loops in an ecosystem. (Reproduced from Wilson, 1980, with permission of the Benjamin/Cummings Publishing Company, Inc.)

the two long positive feedback loops by shorter loops (Fig. 13.5). The perturbation matrix for this system is

$$\mathbf{A} = \begin{vmatrix} a_{11} & a_{12} & 0 & -a_{14} \\ a_{21} & a_{22} & -a_{23} & 0 \\ a_{31} & 0 & -a_{33} & 0 \\ a_{41} & 0 & -a_{43} & -a_{44} \end{vmatrix}. \tag{13.12}$$

where all the a_{ij}'s are positive.

Even for this simplified model, the criteria for stability are complex because positive linear system theory cannot be used and the Routh-Hurwitz criteria would be necessary. It is clear that only if the loop $1 \rightarrow 3 \rightarrow 4 \rightarrow 1$ is more powerful than the loop $1 \rightarrow 3 \rightarrow 2 \rightarrow 1$ will the benefits of toxin production outweigh its costs, although the quantitative condition depends on the relative importance to the plant of the feedback loops $1 \rightarrow 4 \rightarrow 1$ and $1 \rightarrow 2 \rightarrow 1$.

The existence of positive feedback loops in ecosystems can be a latent source of instability when large disturbances upset the balance between positive and negative feedback. A dramatic example of this was investigated in a coastal marine ecological community by Mann (1977, 1982).

Between 1968 and 1976 sea urchins (*Strongylocentrous droebachiensis*) in St. Margaret's Bay, Nova Scotia, were observed to form dense population clusters, which spread and eventually destroyed more than 90 % of the subtidal seaweed beds

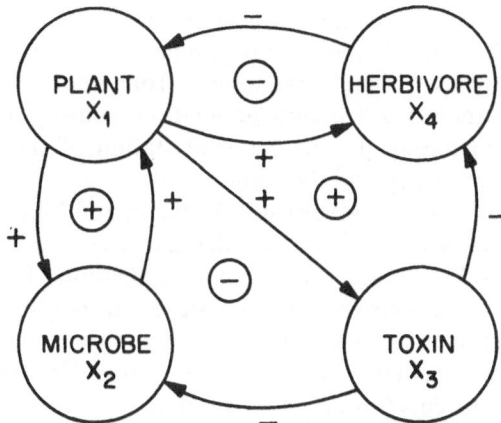

Fig. 13.5. A condensed version of the model in Fig. 13.4

in the bay. Instead of soon declining, as was expected if natural predators increased, the sea urchin population remained at high levels for at least the next two years. A partial explanation for this surprising occurrence (which was subsequently shown to have happened along much of the coastline of Nova Scotia) was attributed to a positive feedback loop involving the kelp-beds, the sea urchins, and the lobster, a primary predator on the urchins (Fig. 13.6). An initial decrease in the lobster population, perhaps due to overfishing, appears to have released the sea urchins from predator regulation, leading to their population increase and their change in habits from detritivory to direct feeding on kelp-beds. Lobsters, which required the type of sheltered habitat provided by the seaweed, dispersed or were removed by

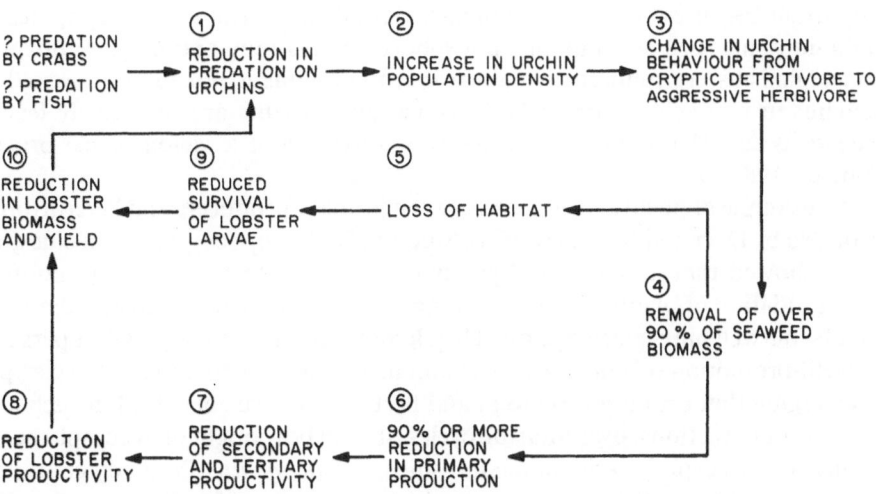

Fig. 13.6. Diagram summarizing the feedbacks involved in the destruction of kelp beds in Nova Scotia. (Reproduced from Mann, 1982, with permission from Blackwell Scientific Publication, Ltd.)

235

predators as the sea urchins destroyed their habitat. The result was a continued increase in the urchin population, which intensified as their natural predators waned. Finally, regulation from another source, disease, caught up with the runaway sea urchin population, causing drastic reductions in numbers, though regulation due to overexploitation of the kelp-beds by the urchin have might eventually had the same result.

As Odum (1969) and Bormann and Likens (1979) point out, the condition of an ecosystem may be considered as a competition between forces tending to increase and decrease ecosystem organization. Loss of some organization through a catastrophe can lead to more loss through positive feedback if the damage is severe enough. This can be the case for erosion, which occurs when there is a loss of biomass cover, as with clear-cutting. Erosion removes further organic matter and nutrients (though there is often a delay following clear-cutting before this starts to happen), leading to lower biomass production and hence lowered resistance to erosion. If the erosion is severe enough to form gullies, the process can accelerate rapidly. The deeper the gullies become, the more water is channeled into them and the faster it flows. Since the theoretical increase in the ability of water to move objects is proportional to the sixth power of its velocity, a strong positive feedback relationship between gully depth and erosional power is generated. Jenny (1980) has alluded to the frequency with which positive feedback forces can overwhelm the homeostatic forces in an ecosystem: "*Soil losses are often belittled and nature's healing power is praised. The forest is put on a par with an equilibrium system and its principle of virtual change: after a minor disturbance the system reverts to its former position of rest and all parts are stabilized again. More appropriate to an open system may be the analogy to biological radiation inputs in which small inputs cause accumulating alterations and eventual system deterioration, especially when erosion rates exceed weathering rates.*"

Another, subtler type of soil deterioration can follow cultivation. Cultivation reduces the teeming microbial life that produces humus. Non-acidic humus is important because it combines with particles of clay, forming soil aggregates that hold nutrients such as ammonia, phosphorus, potassium, and magnesium. When soil organisms are reduced by cultivation, the humus is not fully processed and remains in raw acidic form. The basic cation nutrients are then easily leached. The reduced pH inhibits microorganisms further and a vicious cycle proceeds (Butzer, 1982).

An example of positive feedback between primary producers and heterotrophs is reported by Dyer and Bokhari (1976). Studies of grasshopper grazing on blue grama grass showed that aboveground grazing increases belowground respiration and root exudation of plants, possibly leading to increased metabolism and biomass of plants, outweighing consumption. They hypothesized that the grasshoppers inject growth-promoting substances into a plant metabolic system. In fact, the viewpoint is emerging that primary producers and herbivores have coevolved in such a way that their interactions, over total life cycles, should be thought of as complementary rather than as exploiter-victim relationships. "*Both aquatic* (Pomeroy et al., 1963; *Hargrave and Been, 1970) and terrestrial (Arman et al., 1975; Woodmansee, 1978) herbivores accelerate mineralization rates substantially above those prevailing in their absence. Moreover, remineralization due to zooplankton grazing can exert a positive feedback on phytoplankton populations by stimulating growth of the latter when their density is low (Larow and McNaught, 1978)* (Dyer and McNaughton, unpublished).

Swift et al. (1979, p. 289) sketch a causal sequence that can result from increased grazing in a pasture. The increased grazing can accelerate the decomposition of organic matter. Accumulation of organic matter, and nitrogen in particular, is reduced, decreasing soil acidity. The reduced acidity benefits the earthworm population, which further increases organic decomposition. Douce and Webb (1978, p. 348) show, with reference to a tundra model, that soil invertebrates (microphytophages), by increasing the litter decomposition rate, stimulate microflora, which benefits the microphytophages in a positive feedback cycle.

13.7 Nutrient Cycling

An ecosystem is maintained not only by a flow of energy, but also by flows of an array of chemical elements. In fact, the availability of water and nutrients are much more likely to be limiting factors in ecosystems than the availability of radiant energy. Unlike energy, chemical elements do not degrade, so nutrient limitations can be partly offset by conservation and recycling of essential elements within an ecosystem.

Recycling of elements is particularly feasible on a local scale for minerals like calcium, phosphorus, and potassium, which do not easily volatilize. Even so, complete conservation is almost always impossible, and the unavoidable outflows of nutrients must be balanced by inflows for a system to maintain itself.

Mineral cycling models have been formulated for many ecosystems and elements (e.g., Jordan et al., 1972; Shugart et al., 1976; Sebetich, 1975), and tend to be fairly standardized in form. Let us consider a hypothetical but typical model of the cycling of a given element, say phosphorus. Let the system be divided into three compartments, soil water solution, phosphorus in plant biomass, and phosphorus in litter, with amounts in the three compartments (perhaps in grams per square meter) being represented by X_1, X_2, and X_3, respectively (Fig. 13.7). A simple linear model of this system might take the form

$$\frac{dX_1}{dt} = I_0 + a_{31}X_3 - (a_{12} + \alpha_1)X_1, \qquad (13.13a)$$

$$\frac{dX_2}{dt} = a_{12}X_1 - (a_{23} + \alpha_2)X_2, \qquad (13.13b)$$

$$\frac{dX_3}{dt} = a_{23}X_2 - (a_{31} + \alpha_3)X_3, \qquad (13.13c)$$

where I_0 is a constant external input and the a_{ij}'s (transfer coefficients) and the α_i's (loss rates from the system) are positive constants.

237

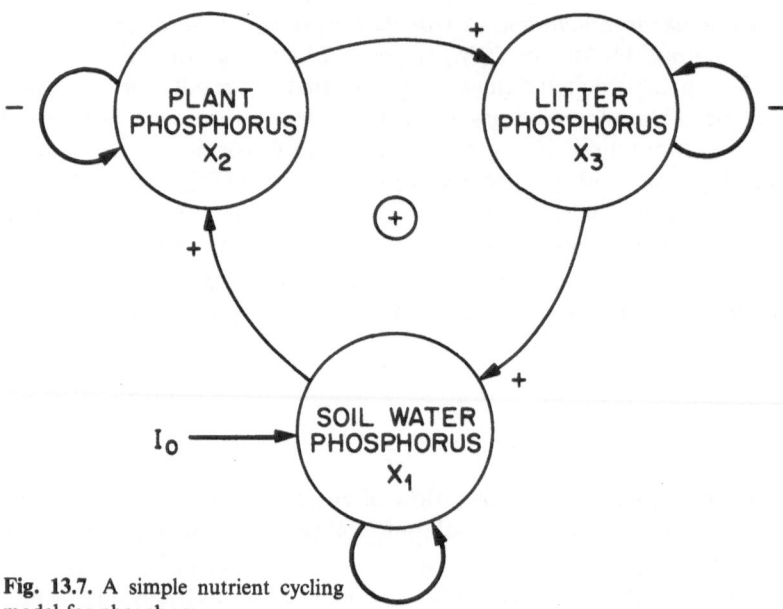

Fig. 13.7. A simple nutrient cycling model for phosphorus

The system (13.13a–c) can be shown to have a single equilibrium point

$$\bar{X}_1 = I_0/(a_{12} + \alpha_1 + \varrho), \tag{13.14a}$$

$$\bar{X}_2 = \bar{X}_1 a_{12}/(a_{23} + \alpha_2), \tag{13.14b}$$

$$\bar{X}_3 = \bar{X}_2 a_{23}/(a_{31} + \alpha_3),$$

where

$$\varrho = (a_{31} a_{23} a_{12})/\{(a_{31} + \alpha_3)(a_{23} + \alpha_2)\}. \tag{13.14c}$$

The stability matrix at this equilibrium point is

$$\mathbf{A} = \begin{vmatrix} -(a_{12} + \alpha_1) & 0 & a_{31} \\ a_{12} & -(a_{23} + \alpha_2) & 0 \\ 0 & a_{23} & -(a_{31} + \alpha_3) \end{vmatrix}. \tag{13.15}$$

Matrix **A** is a positive feedback matrix, and hence it is of the type to which positive linear theory can be applied.

Simple considerations show that matrix **A** is always stable. Because all column sums of **A** are negative, all eigenvalues are negative and the system is stable about the equilibrium point $(\bar{X}_1, \bar{X}_2, \bar{X}_3)$ (see Appendix G). Therefore, persistence and stability are guaranteed. We must remember, however, that the model (13.13) is an oversimplification of real systems. In real systems the possibility of extinction must generally be considered. What makes extinction impossible in the present case is the constant input I_0.

238

In ecological systems, at least in the early stages of succession, there might not be a guaranteed input I_0. Instead, any influx of nutrient is likely to depend on positive feedback linkages between the biota and the edaphic environment. As Alexander (1964) states, "*Webley and collaborators (1963) have demonstrated that, coincidental with the colonization of rock surfaces by lichens of the pioneer community, there is an increase in abundance of bacteria and fungi. Many of these heterotrophs, which undoubtedly participate in the weathering of rocks and pedogenesis, dissolve calcium, magnesium, and zinc silicates. Roy (1962), using samples of rock, weathered rock, and soil materials that resemble the weathering sequence in the conversion of rock to soil, observed that the relative quantity of phosphorus and potassium that becomes biologically available increased with the extent of weathering, despite the quantitative phosphorus and potassium loss from the material as it was subjected to weathering.*" In addition to promoting weathering, the presence of biota may also act as a trap for allochthonous litter.

Weathering by microorganisms can be particularly important at low temperatures when purely chemical processes are extremely slow. Also, as the populations of microorganisms grow, the rate of biological weathering can increase through time in contrast to chemical weathering rates, which, though temperature-dependent and thus seasonally varying, tend to remain constant on the average over a period of many years (Brock, 1966).

Gutierrez and Fey (1980) extended their model of secondary succession to include primary succession processes. The positive feedback effect of rock weathering is embodied in two feedback loops:

1) particulate matter is generated which serves as a habitat for organisms that secrete acids that weather more rock, and

2) plant nutrients are generated by weathering "*which support plants whose dead litter both stimulates soil organisms that secrete acids and produces humus that removes positive ions from the soil solution leaving a weathering acid.*"

In addition to nutrients procured by rock weathering, nitrogen is also obtained from the atmosphere by nitrogen-fixing fungi and bacteria. In Chap. 8 we discussed symbioses between certain autotrophs and mycorrhizal fungi and microorganisms of the genus *Rhizobium*. There are also free-living, non-symbiotic nitrogen fixers, members of the genus *Azotobacter*, common in temperate soils. These bacteria need organic compounds that plants can provide, and they fix nitrogen in return. This feedback cycle builds up the organic biomass in a terrestrial ecosystem.

Hutchinson (1948) has outlined a similar scenario for the build-up of nutrients and biomass in a lake: "*Starting with a barren glacial basin newly filled with water, the concentration of phosphate in solution in the drainage basin will depend on the general geochemistry of the region. At first, the nitrogen available will probably be derived solely from rain water, but nitrogen will tend to be fixed biologically according to the availability of organic matter, the production of which will depend on the availability of phosphorus and other nutrients. The water of the lake will gradually develop a phytoplankton population. The first sediments to be deposited will be almost entirely inorganic, but as soon as remains of organisms are included in the surface layer of these sediments, an internal cycle of the kind already described will be established. Phosphate will be more easily liberated from mineral particles, owing to the production of carbon dioxide by decomposition, and the phosphate of the decomposing organisms will be returned to the lake water. As productivity increases, the sediments will become more and more organic, and thus more and more able to return the nutrients rapidly to the cycle. The process continues until the geochemically determined nutrient potential of the silt and the water of the drainage basin is fully utilized."*

Positive feedback processes in lake sediments can also produce or intensify conditions of anoxia in bodies of water. Anoxia occurs in the deeper waters of the central portions of Chesapeake Bay, for example, due to stratification and the respiration of dead organic matter that sinks down from the photosynthetic zone. Officer et al. (1984) have suggested that conditions of anoxia in bottom waters can cause more rapid mineralization of phosphorus from the sediments, possibly stimulating increased primary production and greater anoxia.

Besides the self-generated input of inorganic nutrients, there is also an increase of organic nutrients during succession: *"Progressive biochemical enrichment of nutritionally impoverished areas parallels the progressive enrichment with additional populations. A widespread characteristic of growing heterotrophs and autotrophs is their ability to excrete certain metabolites, including compounds that are needed by neighboring individuals. The biochemical enrichment acts selectively, allowing for the increase of only particular species, the selectivity depending on, among other things, the molecules liberated and the requirements of those organisms reaching the particular place at the appropriate time" (Alexander, 1971).*

Let us consider the system (13.13) again, this time assuming that $I_0 = 0$, but that there is an allochthonous input of litter proportional to the plant compartment, βX_2. Because $I_0 = 0.0$, the point $(0,0,0)$ is a feasible equilibrium point. Stability of this point corresponds to inability of succession to proceed. The stability matrix at this point is

$$\mathbf{A} = \begin{vmatrix} -(a_{12} + \alpha_1) & 0 & a_{31} \\ a_{12} & -(a_{23} + \alpha_2) & 0 \\ 0 & \beta + a_{23} & -(a_{31} + \alpha_3) \end{vmatrix}. \tag{13.16}$$

We are interested in determining if succession can take place; that is, if $(0,0,0)$ is a repellor. If $\beta + a_{23} > a_{23} + \alpha_2$, there is a possibility that $\mathrm{Re}(\lambda) > 0$. We can determine if $\mathrm{Re}(\lambda) > 0$ by applying the principal minor test. If $(-1)^3 \det(\mathbf{A}) < 0$, or

$$\beta a_{12} a_{31} > \alpha_1 a_{23} a_{31} + \alpha_2 a_{31} a_{12} + \alpha_3 a_{12} a_{23}$$

$$+ \alpha_1 \alpha_3 a_{23} + \alpha_2 \alpha_3 a_{12} + \alpha_1 \alpha_2 a_{31} + \alpha_1 \alpha_2 \alpha_3. \tag{13.17}$$

then the point $(0,0,0)$ is a repellor and the system is persistent. Hence, a strong ability of the pioneer plant biota to trap allochthonous litter, as measured by β, can lead to persistence and growth of the system. We have not included carrying capacity effects, which must ultimately limit growth, in the model.

A related question has to do with the importance of positive feedback effects on nutrient cycling stemming from migration of adult fish to the streams of their origin to spawn and die. Hall (1972) has summarized some of the literature on the subject. Among examples are the observations of Juday et al. (1932) that the carcasses of Pacific salmon might be significant in replenishing the nutrient-poor upstream spawning areas. Donaldson (1967) and Krokhin (1967) quantified these observations, showing that migration plays a role in supplying phosphorous to the upstream areas. It is possible that migration often plays a crucial role in nutrient cycles.

This growth can be studied analytically by assuming that X_1, X_2, and X_3 of Eqs. (13.13) represent aqueous phosphorus, plant phosphorus, and detrital phosphorus in a stream spawning area. We add a compartment X_4 for juvenile salmon, and a compartment, X_5, for adult spawning salmon. The equations might now read

$$\frac{dX_1}{dt} = a_{31}X_3 - (a_{21} + \alpha_1)X_1, \tag{13.18a}$$

$$\frac{dX_2}{dt} = a_{21}X_1 - (a_{32} + a_{42} + \alpha_2)X_2, \tag{13.18b}$$

$$\frac{dX_3}{dt} = a_{32}X_2 + a_{35}X_5 - (a_{31} + \alpha_3)X_3, \tag{13.18c}$$

$$\frac{dX_4}{dt} = a_{42}X_2 - (a_{34} + \alpha_4)X_4, \tag{13.18d}$$

$$X_5 = \gamma\beta X_4. \tag{13.18e}$$

In these equations, $a_{42}X_2$ represents transfer of plant phosphorus (indirectly through the food chain) to juvenile salmon, $a_{34}X_4$ is the rate of migration downstream of juvenile salmon, and $a_{35}X_5$ is the contribution to detrital phosphorus of dead adult spawners. The constant γ is the fraction of fish spawned in the modeled stream that returns as adults to spawn, and β is the ratio of weight of adult salmon to juveniles. The model has assumed the severe condition that there is no allochthonous source of phosphorus to the system other than from migrating adults. If $\gamma\beta a_{35} > a_{34} + \alpha_4$, however, there is a possibility of the system being self-sustaining. To see if it actually is, we must consider the stability matrix, $\mathbf{A}$, at $(0, 0, 0, 0)$, where

$$\mathbf{A} = \begin{vmatrix} -(a_{21} + \alpha_1) & 0 & a_{31} & 0 \\ a_{21} & -(a_{32} + a_{42} + \alpha_2) & 0 & 0 \\ 0 & a_{32} & -(a_{31} + \alpha_3) & \gamma\beta a_{35} \\ 0 & a_{42} & 0 & -(a_{34} + \alpha_4) \end{vmatrix}. \tag{13.19}$$

If $(-1)^4 \det(\mathbf{A}) < 0$, then this system can be maintained purely on the basis of adult migration.

Positive feedback loops exist that could also lead to deteriorating nutrient conditions. A feedback loop that could possibly lead to loss of forest vigor over the long term in certain sites was identified by Gosz (1981) and Vitousek (1982). This conjecture is based on the observation that woody plants in low-nitrogen sites produce litterfall with a much higher ratio of carbon to nitrogen than plants on nitrogen-rich sites. Nitrogen released from the high $C:N$ litter might be monopolized by decomposer organisms in the soil, leading to intensified nitrogen stress on the trees (see also Pastor et al., 1984).

13.8 Selection on the Community or Ecosystem Level

The essence of an ecosystem resides in the almost limitless complexity of its structure and its feedback loops. We have mentioned earlier the fundamental question of whether or not the interrelatedness of the parts of a community, or perhaps a whole ecosystem, make it appropriate to consider it as an integral unit (or perhaps even a *"superorganism"*), which can compete with other ecosystems in the same manner that species can compete. Many ecologists, typified by Williams (1966), have rejected this view, in part because there is no genetic mechanism linking species of an ecosystem as a heritable unit like, say, the brains, liver, kidneys, etc., for an organism.

Nevertheless, while the biological constituents of an ecosystem are genetically distinct, it is still true that many have evolved coadaptations that are mutually beneficial. Even in basically antagonistic relationships, such as predation, the prey may derive some positive benefits; for example, plant seeds are dispersed, as well as consumed, by seed predators. Odum and Biever (1984) picture the ecosystem as a network of "autotroph-heterotroph mutualisms that involve highly specialized, coevolved species." This concept does not presume that the ecosystem functions like a servomechanism or organism, with goal direction toward particular set points. Instead, the ecosystem is viewed as a diffuse web of feedbacks, which, in effect, seem to promote the efficiency and stability of the system as a whole (Patten and Odum, 1981).

A basic question if whether such a network has evolved through pure natural selection at the individual level or whether some organisms in the ecological community can achieve a net positive feedback by rendering some service, at a cost to themselves, to other members of the community.

Wilson (1976, 1980) argued in the affirmative. Before presenting a possible mechanism for such higher order system properties, he described the traditional argument against it by means of an example. Consider an organism, say an earthworm, labelled genotype P. The earthworm is probably a benefactor to many members of the community since it aerates the soil, helps break down litter, and may even secrete some plant growth inducers. These activities doubtless stimulate increased production by the ecosystem as a whole, and provide positive feedback to the earthworm in terms of increased litter, etc. The problem, however, is that another organism, labelled genotype Q, that used the same resources as genotype P without making any of these positive contributions to the ecosystem, could still enjoy the benefits from the indirect positive feedback, engendered by genotype P. The system would in effect, be tolerating *"cheaters"*. Hence, there is no apparent mechanism to integrate a loose collection of species into a superorganism of mutualistic components.

Wilson (1976, 1980) phrased this problem in mathematical form. He considered two species (or two genotypes of the same species) P and Q. Each is assumed to interact, directly or indirectly, with the other biota of the ecosystem. The two genotypes are assumed different from each other in that they affect the rest of the biota to different degrees; let us say that Q affects the rest of the system in a more positive way than P. The question of whether or not Q is selected over genotype P can then be answered by looking at the community matrix,

$$
\begin{vmatrix}
g & g & h & i & j & \cdots \\
g & g & h & i & j & \cdots \\
k & k' & a_{33} & a_{34} & & \\
l & l' & a_{43} & & & \\
\cdot & \cdot & & & & \\
\cdot & \cdot & & & & \\
\cdot & \cdot & & & &
\end{vmatrix},
\qquad (13.20)
$$

with $k' > k$, $l' > l$, etc. The first two columns of the matrix represent the effects of P and Q on other ecosystem constituents, while the first two rows represent the effects of the total ecosystem on genotypes P and Q. Since the two genotypes were assumed to differ only in their contributions to the ecosystem, the first two rows are identical. This implies that the dynamics of P and Q are identical, so Q obtains no relative benefit from its greater positive contribution to the other ecosystem components. Actually, since there presumably has been some cost to Q in helping the other biota, the second term in row 2 should be replaced by g', where $g' < g$. Hence, Q is at a disadvantage relative to P and will be eliminated by natural selection. Thus, the mathematical approach supports the traditional argument that there is no apparent mechanism to transform an ecosystem from a loose assemblage of species into a higher level system of mutualistic components.

Wilson (1976, 1980) was not satisfied with this result and attributed it to an overly simplistic model that implicitly assumed spatial homogeneity. He reasoned that the introduction of spatial structure could produce different results, citing evidence that, at least for short time intervals, indirect effects within a community are often localized to small spatial areas around the individuals that cause the effects (Wiens, 1976; Levin, 1976; Whittaker and Levin, 1977). Because individual genotypes are localized, a genotype Q, for example, that has a positive effect on the rest of the local community is likely to be the recipient of more of the positive feedback from the community than is some other genotype P, that is spatially displaced. If the positive feedback Q receives more than compensates for the cost incurred on itself, Q will grow relative to P.

Let us consider this structured approach from the viewpoint of positive feedback matrices, but otherwise following Wilson's (1976, 1980) approach. Assume replicates of genotype P, say earthworms, are scattered randomly at various spatial sites. Assume also that initially the earthworms are in equilibrium with their local communities. Now assume that a mutant form, genotype Q, appears on one or a few spatial sites, and that it differs from P in having additional positive effects on its local community, while incurring some additional costs to itself. On the other hand, other community members can have various direct effects, either positive or negative, on the earthworms. For the system to be a positive feedback system, however, every loop of length greater than zero must be positive. For example, if the system includes bacteria pathogenic to worms, we assume that the actions of the earthworms stimulate other species that suppress these. Hence, we are excluding, for simplicity, any negative feedback the incremental change on earthworm behavior will have on the earthworm population.

The submatrix of **A** governing the differential growth of genotype Q relative to genotype P will then have the form of the matrix **B**, where

$$
\mathbf{B} = \begin{vmatrix} g & h & i & j & \cdots \\ k & a_{12} & a_{23} & & \\ l & a_{32} & & & \\ m & & & & \\ \cdot & & & & \\ \cdot & & & & \\ \cdot & & & & \end{vmatrix}, \tag{13.21}
$$

where $k > 0$, $l > 0$, $m > 0$, etc., and the matrix is then positive feedback. Because genotype Q incurs additional costs in itself, $g < 0$. Whether or not genotype Q will increase beyond the former equilibrium in its local region depends on the eigenvalues, λ_i, of **B**. If the largest eigenvalue, λ_1, of **B** is positive, then genotype Q will increase. Equivalently, because **B** is a positive feedback matrix, if

$$
(-1)^n \det (\mathbf{B}) < 0, \tag{13.22}
$$

where n is the order of the matrix, genotype Q will increase relative to P. In Wilson's computer model, it was assumed that periods of growth would be divided by periods of dispersal, in which new relative proportions of the genotypes are calculated and spatially relocated. Wilson also built into the model a certain degree of random variation. If, on the average, the community positive feedback on genotype Q outweighs the incurred extra cost on itself, genotype Q increases relative to genotype P over the long term.

The above argument, if correct, would support view that the ecosystem can be usefully viewed as a distinct level of organization, above that of the organisms (Patten and Odum, 1981; Jordan, 1981; McNaughton and Coughenour, 1981; Knight and Swaney, 1981). A key question, however, is what degree of coherence this organization has; that is, whether internal feedbacks dominate in shaping the course of ecosystem evolution, or whether exogenous factors of chance disturbances and invasions are more significant over the long term. Although this question is still an open one, we hope that by drawing attention, in this book, to the ubiquity of positive feedbacks in the ecosystem and in nature in general, we are a little closer to an answer.

13.9 Summary and Conclusions

The ecosystem, a unit of biological organization in a given area, is an intricate network of feedbacks. The degree to which an ecosystem represents an integrated unit is the subject of debate, but is is clear at least that some reciprocal interactions between biota and their environment occur that gradually alter both during ecological succession. In primary succession, at least, the gradual amelioration of the environment by early successional species may pave the way for a more complex community.

Clementsian succession models directly incorporate positive feedback mechanisms. Matrix species-replacement models do not assume positive feedback. However, deviations from the behavior predicted by the matrix models may reflect the fact that certain successional stages may hasten their dominance by creating conditions more favorable for themselves. Fire dominated communities are a possible example of this.

Numerous positive feedback loops exist actively or potentially in ecosystems and can be triggered by certain types of perturbation. Disturbance of soil can cause a cycle of intensifying erosion or nutrient losses. Erosion and nutrient loss can also be stemmed by natural positive feedback processes under proper conditions.

Natural selection is normally assumed to act on the level of individuals or, possibly, species. Most ecologists are sceptical that selection between groups of interacting species, or whole ecosystems, can occur. However, mechanisms have been proposed that involve competition between positive feedback loops in different spatial patches. Determination of whether selection of this sort can take place is a challenge for ecologists.

Appendices

Appendix A: Positive Linear Systems

The subject of positive linear systems is reviewed in many places. A highly readable survey is given by Luenberger (1979). In this appendix we define a positive linear system and in the later appendices we describe some of its properties.

A positive linear system is defined as a system whose state variables always have positive, or at least non-negative, values. This concept applies to both discrete and continuous linear systems. Consider first an n-dimensional discrete system;

$$\mathbf{X}(t+1) = \mathbf{A}\mathbf{X}(t), \tag{A1}$$

where $\mathbf{X}(t)$ is an n-dimensional vector and $\mathbf{A}$ is an $n \times n$ matrix. The constraint that $\mathbf{X}(t)$ remain non-negative through time imposes conditions on the matrix $\mathbf{A} = (a_{ij})$. For every element of $\mathbf{X} = (X_1, X_2, \ldots, X_n)$ to remain non-negative at all times, given that they all start with non-negative values, it is necessary and sufficient that all elements of $\mathbf{A}$ be non-negative ($a_{ij} \geq 0$ for all i and j).

Consider next a continuous system;

$$\frac{d\mathbf{X}(t)}{dt} = \mathbf{A}\mathbf{X}(t). \tag{A2}$$

The conditions on the matrix $\mathbf{A}$ for the linear continuous system to be positive are not as stringent as the conditions on $\mathbf{A}$ for the discrete system. This is because it is not necessary that $dX_i(t)/dt$ be non-negative for $X_i(t)$ to remain positive. The continuous system is positive if and only if $a_{ij} \geq 0$ for all $i \neq j$. Thus, only the off-diagonal elements need to be non-negative. Matrices with this property are called Metzler matrices. A Metzler matrix, $\mathbf{A}$, can be written as $\mathbf{A} = c\mathbf{I} + \mathbf{B}$, where $\mathbf{B}$ is a non-negative matrix, $\mathbf{I}$ is the identity matrix, and c is a scalar (positive, negative, or zero).

We next introduce some notation to be used in later appendices. If $\mathbf{A} = (a_{ij})$ is a matrix, then

1) $\mathbf{A} > 0$ means $a_{ij} > 0$ for all i, j. ($\mathbf{A}$ is termed strictly positive.)
2) $\mathbf{A} \geq 0$ means $a_{ij} \geq 0$ for all i, j and $a_{ij} > 0$ for at least one element. ($\mathbf{A}$ is termed strictly non-negative.)
3) $\mathbf{A} \geq 0$ means $a_{ij} \geq 0$ for all i, j. ($\mathbf{A}$ is termed non-negative.)

247

Appendix B: Stability of Positive Feedback Systems

Criteria for the local stability of mathematical representations of ecological communities ultimately reside in the nature of the eigenvalues of the community matrix. The community matrix is a matrix of first partial derivatives evaluated at equilibrium, often referred to as the Jacobian of the system of equations. If all the eigenvalues have negative real parts then the equilibrium is stable. If at least one eigenvalue has a positive real part then the equilibrium is unstable. Thus determination of the eigenvalues is necessary and sufficient to determine local stability of the equilibrium.

It is our purpose in this appendix to develop criteria for the stability of positive feedback systems that are simpler to compute than eigenvalues and have direct biological interpretations in terms of population parameters. The translation of mathematical abstractions into understandable functions of population parameters is not only helpful in interpreting mathematical concepts in biology, but it is also important in relating theory to observations and experiments.

Let $\mathbf{B} = (b_{ij})$, $1 \leq i, j \leq n$, be a matrix satisfying $b_{ij} \geq 0$ for all i, j. The Perron-Frobenius theorem states that $\mathbf{B}$ has a positive eigenvalue λ^* with the property that $\lambda^* \geq \lambda_i$, where λ_i is any other eigenvalue of $\mathbf{B}$. Thus, all the eigenvalues of the matrix $\mathbf{B}$ lie in a circle in the complex plane with radius λ^* centered at the origin. Because the eigenvalues of any matrix of the form $\mathbf{B} - c\mathbf{I}$ are found by subtracting c from each eigenvalue of the matrix $\mathbf{B}$, the eigenvalues of the matrix $\mathbf{B} - c\mathbf{I}$ lie in a circle in the complex plane centered at $-c$ with radius λ^*. Thus subtracting a sufficiently large constant c from the diagonal of $\mathbf{B}$ (i.e., $c > \lambda^*$) will move all the eigenvalues of $\mathbf{B} - c\mathbf{I}$ into the left half of the complex plane. This condition is illustrated in Fig. B1.

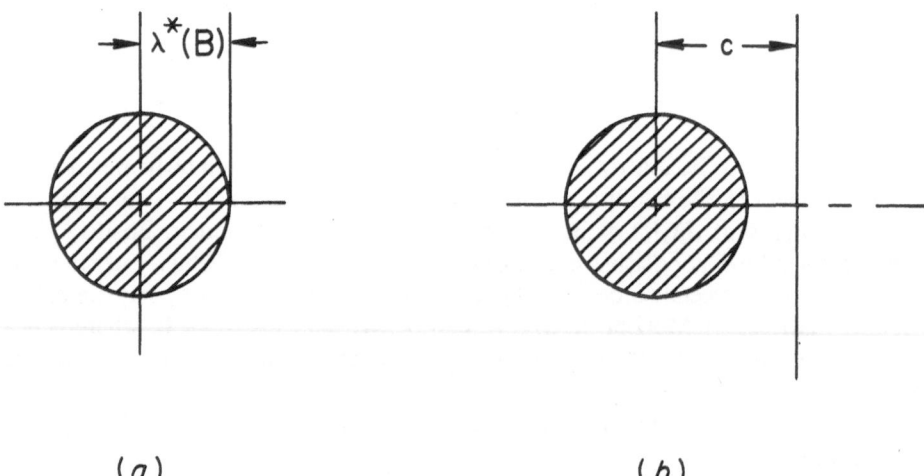

(a)

(b)

Fig. B1a and b. Assume $\mathbf{B}$ is a matrix with nonnegative entries. The hatched circle in **a** represents the region in the complex plane containing the eigenvalues of the matrix $\mathbf{B}$. The hatched circle in **b** represents the region in the complex plane containing the eigenvalues of the matrix $\mathbf{B} - c\mathbf{I}$. (Reproduced from Travis and Post, 1979, with permission of Academic Press, Inc.)

Because the community matrix $\mathbf{E}$ of any mutualistic community may be written as $\mathbf{E} = \mathbf{B} - c\mathbf{I}$, where $\mathbf{B} = (b_{ij})$ has the property $b_{ij} \geq 0$ for all i, j, we see that all of the eigenvalues of the community matrix $\mathbf{E}$ will have negative real parts if and only if $c > \lambda^*(\mathbf{B})$, where $\lambda^*(\mathbf{B})$ denotes the largest positive eigenvalue of $\mathbf{B}$. Thus we have established that stability of a mutualistic community near equilibrium requires that intraspecific regulation of population densities must be stronger in some sense than the benefits gained from interactions with other species in the community. To establish the precise relationship between intraspecific regulation and mutualistic benefit which must be satisfied for stability we introduce the concept of an M-matrix (see the excellent review of M-matrix theory by Plemmons, 1977).

Definition

An $n \times n$ matrix $\mathbf{R} = (r_{ij})$, $1 \leq i, j \leq n$, is said to be an M-matrix if $r_{ij} \leq 0$ for all $i \neq j$ and if any one of the following equivalent statements is true:

1) all the principal minors of $\mathbf{R}$ are positive;
2) all eigenvalues of $\mathbf{R}$ have positive real parts;
3) $\mathbf{R}$ is non-singular and $\mathbf{R}^{-1} > 0$;
4) there is a vector $\mathbf{t} > 0$ such that $\mathbf{R}\mathbf{t} > 0$;
5) there is a vector $\mathbf{u} > 0$ such that $\mathbf{R}^T\mathbf{u} > 0$;
6) there is a positive diagonal matrix $\mathbf{D}$ such that $\mathbf{D}\mathbf{R} + \mathbf{R}^T\mathbf{D}$ is positive definite.

A mutualistic community is stable in the neighborhood of an equilibrium if and only if its community matrix $\mathbf{A}$ is stable. Because the off-diagonal elements of $\mathbf{A}$ are non-negative we can conclude from (2) that a mutualistic community will be stable if and only if $-\mathbf{A}$ is an M-matrix.

The power of relating community stability to the theory of M-matrices lies in the availability of the many useful equivalent conditions that one can employ to determine stability or instability. Properties (4) and (5) imply that necessary and sufficient conditions for a mutualistic community to be stable in the neighborhood of an equilibrium is that there exist positive constants $t_1, t_2, \ldots, t_n$ such that

$$-t_i a_{ii} > \sum_{\substack{j=1 \\ j \neq i}}^{n} t_j a_{ij}, \quad i = 1, 2, \ldots, n,$$

or
$$(B1)$$

$$-t_i a_{ii} > \sum_{\substack{j=1 \\ j \neq i}}^{n} t_j a_{ji}, \quad i = 1, 2, \ldots, n.$$

This specifies precisely the relationship between intraspecific regulation and mutualistic benefit insuring stability. Using property (1) of an M-matrix, we can verify the stability of a mutualistic community in a small number of elementary arithmetical operations. Property (1) implies that the equilibrium of a mutualistic community is locally stable if and only if the community matrix satisfies the conditions

$$(-1)^k \begin{vmatrix} a_{11} & a_{12} & \cdots & a_{1k} \\ a_{21} & a_{22} & \cdots & a_{2k} \\ \cdot & & & \\ \cdot & & & \\ \cdot & & & \\ a_{k1} & a_{k2} & \cdots & a_{kk} \end{vmatrix} > 0 \quad \text{for } k = 1, 2, \ldots, n. \tag{B2}$$

The simplicity of these conditions should be compared with a similar necessary and sufficient condition for the local stability of competitive communities (Strobeck, 1973).

Application to per Capita Population Equations of Chapter 8

In this section we establish the statement in Chap. 8 that the stability of mutualistic communities described by the general population equations

$$\frac{dX_i}{dt} = X_i g_i(X_1, X_2, \ldots, X_n), \quad i = 1, \ldots, n, \tag{B3}$$

can be determined from the interaction matrix defined as $\mathbf{A} = (a_{ij})$ where

$$a_{ij} = \frac{\partial g_i}{\partial X_j} (\bar{X}_1, \ldots, \bar{X}_n) \tag{B4}$$

rather than the community matrix, assuming that all equilibrium population sizes are non-negative. The model mutualistic community (B3) will be stable in a neighborhood of the feasible equilibrium $\bar{X}$ if and only if the matrix $-\mathbf{S} = [-s_{ij}]$ is an M-matrix, where the elements s_{ij} are given by

$$s_{ij} = \frac{\partial}{\partial X_j} [X_i g_i(X_1, X_2, \ldots, X_n)], \tag{B5}$$

and the partial derivatives in (B5) are evaluated at the equilibrium point $\bar{X}$. Because

$$\frac{\partial}{\partial X_j} [\bar{X}_i g_i(\bar{X}_1, X_2, \ldots, \bar{X}_n)] = \begin{cases} \bar{X}_i \dfrac{\partial g_i}{\partial X_j} (\bar{X}_1, \ldots, \bar{X}_n) & \text{if } j \neq i, \\ g_i(\bar{X}_i, \ldots, \bar{X}_n) & \\ + \bar{X}_i \dfrac{\partial g_i}{\partial X_j} (\bar{X}_1, \ldots, \bar{X}_n) & \text{if } j = i, \end{cases} \tag{B6}$$

and $g_i(\bar{X}_1, \bar{X}_2, \ldots, \bar{X}_n) = 0$, $\mathbf{S}$ equals $\mathbf{DA}$ where $\mathbf{D}$ is the diagonal matrix $\mathbf{D} = \text{diag}(\bar{X}_1, \bar{X}_2, \ldots, \bar{X}_n)$ and $\mathbf{A}$ is the interaction matrix defined by (B4). Since $\bar{X}_i$ is positive and $a_{ij} \geq 0$ for $i \neq j$, all the off diagonal elements of the matrix $\mathbf{DA}$ are non-negative. It follows that the community matrix $\mathbf{DA}$ will be stable if and only if $-\mathbf{DA}$ is an M-matrix. By property (5) of an M-matrix, a necessary and sufficient condition for $-\mathbf{DA}$ to be an M-matrix is that there exists a vector $\mathbf{t} > 0$ such that $-(\mathbf{DA})^T\mathbf{t} = -\mathbf{A}^T\mathbf{Dt} > 0$. This last condition is equivalent to the fact that there exists a vector $\mathbf{u} > 0$, $\mathbf{u} = \mathbf{Dt}$, such that $-\mathbf{A}^T\mathbf{u} > 0$. Again, using property (5) of an M-matrix and the fact that $a_{ij} \geq 0$ for $i \neq j$, we conclude that in a mutualistic community, the community matrix $\mathbf{DA}$ will be stable if and only if $-\mathbf{A}$ is an M-matrix, that is, if and only if the interaction matrix $\mathbf{A}$ is stable.

Appendix C: Stability of Discrete-Time Systems

Consider the n-dimensional discrete system,

$$\mathbf{X}(t+1) = \mathbf{AX}(t), \tag{C1}$$

where $\mathbf{X}(t)$ is an n-dimensional vector and $\mathbf{A}$ is an $n \times n$ matrix. A necessary and sufficient condition for an equilibrium point of system (C1) to be asymptotically stable is that the spectral radius of the matrix $\mathbf{A}$ be less than one (that is, the eigenvalues of $\mathbf{A}$ must all be inside the unit circle in the complex plane) (see Luenberger, 1979).

It is our purpose in this appendix to develop criteria for stability of discrete-time systems that are simpler to apply than computation of the eigenvalues of the matrix $\mathbf{A}$. We will show that the eigenvalues spectral radius of $\mathbf{A}$ has magnitude less than one if, and only if there exists a matrix $\mathbf{S}$ such that $\mathbf{G} = \mathbf{I} - \mathbf{SAS}^{-1}$ is an M-matrix. First notice that if $\mathbf{G} = \mathbf{I} - \mathbf{SAS}^{-1}$ is an M-matrix then $\mathbf{SAS}^{-1}$, which is similar to $\mathbf{A}$,

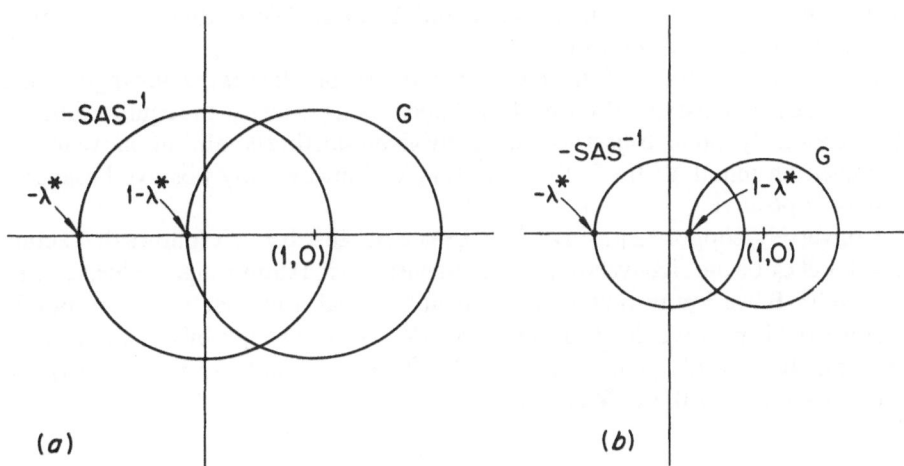

(a) $\qquad\qquad\qquad\qquad\qquad\qquad (b)$

Fig. C1a, b. The eigenvalues of the matrix $-\mathbf{SAS}^{-1}$ lie in the circle centered at the origin of the complex plane. **a** If the radius of the circle described by the Perron-Frobenius eigenvalue is greater than one (i.e. the matrix $\mathbf{A}$ represents an unstable discrete system) then $\mathbf{G} = \mathbf{I} - \mathbf{SAS}^{-1}$ is not an M-matrix. **b** If the circle radius is less than one (A represents a stable discrete system) then $\mathbf{G}$ is an M-matrix. (Reproduced from Travis et al., 1980, with permission from Academic Press) is positive. Therefore, the Perron-Frobenius Theorem guarantees that there exists one eigenvalue λ^* of the matrix $\mathbf{SAS}^{-1}$ which is real, positive, and larger than the absolute value of any other eigenvalue of $\mathbf{SAS}^{-1}$. This implies that the eigenvalues of $\mathbf{SAS}^{-1}$ (and therefore the eigenvalues of $\mathbf{A}$ since similar matrices have the same eigenvalues) lie in a circle centered at the origin with radius λ^* (Fig. C1). The eigenvalues of $-\mathbf{SAS}^{-1}$ lie in the same circle, and $-\lambda^*$, an eigenvalue of $-\mathbf{SAS}^{-1}$, lies at the left extreme of the circle. Adding $\mathbf{I}$ to the matrix $-\mathbf{SAS}^{-1}$ shifts the circle one unit to the right. Now, if the spectral radius of $\mathbf{SAS}^{-1}$ or equivalently of $\mathbf{A}$, is less than 1, the circle will be moved entirely into the right half of the complex plane and $\mathbf{G} = \mathbf{I} - \mathbf{SAS}^{-1}$ will be an M-matrix. If the spectral radius of $\mathbf{A}$ is greater than 1, then the eigenvalue $1 - \lambda^*$ of $\mathbf{G}$ will be in the left half of the complex plane and $\mathbf{G}$ will not be an M-matrix.

Appendix D: Positive Equilibria and Stability

Consider non-homogeneous discrete and continuous system,

$$\mathbf{X}(t+1) = \mathbf{AX}(t) + \mathbf{a}, \tag{D1}$$

$$\frac{d\mathbf{Y}(t)}{dt} = \mathbf{BY}(t) + \mathbf{b}, \tag{D2}$$

where $\mathbf{a} > 0$, $\mathbf{b} > 0$, and $\mathbf{A}$ is non-negative and $\mathbf{B}$ has off-diagonal elements. Such systems have the useful property that the existence of positive equilibria, $\bar{\mathbf{X}}$ and $\bar{\mathbf{Y}}$, automatically implies the stability of these equilibria. Conversely, if all the eigenvalues of $\mathbf{A}$ are within the unit circle and all the eigenvalues of $\mathbf{B}$ have negative real parts, then positive equilibria, $\bar{\mathbf{X}}$ and $\bar{\mathbf{Y}}$, exist. We outline a proof of this assertion for the continuous case.

If all the eigenvalues of $\mathbf{B}$ lie in the left-hand plane, then $\mathbf{B}$ is non-singular and a unique $\bar{\mathbf{Y}}$ exists satisfying $\mathbf{B}\bar{\mathbf{Y}} + \mathbf{b} = 0$ (s. Appendix B). Because the equilibrium point $\bar{\mathbf{Y}}$ is necessarily stable, trajectories with any given starting conditions move towards it. Given an initial $\mathbf{Y}(0) \geq 0$, the trajectory will always stay positive because the system is positive.

Conversely, suppose a positive $\bar{\mathbf{Y}} \geq 0$ exists. Because $\mathbf{b} > 0$, it follows that actually $\bar{\mathbf{Y}} > 0$. Let $\mathbf{e}_0^T$ be the eigenvector associated with the maximum eigenvalue, λ_0. From Appendix B we know that this eigenvalue is real and that the corresponding eigenvector is positive. Multiplication of $\mathbf{B}\bar{\mathbf{Y}} + \mathbf{b} = 0$ through by $\mathbf{e}_0^T$, and using the fact that $\mathbf{Be}_0^T = \lambda_0 \mathbf{e}_0^T$, yields $\lambda_0 \mathbf{e}_0^T \bar{\mathbf{Y}} + \mathbf{e}_0^T \mathbf{b} = 0$. Because both $\mathbf{e}_0^T \bar{\mathbf{Y}} > 0$ and $\mathbf{e}_0^T \mathbf{b} > 0$, it follows that $\lambda_0 < 0$, i.e. $\bar{\mathbf{Y}}$ is stable.

Appendix E: Comparative Statics of Positive Feedback Systems

The study of the comparative statics of ecological communities is concerned with how equilibrium population densities are altered by changes in biological and environmental factors which affect the per capita growth rate of the component species. For a given set of biological or environmental parameters $\mathbf{b}^0 = (b_1^0, b_2^0, \ldots, b_n^0)$, an equilibrium population density is defined as a set of values $\bar{\mathbf{X}} = (\bar{X}_1, \bar{X}_2, \ldots, \bar{X}_n)$ for which

$$g_i(\bar{\mathbf{X}}; b_i^0) = 0 \quad (\text{for } i = 1, 2, \ldots, n). \tag{E1}$$

We wish to determine the qualitative changes that occur in $\bar{\mathbf{X}}$ as the biological or environmental parameter change from $\mathbf{b}^0$ to $\mathbf{b}^1$. We assume that when a mutualistic community is perturbed in such a way as to slightly alter the equilibrium population densities, community stability will be preserved, that is, the community will be stable near its new equilibrium (s., for example, the perturbation theorem of Levin, 1974). Throughout this appendix, we will assume that the vector b_i is a scalar, although, in general, it could be a vector. This assumption is made for simplicity of notation and the results can be extended to any number of parameters. We also assume that the ecological community is indecomposable; that is, we assume that it is impossible to decompose the community into two subcommunities in such a way that one subcommunity has no effect on the other subcommunity.

Let b_k be a parameter with respect to which the per capita growth rate at equilibrium of the k-th species increases; that is assume

$$\frac{\partial g_k}{\partial b_k} (\bar{\mathbf{X}}; b_i) > 0. \tag{E2}$$

If we differentiate the system of algebraic equations

$$g_i(\bar{X}_1, \bar{X}_2, \ldots, \bar{X}_n; b_i) = 0, \quad i = 1, 2, \ldots, n \tag{E3}$$

with respect to b_k, we obtain

$$\sum_{j=1}^{n} \frac{\partial g_i}{\partial \bar{X}_j} \frac{\partial \bar{X}_j}{\partial b_k} = 0 \quad \text{if } k \neq i$$

and

$$\sum_{j=1}^{n} \frac{\partial g_i}{\partial \bar{X}_j} \frac{\partial \bar{X}_j}{\partial b_k} = -\frac{\partial g_k}{\partial b_k} \quad \text{if } k = i. \tag{E4}$$

Solving for $\partial \bar{X}_j / \partial b_k$, we obtain

$$\frac{\partial \bar{X}_j}{\partial b_k} = -\frac{\partial g_k}{\partial b_k} \frac{J_{ki}}{J}, \quad j = 1, 2, \ldots, n, \tag{E5}$$

where J_{ki} is the (k, i)-th cofactor of the interaction matrix $\mathbf{A}$ and J is its determinant. That is, J_{ki}/J is the k, i-th element of the inverse matrix $\mathbf{A}^{-1}$. Since under our assumptions, the interaction matrix $\mathbf{A}$ is stable, it follows from property (3) of an M-matrix (Appendix B) that J_{ki}/J is negative. This establishes that in mutualistic communities at equilibrium, an increase in the per capita growth rate of one species increases the population density of every species.

We now determine the conditions under which an increase in the per capita growth rate of a species will have a greater effect on its own equilibrium population density than that of other species. If the interaction matrix satisfies the condition of diagonal dominance

$$a_{ii} > \sum_{\substack{j=1 \\ j \neq i}}^{n} a_{ij}, \quad i = 1, \ldots, n,$$ (E6)

we call the community strongly self-regulatory. Using a result of Debreu and Herstein (1953, p. 603), one can establish that if a mutualistic community is strongly self-regulatory, then the cofactor J_{ij} of the interaction matrix A satisfies $0 > J_{ij}/J > J_{ii}/J$ for all i,j. This fact, combined with (E5) demonstrates that an increase in the per capita growth rate of the i-th species of a strongly self-regulatory mutualistic community at equilibrium increases the population density of the i-th species proportionately more than that of the j-th species ($j \neq i$).

Appendix F: Similarity Transforms

The question we address here is: Given a system that is not quite a positive feedback or mutualistic system, how do we choose a similarity transformation that will make it one?

The simplest case occurs when the community matrix represents a community with limited competition. Such an community matrix can, by identical row and column permutations, be brought into the form

$$A = \begin{bmatrix} A_{11} & A_{12} \\ A_{21} & A_{22} \end{bmatrix},$$ (F1)

where A_{11}, A_{22} are square blocks and have non-negative off diagonal elements. A_{12} and A_{21} have nonpositive elements. The similarity transformation that transforms A into a positive feedback matrix is

$$S = \begin{bmatrix} I & 0 \\ 0 & -I \end{bmatrix},$$ (F2)

where the dimensions of I and $-I$ agree with the dimensions of A_{11} and A_{22} respectively.

A community with limited competition is one that can be divided into two subcommunities with the property that all interactions within each subcommunity are mutualistic but where interactions between subcommunities are all competitive (Fig. F1). It is a simple matter to arrange the equations into the required order for the interaction matrix to have the form (F1) once this community structure has been identified.

254

Fig. F1. Canonical form of a community with limited competition (adapted from Takeuchi et al., 1978)

There are many communities whose community matrices do not satisfy the restrictions of limited competition. In particular, limited competition allows no predator-prey interactions. We propose a technique here that can be useful in finding a similarity transformation for particular systems that are nearly positive feedback.

The dynamics of a community in the neighborhood of a feasible equilibrium can be described by

$$\frac{d\mathbf{X}}{dt} = \mathbf{A}\mathbf{X}. \tag{F3}$$

Making a similarity transformation on the matrix $\mathbf{A}$ corresponds to a variable substitution

$$\frac{d\mathbf{Y}}{dt} = \mathbf{S}^{-1}\mathbf{A}\mathbf{S}\mathbf{Y}, \tag{F4}$$

where $\mathbf{X} = \mathbf{S}\mathbf{Y}$. For many systems that are nearly in the form required by the theorem of Appendix B the substitution

$$X_i = Y_i + \alpha X_j \tag{F5}$$

is useful. Consider the matrix

$$\mathbf{A} = \begin{vmatrix} a_{11} & a_{12} & a_{13} \\ a_{21} & a_{22} & a_{23} \\ a_{31} & a_{32} & a_{33} \end{vmatrix}. \tag{F6}$$

The transformation $X_1 = Y_1 + \alpha X_2$ will transform it into

$$\mathbf{A}' = \begin{vmatrix} a_{11} - \alpha a_{21} & a_{12} + \alpha a_{11} - \alpha(a_{22} + \alpha a_{21}) & a_{13} - \alpha a_{23} \\ a_{21} & a_{22} + \alpha a_{21} & a_{23} \\ a_{31} & a_{32} + \alpha a_{31} & a_{33} \end{vmatrix}. \tag{F7}$$

255

This is equivalent to a similarity transformation with

$$S = \begin{vmatrix} 1 & \alpha & 0 \\ 0 & 1 & 0 \\ 0 & 0 & 1 \end{vmatrix}.$$ (F8)

This is generalized by the following theorem.

Theorem F1. The substitution

$$X_i = Y_i + \alpha X_j$$

will result in changes to all elements in the i-th row and all the elements in the j-th column of Matrix A. In particular,

1) every element, a_{ik}, in the i-th row will have $-\alpha a_{jk}$ added to it;
2) every element, a_{kj}, in the j-th column will have αa_{ki} added to it;
3) the element a_{ij} will have $\alpha a_{ii} - \alpha a_{jj} - \alpha^2 a_{ji}$ added to it.

The proof of this theorem is a straightforward procedure. Using this theorem, one can often find ways of changing a given matrix into a positive feedback matrix. If necessary, one can apply more than one substitution. The sequence

$$X_1 = Y_1 + \alpha_1 X_j \quad j \neq 1$$
$$X_2 = Y_2 + \alpha_2 X_k \quad k \neq 2$$ (F9)
$$X_i = Y_i \quad i \neq 1, 2$$

is equivalent to

$$A'' = S_2^{-1}(S_1^{-1} A S_1) S_2.$$ (F10)

Example

Consider the system depicted in Fig. F2. This could characterize the competition between two competitors (X_2, X_3) for a limited resource (X_1). The stronger competitor has a negative influence on the resource and the weak competitor has a negligible impact on the resource. The equations near the feasible equilibrium are

$$dX_1/dt = -a_{11}X_1 \qquad\qquad -a_{13}X_3$$
$$dX_2/dt = \quad a_{21}X_1 - a_{22}X_2 - a_{23}X_3$$ (F11)
$$dX_3/dt = \quad a_{31}X_1 - a_{32}X_2 - a_{33}X_3$$

and the corresponding community matrix is, for all $a_{ij} > 0$,

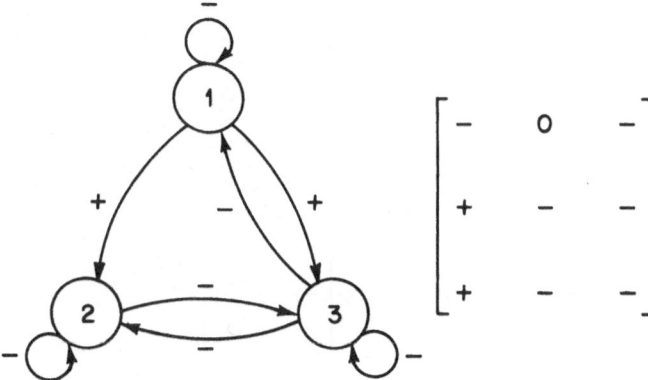

Fig. F2. Model for competition during succession and the sign structure of the corresponding feedback matrix

$$A = \begin{vmatrix} -a_{11} & 0 & -a_{13} \\ a_{21} & -a_{22} & -a_{23} \\ a_{31} & -a_{32} & -a_{33} \end{vmatrix}. \qquad (F12)$$

This would represent a community with limited competition if $a_{31} \leq 0$. With the substitution

$$X_3 = Y_3 + \alpha X_1 \qquad (F13)$$

the transformed community matrix A' becomes

$$A' = \begin{vmatrix} -a_{11} + \alpha a_{13} & 0 & -a_{13} \\ a_{21} + \alpha a_{23} & -a_{22} & -a_{23} \\ a_{31} + \alpha a_{33} + \alpha^2 a_{13} - \alpha a_{11} & -a_{32} & -a_{33} - \alpha a_{13} \end{vmatrix}. \qquad (F14)$$

This will be stable provided $a'_{31} \leq 0$ and A' satisfies the conditions on the principal minors described in Appendix B (note: the condition $a'_{31} \leq 0$ implies that $a'_{11} < 0$). The principal minor condition yields the well known necessary condition for the stability of A (also A'), namely

$$\det A = -a_{11}a_{22}a_{33} + a_{11}a_{23}a_{32} + a_{21}a_{32}a_{13} - a_{22}a_{31}a_{13} < 0. \qquad (F15)$$

This, combined with the condition that

$$a'_{31} = a_{31} + \alpha a_{33} + \alpha^2 a_{13} - \alpha a_{11} < 0 \qquad (F16)$$

is sufficient for the stability of A' (and therefore A). From Fig. F3 we see that (F16) is satisfied if

$$a_{11} > \frac{a_{31}}{\sqrt{a_{31}/a_{13}}} + a_{33} + a_{13}\sqrt{a_{31}/a_{13}}, \qquad (F17)$$

since we can choose $\alpha = \sqrt{a_{31}/a_{13}}$.

257

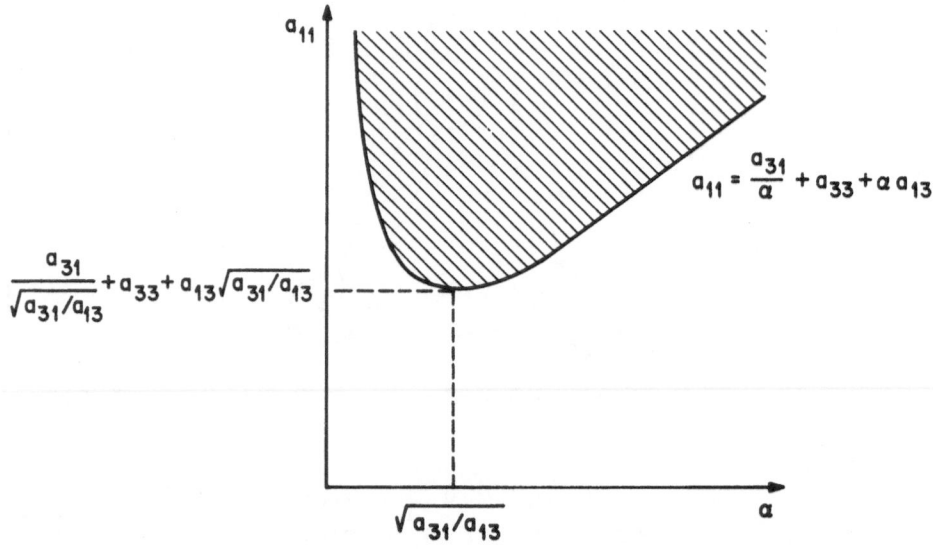

Fig. F3. Hatched region is where it is possible for $a'_{31} < 0$ for fixed values of a_{31}, a_{13}, a_{33}

Appendix G: Bounds on the Roots of a Positive Linear System

Theorem. If the matrix **A** is a Metzler matrix (see Appendix A), then either all the column sums equal the largest eigenvalue, λ_0 (s. Appendix B), or λ_0 lies between the maximum and minimum column sums of **A**; that is,

$$\min_{1 \leq j \leq n} \left\{ \sum_{i=1}^n a_{ij} \right\} < \lambda_0 < \max_{1 \leq j \leq n} \left\{ \sum_{i=1}^n a_{ij} \right\}. \tag{G1}$$

Proof (adapted from Luenberger, 1979). Let $\mathbf{X}_0 = (X_1, X_2, \ldots, X_n)$ be the positive eigenvector corresponding to λ_0 and assume that it is normalized so that

$$\sum_{i=1}^n X_i = 1. \tag{G2}$$

Writing out $\mathbf{A}\mathbf{X}_0 = \lambda_0 \mathbf{X}_0$ in detail, we have

$$a_{11}X_1 + a_{12}X_2 + \ldots + a_{1n}X_n = \lambda_0 X_1$$
$$a_{21}X_1 + a_{22}X_2 + \ldots + a_{2n}X_n = \lambda_0 X_2$$
$$\phantom{a_{21}X_1 + a_{22}X_2}\cdot \tag{G3}$$
$$\phantom{a_{21}X_1 + a_{22}X_2}\cdot$$
$$\phantom{a_{21}X_1 + a_{22}X_2}\cdot$$
$$a_{n1}X_1 + a_{n2}X_2 + \ldots + a_{nn}X_n = \lambda_0 X_n$$

258

Summing these equations, we obtain

$$\Delta_1 X_1 + \Delta_2 X_2 + \ldots + \Delta_n X_n = \lambda_0 (X_1 + X_2 + \ldots + X_n) = \lambda_0, \qquad \text{(G4)}$$

where

$$\Delta_i = \sum_{j=1}^{n} a_{ji}. \qquad \text{(G5)}$$

Hence, λ_0 is a weighted average of the column sums. It must lie between $\min (\Delta_i)$ and $\max (\Delta_i)$, or from (G5), it must obey (G1).

Appendix H: Relationship Between Positive Linear System Stability Criteria and the Routh-Hurwitz Criteria

The Routh-Hurwitz Theorem provides a way of determining the signs of the eigenvalues, and hence stability of an arbitrary matrix, $\mathbf{A}$. For all eigenvalues to have negative real parts, two conditions must be true:

1) All principle minors of $\mathbf{A}$ must obey the rule,

$$(-1)^i \det (\mathbf{A}_i) > 0 \qquad (i = 1, 2, \ldots, n). \qquad \text{(H1)}$$

2) A sequence of inequalities must hold that in essence require that negative feedbacks coming from short feedback loops dominate in magnitude those coming from long loops.

Since positive linear systems have no negative feedback loops longer than unity, the second condition can be dispensed with in these cases. Therefore, the Routh-Hurwitz criteria reduce to the criterion introduced in Appendix B for positive linear systems.

As a corollary of the Routh-Hurwitz Theorem, we can state that a system that has a positive feedback loop that is longer than any of the negative feedback loops in the system will always be unstable.

References

Abraham, R. H. and C. D. Shaw. 1983. Dynamics: The Geometry of Behavior. Part One: Periodic Behavior. Aerial Press, Inc., Santa Cruz, California. 220 pp.

Abugov, R. 1982. Species diversity and phasing of disturbance. Ecology 63:289–293.

Adams, H. 1918. The Education of Henry Adams. Modern Library (1931), New York. 517 pp.

Addicot, J. F. 1979. A multispecies aphid-ant association. Density-dependence and species-specific effects. Can. J. Zool. 57:558–569.

Ahmadjian, V. 1966. Lichens. IN Symbiosis (S. M. Henry, ed.) Academic Press, New York. pp. 35–97.

Aidley, D. J. 1978. The Physiology of Excitable Cells. Cambridge University Press, Cambridge Massachusetts. 239 pp.

Albrecht, F., H. Gatzke, A. Haddad, and N. Wax. 1974. The dynamics of two interacting populations. J. Math. Anal. Appl. 46:658–670.

Alexander, M. 1964. Biochemical ecology of soil microorganisms. Ann. Rev. Microbiol. 18:217–252.

Alexander, M. 1971. Microbial Ecology. John Wiley and Sons, Inc. New York. 511 pp.

Alexander, R. D. 1979. Darwinism and Human Affairs. University of Washington Press, Seattle, Washington. 317 pp.

Allee, W. C. 1938. The Social Life of Animals. W. W. Norton. New York. 293 pp.

Allen, P. M. 1976. Evolution, population dynamics and stability. Proc., Nat. Acad. Sci., USA 73: 665–668.

Anderson, R. M. 1979. The influence of parasitic infection on the dynamics of host population growth. IN (R. M. Anderson, B. D. Turner, and L. R. Taylor, eds.) Population Dynamics. Blackwell Scientific Publication, Oxford, England.

Anderson, R. M. 1980. Depression of host population abundance by direct life cycle macroparasites. J. Theor. Biol. 82:283–311.

Anderson, R. M. and R. M. May. 1979. Population biology of infectious diseases: Part 1. Nature 280:361–367.

Archer, J. 1975. The organization of aggression and fear in vertebrates. pp. 231–298. IN Perspectives in Ethology, Vol. 2 (P. P. G. Bateson and P. Klopfer, eds.),. Plenum Press. New York.

Arman, P., D. Hopcraft, and I. McDonald. 1975. Nutritional studies of East African herbivores. 2. Losses of nitrogen in the feces. Br. J. Nutrition 33:265–276.

Armstrong, R. M. 1976. Fugitive species: Experiments with fungi and some theoretical considerations. Ecology 57:953–963.

Arndt, W. F. and R. E. Ritts 1961. Synergism between *Staphylocci* and *Proteus* in a mixed infection. Proc. Soc. Exptl. Biol. Med. 108:166–169.

Aron, J. L. and R. M. May. 1982. The population dynamics of malaria. IN Population Dynamics of Infectious Diseases: Theory and Applications. (R. M. Anderson, ed.). Chapman and Hall, London. 368 pp.

Ashby, W. R. 1952. Design for a Brain. Science Paperbacks (1965), Chapman and Hall, Ltd., London, England. 286 pp.

Atsatt, P. R. and D. J. O'Dowd. 1976. Plant defence guilds. Science 193:24–29.

Axelrod, R. 1984. The Evolution of Cooperation. Basic Books, Inc. New York. 241 pp.

Axelrod, R. and W. D. Hamilton. 1981. The evolution of cooperation. Science 211:1390–1396.

Bailey, N. T. J. 1975. The Mathematical Theory of Infectious Diseases and its Applications. Hafner, New York, 413 pp.

Bajzer, Z., K. Pavelic, and S. Vuk-Pavlovic. 1984. Growth self-incitement in murine melanoma B12: A phenomenological model. Science 225:930–932.

261

Bakker, R. T. 1983. The deer flees, the wolf pursues: Incongruencies in predator-prey coevolution, pp. 350–382. IN Coevolution (Futuyma, D. J. and M. Slatkin, eds.). Sinauer Associates, Inc., Sunderland, Massachusetts. 555 pp.

Barash, D. 1977. Sociobiology and Behavior. Elsevier, New York. 378 pp.

Barash, D. 1979. The Whisperings Within: Evolution and the Origin of Human Nature. Harper and Row, Publishers, New York, New York. 274 pp.

Barbour, C. D. and J. H. Brown. 1974. Fish species diversity in lakes. Amer. Natur. 108:473–478.

Barlow, G. W. 1974. Hexagonal territories. Anim. Behav. 22:876–878.

Bartlett, M. S. 1957. Measles periodicity and community size. J. R. Statist. Soc., Ser. A, 120:48–70.

Bartlett, M. S. 1960. The critical community size for measles in the United States. J. R. Statist. Soc., Ser. A, 123:37–44.

Bateson, G. 1972. Steps to an Ecology of Mind. Ballentine Books, New York. 541 pp.

Batra, L. R. 1966. Ambrosia fungi: extent of specificity to ambrosia beetles. Science 153:193–195.

Batra, S. W. T. and L. R. Batra. 1967. The fungus gardens of insects. Scientific American 217:112–120.

Beattie, A. J. and D. C. Culver. 1981. The guild of myrmechores in the herbaceous flora of West Virginia forests. Ecology 62:107–115.

Bender, E. A. 1978. An Introduction of Mathematical Modeling. John Wiley and Sons, Inc., New York. 256 pp.

Bently, B. L. 1976. Plants bearing extrafloral nectaries and the associated ant community: interhabitat differences in the reduction of herbivore damage. Ecology 57:815–820.

Bently, B. L. 1977. Extrafloral nectaries and protection by pugnacious bodyguards. Ann. Rev. Ecol. Syst. 8:407–427.

Berg, R. Y. 1972. Dispersal ecology of *Vancouveria* (Berberidaceae). Amer. J. Bot. 59:109–122.

Bernadelli, H. 1941. Population waves. J. Burma Res. Soc. 31:1–18.

Berry, R. S., S. A. Rice, and J. Ross. 1980. Physical Chemistry. Part. 3. Physical and Chemical Kinetics. John Wiley and Sons, Inc., New York. 1264 pp.

Berryman, A. A. 1981. Population Systems: A General Introduction. Plenum Press, New York. 222 pp.

Beverton, R. J. and S. J. Holt. 1957. On the dynamics of exploited fish populations, Fish. Invest. Min. Agric. Fish. Food (Gt. Brit.) Ser. II Salmon Freshwater Fish 19:533–545.

Bishop, R. E. D. 1979 (second edition). Vibrations. Cambridge University Press, Cambridge, England. 164 pp.

Black, F. L. 1966. Measles endemicity in insular populations: Critical community size and its evolutionary implications. J. Theor. Biol. 11:207–211.

Blackman, V. H. 1919. The compound interest law and plant growth. Ann. Bot. 33:353–360.

Blagden, C. 1775. Experiments and observations in a heated room. Phil. Trans. R. Soc. London 65: 111–123.

Bobisud, L. E. and R. J. Neuhaus. 1975. Pollinator constancy and survival of rare species. Oecologia 21:263–272.

Bock, K. 1980. Human Nature and History: A Response to Sociobiology. Columbia University Press, New York. 241 pp.

Bonner, J. T. 1974. On Development: The Biology of Form. Harvard University Press, Cambridge, Massachussetts.

Bonner, J. T. 1980. Evolution of Culture in Animals. Princeton University Press, Princeton, New Jersey. 216 pp.

Booth, D. A. and P. C. Simpson. 1971. Food preferences acquired by association with variations in amino acid nutrition. Quart. J. Exper. Psych. 23:135–145.

Bormann, F. H. and G. E. Likens. 1979. Pattern and Process in a Forested Ecosystem. Springer-Verlag. New York. 253 pp.

Botkin, D. B., J. F. Janak, and J. R. Wallis. 1972. Some ecological consequences of a computer model of forest growth. J. Ecol. 60:849–873.

Boucher, D. H., S. James, and K. Keeler. 1982. The ecology of mutualism. Ann. Rev. Ecol. Syst. 13: 315–347.

Bourgeois-Pichat, J. 1968. The concept of a stable population: application to the study of populations of countries with incomplete demographic statistics. United Nations, New York. ST1SOA1, SER.A139.

Bowley, A. L. 1924. Births and population of Great Britain, J. Roy. Econom. Soc. 34:188–192.

Boyce, M. S. 1979. Seasonality and patterns of natural selection for life histories. Am. Nat. 114:569–583.

262

Briand, F. and P. Yodzis. 1982. The phylogenetic distribution of obligate mutualism-evidence of limiting similarity and global instability. Oikos 39:273–275.

Brock, E. M. 1960. Mutualism between the midge *Cricotopus* and the alga *Nostoc*. Ecology 41:474–483.

Brock, T. D. 1966. Principles of Microbial Ecology. Prentice-Hall, Inc. Englewood Cliffs, New Jersey.

Brown, J. L. 1974. Alternate routes to sociality in jays – with a theory for the evolution of altruism and communal breeding. Amer. Zool. 14:63–80.

Brown, J. L. 1978. Avian communal breeding systems. Ann. Rev. Ecol. Syst, 9:123–156.

Brush, S. G. 1983. Statistical Physics and the Atomic Theory of Matter from Boyle and Newton to Landau and Onsager. Princeton University Press, Princeton, New Jersey. 356 pp.

Buchsbaum, R. 1948. Animals Without Backbones. University of Chicago Press, Chicago.

Bullard, E. C. 1949. The magnetic field within the earth. Proc. Roy. Soc. A 197:433–453.

Bullard, E. C. 1955. The stability of a homopolar dynamo. Proc. Camb. Phil. Soc. 51:744–760.

Bush, G. L. 1975. Sympatric speciation in phytophagous parasitic insects. pp. 187–206. IN Evolutionary Strategies of Insect, and Mites (P. W. Price, ed.) Plenum Press, New York.

Butler, J. H. 1980. Economic Geography: Spatial and Environmental Aspects of Economic Activity. John Wiley and Sons. New York. 402 pp.

Butzer, K. W. 1982. Archeology as Human Ecology: Method and Theory for a Contextual Approach. Cambridge University Press, Cambridge, England. 364 pp.

Cairns-Smith, A. G. 1971. The Life Puzzle. University of Toronto Press, Toronto, Canada. 165 pp.

Calow, P. 1978. Life Cycles. Chapman and Hall, Ltd., London, England. 164 pp.

Calvin, M. 1956. Chemical evolution and the origin of life. Am. Sci. 44:248–263.

Cannan, E. 1895. The probability of a cessation of the growth of population in England and Wales during the next century. The Economic Journal 5:505–515.

Cannon, W. A. 1929. Organization for physiological homeostasis. Physiol. Rev. 9:399–431.

Caraco, T. and L. L. Wolf. 1975. Ecological determinants of groups sizes of foraging lions. Amer. Natur. 109:343–352.

Careri, G. 1984. Order and Disorder in Matter. The Benjamin/Cummings Publishing Company, Inc., Menlo Park, California. 162 pp.

Cates, R. G. 1975. The interface between slugs and wild ginger: Some evolutionary aspects. Ecology 56:391–400.

Cavanaugh, C. M. 1983. Symbiotic chemotrophic bacteria in marine invertebrates from sulphide-rich habitats. Nature 302:58–61.

Charlesworth, B. 1973. Selection in populations with overlapping generations V. Natural selection and life histories. Am. Nat. 107:303–311.

Charlesworth, B. 1980. Evolution in Age-Structured Populations, Cambridge University Press, Cambridge. 300 pp.

Charlesworth, B. and J. A. Leon. 1976. The relation of reproductive effort to age. Am. Nat. 110:449–459.

Chase, I. V. 1974. Models of hierarchy formation in animal societies. Behavioral and Brain Sciences 19:374.

Chase, I. V. 1980. Cooperative and noncooperative behavior in animals. Amer. Natur. 115:827–857.

Chase, I. V. 1982. Behavioral sequences during dominance hierarchy formation in chickens. Science 216:439–440.

Chiang, H. C. 1961. Fringe populations of the European corn borer, *Pyrausta mubilalis*: Their characteristics and problems. Ann. Entomol. Soc. Amer. 54(3):378–387.

Clark, C. W. 1976. Mathematical Bioeconomics: The Optimal Management of Renewable Resources. John Wiley and Sons. New York. 352 pp.

Clark, C. W. and M. Mangel. 1979. Aggregation and fishing dynamics: A theoretical study of schooling and the purse seine tuna fisheries. Fishery Bulletin 77:317–337.

Clark, L. R., P. W. Geier, R. D. Hughes, and R. F. Morris. 1967. The Ecology of Insect Populations in Theory and Practice. John Wiley and Sons, New York.

Clements, F. E. 1916. Plant Succession: An Analysis of the Development of Vegetation. Carnegie Inst., Publ. 242. Washington, D.C. pp. 1–512.

Cohen, D. 1971. Maximizing final yield when growth is limited by time or by limiting resources. J. Theor. Biol. 33:299–307.

Cohen, J. E. 1969. Natural primate troops and a stochastic population model. Amer. Natur. 103:455–477.

Cohen, J. E. 1970. A Markov contingency table model for replicated Lotka-Volterra systems near equilibrium. Amer. Natur. 104:547–559.

Cohen, J. E. 1979. Ergodic theorems in demography. Bulletin of the American Mathematical Society 1:275–295.

Cohn, R. M., M. Yudkoff, and P. D. McNamara. 1980. Servomechanisms and oscillatory phenomena. pp. 295–312. IN Principles of Metabolic Control in Mammalian Systems (R. H. Herman, R. M. Cohn, and P. D. McNamara, eds.) Plenum Press, New York.

Colgan, P. 1983. Comparative Social Recognition. John Wiley and Sons, Inc., New York. 281 pp.

Colwell, R. K. 1973. Competition and coexistence in a simple tropical community. Amer. Natur. 107:737–760.

Connell, J. H. 1979. Tropical rain forests and coral reefs as open, non-equilibrium systems. IN Population Dynamics (R. M. Anderson and B. D. Turner, eds.) 20th Symp. Brit. Ecol. Soc. Blackwell Scientific Publications, Oxford. pp. 141–165.

Connell, J. H. and E. Orias. 1964. The ecological regulation of species diversity. Amer. Natur. 98:399–414.

Copasso, V. and G. Serio. 1978. A generalization of the Kermack-McKendrick deterministic model. Math. Biosci. 42:43–61.

Corning, P. A. 1983. The Synergism Hypothesis. McGraw Hill Book Company. New York. 492 pp.

Cowles, H. C. 1911. The causes of vegetative cycles. Botanical Gazette 51:161–183.

Culver, D. C. and A. J. Beattie. 1978. Mymecochory in *Viola*: dynamics of seed-ant interactions in some West Virginia species. J. Ecol. 66:53–72.

Darling, F. F. 1938. A Herd of Red Deer. Oxford University Press. London, England.

Darlington, P. J., Jr. 1980. Evolution for Naturalists: The Simple Principles and Complex Reality. John Wiley and Sons, Inc., New York. 262 pp.

Darwin, C. 1871. The Descent of Man. Modern Library (1949), New York, New York.

Darwin, C. R. 1859. On the Origin of Species by Means of Natural Selection, or the Preservation of Favoured Races in the Struggle for Life. John Murray, London.

Davidson, D. W. 1980. Some consequences of diffuse competition in a desert ant community. Amer. Natur. 116:92–105.

Dawkins, R. 1976. The Selfish Gene. Oxford University Press. New York, New York. 224 pp.

Dawkins, R. 1982. The Extended Phenotype. W. H. Freeman and Co., San Francisco, California. 307 pp.

Dawkins, R. and J. R. Krebs. 1979. Arms races between and within species. Proc. Roy. Soc. London B 205:489–511.

Dean, A. M. 1983. A simple model of mutualism. Amer. Natur. 121:409–417.

DeAngelis, D. L., C. C. Travis, and W. M. Post. 1979. Persistence and stability of seed-dispersal species in a patchy environment. Theor. Pop. Biol. 16:107–125.

Debreu, G. and I. N. Herstein. 1953. Nonnegative square matrices. Econometrica 21:597–607.

Deneubourg, J. L. 1977. Application to order by fluctuations to description of some stages in building of termite nest. (French). Insectes Sociaux 23:329.

Desowitz, R. S. 1977. Harmonious parasites. Nat. History 86:34–38.

DeVay, J. E. 1956. Mutual relationships in fungi. Ann. Rev. Microbiol. 10:115–140.

Diamond, J. M. 1972. Biogeographic kinetics: Estimation of relaxation times for avifaunas of southwest Pacific islands. Proc. Nat. Acad. Sci. USA 69:3199–3203.

Diamond, J. M. 1974. Colonization of exploded volcanic islands by birds: The supertramp strategy. Science 184:803–806.

Diamond, J. M. 1975. The island dilemma: Lessons of modern biogeographic studies for the design of natural reserves. Biol. Cons. 7:129–146.

Diamond, J. M. 1976. Island biogeography and conservation: Strategy and limitations. Science 193:1027–1029.

Diamond, J. M. 1980. Patchy distributions of tropical birds. IN Conservation Biology (M. E. Soule and B. A. Wilcox, eds.) Sinauer Associates, Inc. Sunderland, Massachusetts. pp. 75–92.

Diamond, J. M. and R. M. May. 1976. Island biogeography and the design of nature reserves. IN Theoretical Ecology (R. M. May, ed.) Saunders Publ. Co., Philadelphia, Pennsylvania. pp. 163–186.

Dinsmore, J. J. 1970. History and natural history of *Paradisaea apoda* on Little Tobago Island, West Indies. Caribbean Journal of Science 10:93–100.

Dodson, S. I. 1970. Complementary feeding niches sustained by size-selective predation. Limnology and Oceanography 15:131–137.

Donaldson, J. R. 1967. Phosphorus budget of Iliamna Lake, Alaska, as related to the cyclic abundance of sockeye salmon. Ph. D. Thesis, Univ. Washington, Seattle.

Douce, G. K. and D. P. Webb. 1978. Indirect effects of soil invertebrates on litter decomposition: Elaboration via analysis of a tundra model. Ecol. Model. 4:339–359.

Doutt, R. L. and J. Nakada. 1973. The *Rubus* Leafhopper and its egg parasitoid: An endemic biotic system useful in grape-pest management. Environmental Entomol. 2:381–386.

Drury, W. H. and I. C. T. Nisbet. 1973. Succession. J. Arnold Arboretum 54:331–368.

Duncan, N. 1978. The effects of culling herring gulls *(Larus argentatus)* on recruitment and population dynamics. J. Appl. Ecol. 15:697–713.

Dworkin, M. 1972. The Myxobacteria: New directions in the study of procaryotic development. Crit. Rev. Microbiol. 2:435–452.

Dyer, M. I. and S. McNaughton. Soft defenses and the evolution of plants and herbivores (unpublished manuscript).

Dyer, M. I. and U. G. Bokhari. 1976. Plant-animal interaction: studies of the effects of grasshopper grazing on blue grama grass. Ecology 57:762–772.

Eastop, V. F. 1977. Worldwide importance of aphids as virus vectors. pp. 413–431. IN (K. F. Harris and K. Maramorosch, eds.) Aphids as Virus Vectors. Academic Press, New York.

Edlin, H. L. 1976. Trees and Man. Columbia University Press, New York. 269 pp.

Ehrlich, P. R. and L. C. Birch. 1967. The "balance of nature" and "population control." Am. Nat. 101:97–107.

Ehrlich, P. and A. Ehrlich. 1981. Extinction: The Causes and Consequences of the Disappearance of Species. Random House, New York, New York. 305 pp.

Eigen, M. 1971. Self-organization of matter and the evolution of biological macromolecules. Naturwissenschaften 58:465–522.

Eigen, M. and P. Schuster. 1979. The Hypercycle: A Principle of natural Self-Organization. Springer-Verlag, New York. 92 pp.

Eigen, M., W. Gardiner, P. Schuster, and R. Winkler-Oswatitch. 1981. The origin of genetic information. Sci. Am. 244(4):88–118.

Eisenberg, J. F. 1981. The Mammalian Radiations. University of Chicago Press, Chicago, Illinois. 610 pp.

Ek, A. R. and R. A. Monserud. 1974. FOREST: A computer model for the growth and reproduction of mixed species forest stands. Research Report A2635. College of Agriculture and Life Sciences, University of Wisconsin, Madison. 90 pp.

Ellul, J. 1980. The Technological System. Continuum Publishing Corporation, New York. 362 pp.

Elsasser, W. M. 1939. On the origin of the earth's magnetic field. Phys. Rev. 55:489–498.

Elsasser, W. M. 1946a. Induction effects in terrestrial magnetism. Part I: Theory. Phys. Rev. 69:106–116.

Elsasser, W. M. 1946b. Induction effects in terrestrial magnetism. Part II: The secular variation. Phys. Rev. 72:821–833.

Emlen, J. M. 1972. Ecology: An Evolutionary Approach. Addison-Wesley. Reading, Massachusetts.

Esnouf, M. P. and R. G. MacFarlane. 1968. Enzymology and the blood clotting mechanism. Adv. Enzymol. 30:255–315.

Ewing, M. S., S. A. Ewing, M. S. Keener, and R. J. Mulholland. 1981. Mutualism among parasitic nematodes: a population model. Ecological Modelling 15:353–366.

Ewing, S. A. and A. C. Todd. 1961. Association among members of the genus *Metastrongylus* Molin, 1861 (Nematoda: Metastrongylidae). Amer. J. of Vet. Res. 22:1077–1080.

Faegri, K. and L. van der Pijl. 1966. The Principles of Pollination Ecology. Pergamon Press, Toronto. 248 pp.

Fagen, R. M. 1972. An optimal life history in which reproductive effort decreases with age. Amer. Natur. 104:258–261.

Faulkner, P. 1982. A novel class of wasp viruses and insect immunity. Nature 299:489–490.

Feder, H. M. 1966. Cleaning symbiosis in the marine environment. IN Symbiosis (S. M. Henry, ed.) Academic Press, New York. pp. 327–380.

Finerty, J. P. 1981. The Population Ecology of Cycles in Small Mammals: Mathematical Theory and Biological Fact. Yale University Press, New Haven, Connecticut. 234 pp.

Fisher, R. A. 1930. The evolution of dominance in certain polymorphic species. Amer. Natur. 64: 385–406.

Fisher, R. A. 1941. Average excess and average effect of a gene substitution. Ann. Eugen. 11:53–63.

Flannery, K. V. 1972. The cultural evolution of civilizations. Annu. Rev. Ecol. Syst. 3:399–426.

Fraser, H. R. 1979. Releasing hormones. IN Reproduction in Mammals. Book 7. Mechanisms of Hormone Action. Cambridge University Press, Cambridge. 239 pp.

Fritz, R. S. 1982. An ant-treehopper mutualism: effects of *Formica subsericea* on the survival of *Vanduzea arquata*. Ecol. Entom. 7, 267–276.

Futuyma, D. J. 1983. Evolutionary interactions among herbivorous insects and plants, pp. 207–231. IN Coevolution (D. J. Futuyma and M. Slatkin, eds.) Sinauer Associates, Inc., Sunderland, Massachusetts.

Futuyma, D. J. and M. Slatkin (eds.) 1983. Coevolution. Sinauer Associates, Inc., Sunderland, Massachusetts. 555 pp.

Gadgil, M. and O. T. Solbrig. 1972. The concept of "r" and "K" selection: Evidence from wild flowers and some theoretical considerations. Amer. Natur. 106:14–31.

Gadgil, M. and W. H. Bossert. 1970. Life history consequences of natural selection. Amer. Natur. 102:52–64.

Galil, J. and D. Eisikowitch. 1971. Studies on mutualistic symbiosis between synconia and sycophilous wasps in monecious figs. New Phytol. 70:773–787.

Galston, A. W. 1981. Green Wisdom: The Inside Story of Plant Life. Basic Books, Inc. New York. 217 pp.

Gehrz, R. D., D. C. Black, and P. M. Solomon. 1984. The formation of stellar systems from interstellar molecular clouds. Science 224:823–830.

Geiselman, P. J. and D. Novin. 1982. Sugar infusion can enhance feeding. Science 218:491–492.

Geist, V. 1971. Mountain Sheep: A Study in Behavior and Evolution. University of Chicago Press. Chicago, Illinois.

Ghiselin, M. T. 1974. The Economy of Nature and the Evolution of Sex. University of California Press, Berkeley, California. 346 pp.

Gilbert, F. S. 1980. The equilibrium theory of island biogeography – fact or fiction? J. of Biogeography 7:209–235.

Gilbert, L. E. 1975. Ecological consequences of a coevolved mutualism between butterflies and plants, pp. 210–240. IN Coevolution of Animals and Plants (Gilbert, L. E. and P. Raven, eds.) Univ. of Texas Press, Austin.

Gilbert, L. E. 1980. Food web organization and the conservation of neotropical diversity. pp. 11–33. IN Conservation Biology (M. E. Soule and B. A. Wilcox, eds.) Sinauer Associates, Inc., Sunderland, Maryland.

Gilbert, L. E. 1983. Coevolution and mimicry, pp. 263–281. IN Coevolution (D. J. Futuyma and M. Slatkin, eds.) Sinauer Associates, Inc., Sunderland, Massachusetts.

Gilpin, M. E. 1975a. Group Selection in Predator-Prey Communities. Princeton University Press. Princeton, New Jersey. 108 pp.

Gilpin, M. E. 1975b. Limit cycles in competition communities. Amer. Natur. 109:51–60.

Givnish, T. J. 1982. On the adaptive significance of leaf height in forest herbs. Amer. Natur. 120: 353–381.

Glynn, P. W. 1980. Defense by symbiotic crustacea of host corals elicited by chemical cues from predators. Oecologia 47:287–290.

Goh, B. S. 1977. Global stability in many-species systems. Amer. Natur. 111:135–143.

Goh, B. S. 1979. Stability in models of mutualism. Amer. Natur. 113:261–271.

Goodman, D. 1974. Natural selection and a cost on reproduction effort. Amer. Natur. 108:247–268.

Goodwin, D. 1967. Extinct and Vanishing Birds of the World. Dover Press. New York.

Gordon, J. E. 1978. Structures, or Why Things Don't Fall Down. Plenum Press, New York. 395 pp.

Gosz, J. R. 1981. Nitrogen cycling in coniferous ecosystems. In: F. E. Clark and T. H. Rosswall (eds.), Nitrogen Cycling in Terrestrial Ecosystems: Processes, Ecosystem Strategies, and Management Implications. Ecol. Bull. (Stockholm) 33:405–426.

Gould, J. L. 1982. Ethology: The Mechanisms and Evolution of Behavior. W. W. Norton and Company, New York. 544 pp.

Gould, S. J. 1977. Ever Since Darwin: Reflections in Natural History. W. W. Norton and Company, Inc., New York, New York. 285 pp.

266

Gourou, P. 1947. Les pays tropicaux: principes d'une geographie humaine et economique. Presses Universitaires de France, Paris. 196 pp.

Graham, K. 1967. Fungal-insect mutualism in trees and timber. Ann. Rev. Entom. 12:105–126.

Grant, K. A. and V. Grant. 1968. Hummingbirds and their Flowers. Columbia University Press, New York.

Grant, V. 1963. The Origin of Adaptations. Columbia University Press, New York, New York. 606 pp.

Grasse, P.-P. 1959. La reconstruction du nid et les coordinations inter-individuelles chez *Bellicositermes natalensis* et *Cubitermes* sp. La theorie de la stigmergie: essai d'interpellation du comportement des termites constructeurs. Insectes Sociaux 6:41–83.

Grime, J. P. 1979. Plant Strategies and Vegetation Processes. John Wiley and Sons, New York. 222 pp.

Guckenheimer, J., G. F. Oster, and A. Ipaktchi. 1977. The dynamics of density-dependent population models. J. Math. Biol. 4:101–147.

Gutierrez, L. T. and W. R. Fey. 1975. Feedback dynamics analysis of secondary successional transients in ecosystems. Proc. Nat. Acad. Sci. USA 72:2733–2737.

Gutierrez, L. T. and W. R. Fey. 1980. Ecosystem Succession: A General Hypothesis and a Test Model of a Grassland. The MIT Press. Cambridge, Massachusetts. 231 pp.

Hairston, N. G., F. E. Smith, and L. B. Slobodkin. 1960. Community structure, population control, and competition. Amer. Natur. 94:421–425.

Haken, H. 1984. The Science of Structure: Synergetics. Van Nostrand Reinhold Company, New York. 255 pp.

Hall, C. A. S. 1972. Migration and metabolism in a temperate stream ecosystem. Ecology 53:585–604.

Hallam, T. G. 1980. Effects of cooperation on competitive systems. J. Theor. Biol. 82:415–423.

Halle, L. J. 1977. Out of Chaos. Houghton Mifflin Company, Boston, Massachusetts. 657 pp.

Halliday, T. R. 1974. The sexual behavior of the smooth newt, *Triturus vulgaris* (Urodela, Salamandridae). J. Herpetol. 8:277–292.

Halliday, T. R. 1980. The extinction of the passenger pigeon *Ectopistes migratorius* and its relevance to contemporary conservation. Biol. Conserv. 17:157–162.

Hamilton, T. H., R. H. Barth, and I. Rubinoff. 1964. The environmental control of insular variation in bird species abundance. Proc. Nat. Acad. Sci. USA 52:132–140.

Hamilton, W. D. 1967. Extraordinary sex ratios. Science 156:477–488.

Hamilton, W. D. 1971. Geometry of the selfish herd. J. Theor. Biol. 31:295–311.

Hanson, F. B. and H. C. Tuckwell. 1978. Persistence time of populations with large random fluctuations. Theor. Pop. Biol. 14:46–61.

Hardy, H. C. 1959. Cocktail party acoustics. J. Acoust. Soc. Amer. 31:535.

Hargrave, B. T. and G. H. Been. 1970. Effects of copepod grazing on two natural phytoplankton populations. J. Fish. Res. Bd. Canada 27:1395–1405.

Harth, E. 1982. Windows on the Mind. William Morrow and Co., Inc., New York, New York. 285 pp.

Haselwandter, K. and D. J. Read. 1982. The significance of a root-fungus association in two *Carex* species of high-alpine plant communities. Oecologia 53:352–354.

Hassell, M. P. and H. W. Comins. 1976. Discrete time models for two-species competition. Theor. Pop. Biol. 9:202–221.

Hastings, A. 1977. Spatial heterogeneity and the stability of predator-prey systems. Theor. Pop. Biol. 12:37–48.

Heatwole, H. 1965. Some aspects of the association of cattle egrets with cattle. Anim. Behav. 13: 79–83.

Heinrich, B. 1971. Temperature regulation in the sphinx moth *Manduca sexta*. J. Exp. Biol. 54:141–152.

Heinrich, B. 1979. Bumblebee Economics. Harvard University Press. Cambridge, Massachusetts. 245 pp.

Heinrich, B. 1984. In a Patch of Fireweed. Harvard University Press, Cambridge, Massachusetts. 194 pp.

Heinrich, B. and P. H. Raven. 1972. Energetics and pollination ecology. Science 176:597–602.

Heithaus, E. R., D. C. Culver, and A. J. Beattie. 1980. Models of some ant-plant mutualisms. Am. Nat. 116:347–361.

Hess, K. W., M. P. Sissenwine, and S. B. Saila. 1975. Simulating the impact of entrainment of winter flounder larvae. pp. 1–30. IN Fisheries and Energy Production (S. B. Saila, ed.). D. C. Health and Co., Lexington, Mass.

Hethcote, H. W. 1976. Qualitative analysis of communicable disease models. Math. Biosci. 28:335–356.

Hethcote, H. W. 1978. An immunization model for a heterogeneous population. Theor. Pop. Biol. 14:338–349.

Hirsch, M. W. and S. Smale. 1974. Differential Equations, Dynamical Systems and Linear Algebra. Academic Press, New York.

Hocking, B. 1975. Ant-plant mutualisms: evolution and energy. Ann. Rev. Ecol. Syst. 4:78–90.

Homans, G. C. 1950. The Human Group. Harcourt Brace. New York.

Horn, H. S. 1974. The ecology of secondary succession. Ann. Rev. Ecol. Syst. 5:25–37.

Horn, H. S. 1975a. Forest succession. Scientific American 232:90–98.

Horn, H. S. 1975b. Markovian properties of forest succession. pp. 196–211. IN Ecology and Evolution of Communities (M. L. Cody and J. M. Diamond, eds.) The Belknap Press of Harvard University Press. Cambridge, Massachusetts.

Horn, H. S. 1978. Optimal tactics of reproduction and life history. pp. 411–429. IN Behavior Ecology: An Evolutionary Approach (J. R. Krebs and N. B. Davies, eds.) Sinauer Associates, Inc. Sunderland, Massachusetts.

Horn, H. S. and R. H. MacArthur. 1972. Competition among fugitive species in a harlequin environment. Ecology 53:749–752.

Horst, T. J. 1977. Effects of power station mortality on fish population stability in relationship to life history strategy. pp. 297–310. IN Proceedings of the Conference on Assessing the Effects of Power-Plant-Induced Mortality on Fish Populations (W. Van Winkle, 1977). Pergamon Press, New York.

Howard, B. H. 1966. Intestinal microorganisms of ruminants and other vertebrates. pp. 317–385. IN Symbiosis (S. M. Henry, ed.) Academic Press, New York.

Howe, H. F. 1977. Bird activity and seed dispersal of a tropical wet forest tree. Ecology 58:539–550.

Howe, H. F. and G. A. Vande Kerchove. 1979. Fecundity and seed dispersal of a tropical tree. Ecology 60:180–189.

Hutchinson, G. E. 1948. Circular causal systems in ecology. Ann. N. Y. Acad. Sci. 50:221–246.

Hutchinson, G. E. 1961. The paradox of the plankton. Amer. Natur. 95:137–145.

Jackson, J. B. C. and L. Buss. 1975. Allelopathy and spatial competition among coral reef invertebrates. Proc. Nat. Acad. Sci. USA 72:5160–5163.

Jacob, F. 1976. The Logic of Life: A History of Heredity. Vintage Books, New York. 348 pp.

Jacobs, J. A. 1963. The Earth's Core and Geomagnetism. Pergamon Press, Oxford, England. 137 pp.

Janson, C. H. 1983. Adaptation of fruit morphology to dispersal agents in a neotropical forest. Science 219:187–188.

Janzen, D. H. 1966. Coevolution of mutualism between ants and acacias in Central America. Evolution 20:249–275.

Janzen, D. H. 1967. Interaction of the bull's horn Acacia (*Acacia cornigera* L.) with an ant inhabitant (*Pseudomyrmex ferruginea* F. Smith) in eastern Mexico. U. Kans. Sci. Bull. 47:315–558.

Janzen, D. H. 1969. Allelopathy by myrmecophytes: The ant *Azteca* as an allelopathic agent of *Cecropia*. Ecology 50:146–153.

Janzen, D. H. 1972. Protection of *Barteria* (Passifloraceae) by Pachysima ants (Pseudomyrmecinae) in a Nigerian rain forest. Ecology 53:885–892.

Janzen, D. H. 1974. Epiphytic myrmecopytes in Sarawak: mutualism through the feeding of plants by ants. Biotropica 6:237–259.

Janzen, D. H. 1975. What are dandelions and aphids? Am. Nat. 111:586–589.

Janzen, D. H. 1979. How to be a fig. Ann. Rev. Ecol. Syst. 10:13–51.

Janzen, D. H. 1983. Dispersal of seeds by vertebrate guts, pp. 232–262. IN Coevolution (D. J. Futuyma and M. Slatkin, eds.) Sinauer Associates, Inc., Sunderland, Massachusetts.

Jeffers, J. N. R. 1978. An Introduction to Systems Analysis: with Ecological Applications. Edward Arnold. London, England. 198 pp.

Jenny, H. 1980. The Soil Resource: Origin and Behavior. Springer-Verlag. New York. 377 pp.

Johnson, M. P. and P. H. Raven. 1973. Species number and endemism: The Galapagos Islands revisited. Science 179:893–895.

Jordan, C. F. 1981. Do ecosystems exist? Amer. Natur. 118:284–287.

Jordan, C. F., J. R. Kline, and D. S. Sasscer. 1972. Relative stability of mineral cycles in forest ecosystems. Amer. Natur. 106:237–253.

Juday, C., W. H. Rich, G. I. Kemmerer, and A. Mann. 1932. Limnological studies of Karluk Lake, Alaska, 1926–1930. U.S. Bur. Fish. Bull. 57:407–436.

Judson, H. F. 1980. The Search for Solutions. Holt, Rinehart, and Winston, New York. 211 pp.

Karakashian, S. J. and M. W. Karakashian. 1965. Evolution and symbiosis in the genus *Chlorella* and related algae. Evolution 19:368–377.

Karlin, S. and J. McGregor. 1972. Polymorphisms for genetic and ecological systems with weak coupling. Theor. Pop. Biol. 3:210–238.

Kassanis, B. 1963. Interactions of viruses in plants. Advan. Virus Res. 10:219–255.

Katchalsky, A. K., V. Rowland, and R. Blumenthal. 1974. Dynamic Patterns of Brain Cell Assemblies. MIT Press. 187 pp.

Kay, D., J. F. Longworth, and J. M. Thresh. 1961. The interaction between swollen shoot disease and mirids on cocoa in Nigeria. Proc. Inter.-Am. Cacao Conf. 8 Trin. Tobago, 1960:224–235.

Keast, A. 1965. Resource subdivision amongst cohabiting fish species in a bay, Lake Opinicon, Ontario, pp. 106–132. Publ. No. 13. Great Lakes Research Division, University of Michigan.

Keeler, K. H. 1979. Species with extrafloral nectaries in a temperate flora (Nebraska). Prairie Nat. 11:33–38.

Keeler, K. H. 1981a. Cover of plants with extrafloral nectaries in four northern California habitats. Madrono 28:26–29.

Keeler, K. H. 1981b. A model of selection for facultative nonsymbiotic mutualism. Am. Nat. 118:488–498.

Keller, E. F. and L. A. Segel. 1970. The initiation of slime mold aggregation viewed as an instability. J. Theor. Biol. 26:399–415.

Kellogg, R. B. 1971. Matrices similar to a positive or essentially positive matrix. Lin. Alg. and Appl. 4:191–204.

Kierstad, H. and L. B. Slobodkin. 1953. The size of water masses containing plankton bloom. J. Mar. Res. 12:141–147.

Kippenhahn, R. 1983. 100 Billion Suns: The Birth, Life, Death of the Stars. Basic Books, Inc., New York, New York. 264 pp.

Kleinfeldt, S. E. 1978. Ant gardens: the interaction of *Codonanthe crassifolia* (Gesneriaceae) and *Crematagastor longispina* (Formicidae). Ecology 59:449–456.

Kline, M. 1980. Mathematics: The Loss of Certainty. Oxford University Press, New York. 366 pp.

Knight, R. L. and D. P. Swaney. 1981. In defense of ecosystems. Amer. Natur. 117:991–992.

Koch, A. 1966. Insects and their endosymbionts. pp. 1–106. IN Symbiosis (S. M. Henry, ed.) Academic Press, New York.

Koshland, D. E. 1977. A response regulator model in a simple sensory system. Science 196:1055–1063.

Krebs, C. J. 1972. Ecology. Harper and Row, New York.

Krebs, J. R. and N. B. Davies. 1981. An Introduction to Behavioral Ecology. Sinauer Associates, Inc. Sunderland, Massachusetts. 292 pp.

Krokhin, E. M. 1967. Effect of size of escapement of sockeye salmon spawners on the phosphate content of a nursery lake. Izvestiya Tikhookeanskogo Nauchno-Issledovatel'skogo. Instituta Rybnogo Kjozyaistva i Okeanografii. 57. Fish. Res. Bd. Canada. Translation Series No. 1186:31–54.

Kropotkin, P. A. 1902. Mutual Aid: A Factor of Evolution. (1972 edition.) New York University Press, New York.

Kushlan, J. A. 1979. Design and management of continental wildlife reserves: Lessons from the everglades. Biol. Cons. 15:281–290.

Lajmanovich, A. and J. A. Yorke. 1976. A deterministic model for gonorrhea in a nonhomogeneous population. Math Biosci. 28:221–236.

Land, G. T. L. 1973. Grow or Die: The Unifying Principle of Transformation. Dell Publishing Company, Inc. 265 pp.

Landau, H. G. 1965. Development of structure in a society with a dominance relation when new members are added successively. Bull. Math. Biophys., special issue 27:151–160.

Lande, R. 1981. Models of speciation by sexual selection in polygenic traits. Proc. Natl. Acad. Sci. USA 78:3721–3725.

Lange, R. T. 1966. Bacterial symbiosis with plants. pp. 99–170. IN Symbiosis (S. M. Henry, ed.) Academic Press, New York.

Langley, L. L. (ed.) 1973. Homeostasis: Origins of the Concept. Dowden, Hutchinson, and Ross, Stroudsburg, Pennsylvania. 362 pp.

Larow, E. J. and D. C. McNaught. 1978. Systems and organismal aspects of phosphorus remineralization. Hydrobiologia 59:151–154.

Leggett, W. C. 1977. Density dependence, density independence, and recruitment in the American Shad *(Alosa sapidissima)* population of the Connecticut River. In Assessing the Effects of Power-Plant Induced Mortality on Fish Populations (W. Van Winkle, ed.), Pergamon Press, New York.

Leon, J. A. 1976. Life histories as adaptive strategies. J. Theor. Biol. 60:301–336.

Leontief, W. 1966. Input-Output Economics. Oxford University Press, New York. 257 pp.

Leshner, A. I. 1975. A model of hormones and agonistic behavior. Physiology and Behavior 15:225–235.

Leslie, P. H. 1945. On the use of matrices in certain population mathematics. Biometrika 33:183–212.

Leslie, P. H. 1948. Some further remarks on the use of matrices in population mathematics. Biometrika 35:213–245.

Levandowsky, M., W. S. Childress, E. A. Spiegel, and S. H. Hutner. 1975. A mathematical model of pattern formation by swimming microorganisms. J. Protozool. 22:296–306.

Levin, B. R. and R. E. Lenski. 1983. Coevolution in bacteria and their viruses and plasmids, pp. 99–127. IN Coevolution (D. J. Futuyma and M. Slatkin, eds.) Sinauer Associates, Inc., Sunderland, Massachusetts.

Levin, D. A. 1973. The age structure of a hybrid swarm of *Liatris* (Compositae). Evolution 27:532–535.

Levin, S. A. 1974. Dispersion and population interactions. Amer. Natur. 108:207–228.

Levin, S. A. 1976. Population dynamic models in heterogeneous environments. Ann. Rev. Ecol. Systematics 7:287–310.

Levin, S. A. 1978a. Population models and community structure in heterogeneous environments. pp. 439–476. IN Mathematical Association of America, Study in Mathematical Ecology. Vol. II. Populations and Communities (S. A. Levin, ed.) Math. Assoc. Amer., Washington, D.C.

Levin, S. A. 1978b. Pattern formation in ecological communities. pp. 433–465. IN Spatial Patterns in Plankton Communties (J. H. Steele, ed.) Plenum Press. New York.

Levin, S. A. 1979. Non-uniform stable solutions to reaction-diffusion equations: Application to ecological pattern formation. Proc. Intern. Symp. Synergetics, Schloss Elmau, West Germany. April 30–May 5, 1979. Springer-Verlag, New York.

Levin, S. A. 1983. Some approaches to the modelling of coevolutionary interactions, pp. 21–65. IN Coevolution (M. Nitecki, ed.) University of Chicago Press, Chicago, Illinois.

Levin, S. A. and L. A. Segel. 1976. Hypothesis for origin of planktonic patchiness. Nature 259:659.

Levins, R. 1975. Evolution in communities near equilibrium. pp. 16–55. In Ecology and Evolution of Communities (M. L. Cody and J. M. Diamond, eds.) Belknap Press, Cambridge, Massachusetts.

Levins, R. and D. Culver. 1971. Regional co-existence of species and competition between rare species. Proc. Nat. Acad. Sci. 68:1246–1248.

Lewis, E. G. 1942. On the generation and growth of a population. Sankhya 6:93–96.

Limbaugh, C. 1961. Cleaning symbiosis. Sci. Amer. 205:42–49.

Lindeman, R. L. 1942. The trophic-dynamic aspect of ecology. Ecology 23:399–418.

Loefer, J. B. and R. B. Mefferd. 1952. Concerning pattern formation by free-swimming microorganisms. Amer. Natur. 86:325–329.

Loosli, C. G. 1968. Synergism between respiratory viruses and bacteria. Yale J. Biol. Med. 40:522–540.

Lovelock, J. E. 1979. Gaia: A New Look at Life on earth. Oxford University Press, New York, New York. 154 pp.

Ludlow, A. R. 1982. Towards a theory of thresholds. Anim. Behavior. 30:253–267.

Ludwig, D., D. D. Jones, and C. S. Holling. 1978. Qualitative analysis of insect outbreak systems: the spruce budworm and forest. J. Anim. Ecol. 47:315–332.

Luenberger, D. G. 1979. Introduction to dynamic systems. Theory, models, and applications. John Wiley and Sons, Inc., New York. 446 pp.

Lugt, H. J. 1983. Vortex Flow in Nature and Technology. John Wiley and Sons, Inc., New York. 297 pp.

Lumsden, C. J. and E. O. Wilson. 1981. Genes, Mind, and Culture: The Coevolutionary Process. Harvard University Press, Cambridge, Massachusetts. 428 pp.

Lyttle, T. W. 1977. Experimental population genetics of meiotic drive systems. I. Pseudo-Y chromosomal drive as a means of eliminating cage populations of *Drosophila melanogaster*. Genetics 86:413–445.

MacArthur, R. H. and E. O. Wilson. 1963. An equilibrium theory of island biogeography. Evolution 17:373–387.

MacArthur, R. H. and E. O. Wilson. 1967. The Theory of Island Biogeography. Princeton University Press. Princeton, New Jersey.

MacLean, W. R. 1959. On the acoustics of cocktail parties. J. Acoust. Soc. Amer. 31:79–80.

Majer, J. D. 1976. The maintenance of the mosaic in Ghana cocoa farms. J. Appl. Ecol. 13:123–143.

Mann, K. H. 1977. Destruction of kelp beds by sea-urchins: Cyclic phenomenon or irreversible degradation? Helgo. Wiss. Meeresunders. 30:455–467.

Mann, K. H. 1982. Ecology of Coastal Waters: A Systems Approach. University of California Press. Berkeley and Los Angeles. 322 pp.

Margalef, R. 1963. On certain unifying principles in ecology. Am. Nat. 97:357–374.

Margalef, R. 1968. Perspectives in Ecological Theory. The University of Chicago Press. Chicago, Illinois. 111 pp.

Margulis, L. 1970. Symbiosis and evolution. Sci Amer. 58:49–57.

Marler, P. and W. J. Hamilton, III. 1966. Mechanisms of Animal Behavior. John Wiley and Sons, Inc., New York, New York. 771 pp.

Martinez, H. 1972. Morphogenesis and chemical dissipative structures: A computer simulated case study. J. theor. Biol. 36:479–501.

Maruyama, M. 1963. The second cybernetics: Deviation-amplifying mutual causal processes. Am. Sci. 51:164–179.

May, R. M. 1973a. Qualitative stability in model ecosystems. Ecology 54:638–641.

May, R. M. 1973b. Stability and Complexity in Model Ecosystems. Princeton University Press. Princeton, New Jersey. 235 pp.

May, R. M. 1976a. Mathematical aspects of the dynamics of animal populations. IN Studies in Mathematical Biology (S. A. Levin, ed.) American Mathematical Society, Providence, RI.

May, R. M. 1976b. Models for two interacting populations. IN Theoretical Ecology: Principles and Applications (R. M. May, ed.) Saunders, Philadelphia. 317 pp.

May, R. M. and R. M. Anderson. 1979. Population biology of infectious diseases: Part II. Nature 280:455–461.

May, R. M. and R. M. Anderson. 1984. Spatial heterogeneity and the design of immunization programs. manuscript.

Maynard Smith, J. 1976. Evolution and the theory of games. Am. Sci. 64:41–45.

Maynard Smith, J. 1979. Game theory and the evolution of behavior. Proc. Roy. Soc. Lond. B 205:475–488.

Maynard Smith, J. and G. R. Price. 1973. The logic of animal conflicts. Nature 246:15–18.

Mayo, O. 1983. Natural Selection and Its Constraints. Academic Press, New York. 145 pp.

Mayr, E. 1960. The emergence of evolutionary novelties, pp. 349–380. IN Evolution after Darwin. Vol. I. The Evolution of Life (S. Tax, ed.) The University of Chicago Press, Chicago, Illinois.

McCleerly, R. H. 1978. Optimal behavior sequences and decision making. pp. 377–410. IN Behavioral Ecology: An Evolutionary Approach (J. R. Krebs and N. B. Davies, eds.) Sinauer Associates, Inc. Sunderland, Massachusetts.

McClendon, J. H. 1980. The evolution of the chemical isotopes as an analog of biological evolution. J. Theor. Biol. 87:113–128.

McKey, D. 1975. The ecology of coevolved seed dispersal systems. pp. 159–191. IN Coevolution of Plants and Animals (L. E. Gilbert and P. H. Raven, eds.) Univ. Texas Press, Austin.

McLaren, I. A. (ed.) 1971. Natural Regulation of Animal Populations. Atherton Press, New York. 195 pp.

McLaughlin, J. A. and P. A. Zahl. 1966. Endozoic algae. pp. 257–297. IN Symbiosis (S. M. Henry, ed.) Academic Press, New York.

McMahon, T. A. 1973. Size and shape in biology. Science 179:1201–1204.

McNaughton, S. J. 1978. Serengeti ungulates: Feeding selectivity influences the effectiveness of plant defense guilds. Science 199:806–807.

McNaughton, S. J. and M. B. Coughenour. 1981. The cybernetic nature of ecosystems. Amer. Natur. 117:985–990.

McNew, G. L. 1971. The Boyce Thompson Institute Program in Forest Entomology that led to the discovery of pheromones in bark beetles: Symposium on population attractants. Contr. Boyce Thompson Inst. Pl. Res. 24:251–262.

Mech, L. D. 1977. Wolf-pack buffer zones as prey reservoirs. Science 198:320–321.

Medawar, P. B. 1946. Old age and natural death. Modern Quarterly 1:30–56.

Medawar, P. B. 1952. An unsolved problem in biology. London, H. K. Lewis. [Reprinted in the Uniqueness of the Individual (1957), pp. 44–70, London, Methuen.]

Meijer, W. 1973. Devastation and regeneration of lowland diptocarp forests in south-east Asia. BioScience 23:528–533.

Messina, F. J. 1981. Plant protection as a consequence of ant-membracid mutualism: Interactions on goldenrod (*Solidago* sp.). Ecology 62:1433–1440.

Meyer, F. H. 1966. Mycorrhiza and other plant symbioses. pp. 171–255. IN Symbiosis (S. M. Henry, ed.) Academic Press, New York.

Meyer, F. H. 1974. Physiology of mycorrhiza. Ann. Rev. Plant Physiol. 25:567–586.

Michod, R. E. 1979. Evolution of life histories in response to age-specific morality factors. Am. Nat. 113:531–550.

Miles, J. 1979. Vegetation Dynamics. Chapman and Hall, Inc., London, England. 80 pp.

Milne, A. 1957. Theories of natural control of insect populations. Cold Spring Harbor Symp. Quant. Biol. 22:253–271.

Milsum, J. H. (ed.). 1968. Positive Feedback. Pergamon Press, Oxford. 169 pp.

Mimura, M. 1979. Asymptotic behavior of a parabolic system related to a planktonic prey and predatory system. SIAM J. Appl. Math. 37:499–512.

Mimura, M., Y. Nishiura, and M. Yamaguti. 1979. Some diffusive prey and predator systems and their bifurcation problem. Paper presented at Internat. Conf. Bifurcation Theory and its Application to Scientific Disciplines, New York, Oct. 1977. Annals of the New York Academy of Sciences 316:490–510.

Minorsky, N. 1962. Nonlinear Oscillations. D. Van Nostrand Company, Inc., Princeton, New Jersey. 714 pp.

Monk, C. D. 1966. An ecological significance of evergreenness. Ecology 47:504–505.

Monod, J. 1972. Chance and Necessity. Vintage Books, Random House, Inc., New York. 199 pp.

Mooney, H. A. 1972. The carbon balance of plants. Ann. Rev. Ecol. Syst. 3:315–346.

Murchie, G. 1978. The Seven Mysteries of Life: An Exploration in Science and Philosophy. (1st Paperback Edition, 1981). Houghton Mifflin Company, Boston, Massachusetts. 690 pp.

Murdoch, W. W. 1966. "Community structure, population control, and competition" – a critique. Am. Nat. 100:219–226.

Murton, R. K. and N. J. Westwood. 1976. Birds as pests. pp. 89–181. IN Applied Biology, 1 (T. H. Coaker, ed.) Academic Press. London, England.

Muscatine, L. and J. W. Porter. 1977. Reef corals: Mutualistic symbioses adapted to nutrient-poor environments. BioScience 27:454–460.

Mutch, R. W. 1970. Wildland fires and ecosystems – an hypothesis. Ecology 51:1046–1051.

Neill, W. E. 1974. The community matrix and interdependence of the competition coefficients. Amer. Natur. 108:399–408.

Nestler, E. J., S. I. Walaas, and P. Greengard. 1984. Neuronal phosphoproteins: Physiological and clinical implications. Science 225:1357–1364.

Nicholson, A. J. 1957. The self-adjustment of populations to change. Cold Spring Harbor Symp. Quant. Biol. 22:153–172.

Nicolis, J. S. 1980. Bifurcation in cognitive networks: A paradigm of self-organization via de-synchronization. pp. 220–214. IN Dynamics of Synergetic Systems (H. Haken, ed.) Springer-Verlag, Berlin.

Nicolis, G. and I. Prigogine. 1977. Self-Organization in Non-Equilibrium Systems. John Wiley and Sons, New York.

Nitecki, M. H. (ed.). 1983. Coevolution. The University of Chicago Press. Chicago, Illinois. 392 pp.

Noble, J. V. 1974. Geographic and temporal development of plagues. Nature 250:726–728.

Noy-Meir, I. 1973. Desert ecosystems: Environment and producers. Ann. Rev. Ecol. Syst. 4:25–52.

O'Donald, P. 1980. Genetical Models of Sexual Selection. Cambridge University Press, London and Cambridge, England.

O'Dowd, D. J. 1980. Pearl bodies of a neotropical tree, *Ochroma pyramidale*: ecological implications. Amer. J. Bot. 67:543–549.

O'Dowd, D. J. and M. E. Hay. 1980. Mutualism between harvester ants and a desert ephemeral: Seed escape from rodents. Ecology 61:531–540.

Odum, E. P. 1969. The strategy of ecosystem development. Science 164:262–270.

Odum, E. P. 1971. Fundamentals of Ecology (3rd Edition). W. B. Saunders Company. 574 pp.

Odum, E. P. and L. J. Biever. 1984. Resource quality, mutualism, and energy partitioning in food chains. Amer. Natur. 124:360–376.

Officer, C. B., R. B. Biggs, J. L. Taft, L. E. Cronin, M. A. Tyler, and W. R. Boynton. 1984. Chesapeake Bay anoxia: Origin, development, and significance. Science 223:22–27.

Okubo, A. 1974. Diffusion-induced instability in model ecosystems. Chesapeake Bay Institute, The Johns Hopkins University, Baltimore, Tech. Rept. No. 86.

Okubo, A. 1978. Horizontal dispersion and critical scales for phytoplankton patches. pp. 21–42. IN Spatial Patterns in Plankton Communities (J. H. Steele, ed.) Plenum Press. New York.

Okubo, A. 1980. Diffusion and Ecological Problems: Mathematical Models. Springer-Verlag. New York. 254 pp.

Olson, J. S. 1958. Rates of succession and soil changes on southern Lake Michigan sand dunes. Botanical Gazette 119:125–170.

Oosting, H. J. 1955. The Study of Plant Communities: An Introduction to Plant Ecology (2nd ed.). W. H. Freeman and Company. San Francisco, California. 440 pp.

Orgel, L. E. 1973. Ageing of clones of mammalian cells. Nature, (Lond.). 243:441–445.

Osman, R. W. and J. A. Haugsness. 1981. Mutualism among sessile invertebrates: A mediator of competition and predation. Science 211:846–848.

Oster, G. F. and E. O. Wilson. 1978. Caste and Ecology in the Social Insects. Princeton University Press. Princeton, New Jersey. 352 pp.

Owen, D. F. 1978a. The effect of a consumer, *Phyomyza ilicis*, on a seasonal leaf-fall in the holly *Ilex aquifolium*. Oikos 31:268–271.

Owen, D. F. 1978b. Why do aphid synthesize melezitose? Oikos 31, 264–267.

Parsegian, V. L. 1972. This Cybernetic World of Men, Machines, and Earth Systems. Doubleday and Company, Inc., Garden City, New York. 217 pp.

Pasineti, L. L. 1977. Lectures on the Theory of Production. Columbia University Press, New York. 285 pp.

Pastor, J., J. D. Aber, C. A. McClaugherty, and J. M. Melillo. 1984. Aboveground production and N and P cycling along a nitrogen mineralization gradient on Blackhawk Island, Wisconsin. Ecology 65:256–268.

Patten, B. C. and E. P. Odum. 1981. The cybernetic nature of ecosystems. Amer. Natur. 118:886–895.

Peacocke, A. R. 1983. An Introduction to the Physical Chemistry of Biological Organization. Clarendon Press, Oxford, England. 302 pp.

Pearson, K. 1976. The control of walking. Sci. Am. 239(12):72–86.

Pennycuick, C. J., R. M. Compton, and L. Beckingham. 1968. A computer model for simulating the growth of a population or two populations. J. Theor. Biol. 18:316–329.

Perkowski, J. J. 1979. A compensatory Leslie matrix model for two interacting populations. Masters Thesis, University of Tennessee.

Peterken, G. F. 1974. A method of assessing woodland flora for conservation using indicator species. Biol. Cons. 6:239–245.

Phillips, B. P. 1968. pp. 279–286. IN Advances in Germfree Research and Gnotobiology (M. Miyakawa and T. D. Luckey, eds.) Iliffe, London.

Pianka, E. R. 1978. Evolutionary Ecology (second edition), Harper and Row, New York. 397 pp.

Pickett, S. T. A. and J. N. Thompson. 1978. Patch dynamics and the design of nature reserves. Biol. Conserv. 13:27–37.

Picton, H. D. 1979. The application of insular biogeographic theory to the conservation of large mammals in the northern Rocky Mountains. Biol. Conserv. 15:73–79.

Pielou, E. C. 1977. Mathematical Ecology. John Wiley and Sons, Inc. 385 pp.

Pielou, E. C. 1979. Biogeography. John Wiley and Sons, Inc. New York. 351 pp.

Pierce, N. E. P. S. Meade. 1981. Parasitoids as selective agents in symbiosis between lycaenid butterfly larvae and ants. Science 211:1185–1187.

Platt, J. R. 1961. "Bioconvection patterns" in cultures of free-swimming organisms. Science 133:1766–1767.

Plemmons, R. V. 1977. M-matrix characteristics. I. Nonsingular M-matrices. Linear Algebra and Applications 18:175–188.

Plowright, R. C. and L. K. Hartling. 1981. Red clover pollination by bumble bees: a study of the dynamics of a plant-pollinator relationship. J. Appl. Ecol. 18:639–647.

Pomeroy, L. R., H. M. Mathews, and H. S. Min. 1963. Excretion of phosphate and soluble organic phosphorus compounds by zooplankton. Limnol. Oceanogr. 8:50–55.

Porter, R. W. 1952. Alterations in electrical activity of the hypothalmus induced by stress stimuli. Am. J. Physio. 169:629–637.

Post, W. M., D. L. DeAngelis, and C. C. Travis. 1983. Endemic disease in environments with spatially heterogeneous host populations. Math. Biosc. 63:289–302.

Poundstone, W. 1985. The Recursive Universe. Cosmic Complexity and the Limits of Scientific Knowledge. William Morrow and Company, Inc. New York.

Powell, J. A. and R. A. Mackie. 1966. Biological interrelationships of moths and *Yucca whipplei*. University of California Publications in Entomology 42:1–59.

Preston, F. W. 1962. The canonical distribution of commoness and rarity. Ecology 43:185–215.

Price, P. W. 1977. General concepts on the evolutionary biology of parasites. Evolution 31:405–420.

Prigogine, I. and I. Stengers. 1984. Order Out of Chaos: Man's New Dialogue with Nature. Bantam Books, Toronto. 349 pp.

Prigogine, I., G. Nicolis, and A. Babloyantz. 1972a. Thermodynamics and evolution. Part I. Phys. Today. November. pp. 23–28.

Prigogine, I., G. Nicolis, and A. Babloyantz. 1972b. Thermodynamics and evolution. Part II. Phys. Today. December. pp. 38–44.

Prigogine, I. and R. Lefever. 1975. Stability and thermodynamic properties of dissipative structures in biological systems. IN Stability and Origin of Biological Information (I. R. Miller, ed.) First A. Katchalsky-Katzir Conf. Proc. Wiley, New York.

Quinlan, R. J. and J. M. Cherret. 1978. Aspects of the symbiosis of the leaf cutting ant *Acromyrmex actospinosus* (Reich.) and its fungus food. Ecol. Entomol. 3:221–230.

Rai, B., H. I. Freedman, and J. F. Addicot. 1983. Analysis of models of mutualism in predator-prey and competitive systems. Math. Biosci. 65:13–50.

Rapoport, A. 1974. Conflict in Man-Made Environment. Penguin Books, Inc. , Baltimore, Maryland. 272 pp.

Rapp, P. E. 1980. Biological applications of control theory, pp. 146–247. IN Mathematical Models in Molecular and Cellular Biology (L. A. Segel, ed.) Cambridge University Press, Cambridge, England.

Ratcliffe, D. A. 1970. Changes attributable to pesticides in egg breakage frequency and eggshell thickness in some British birds. J. Appl. Ecol. 7:67–107.

Rehr, S. S., P. P. Feeny, and D. H. Janzen. 1973. Chemical defense in Central American non-ant-acacias. J. Anim. Ecol. 42:405–416.

Rensch, B. 1959. Evolution Above the Species Level. Columbia Univesity Press. New York, New York. 419 pp.

Ricker, W. E. 1954. Stock and recruitment. J. Fish. Res. Board Canada. 1:559–623.

Ricklefs, R. E. 1976. The Economy of Nature. Chiron Press, Inc., Portland, Oregon. 455 pp.

Rickson, F. R. 1979a. Absorption of animal tissue breakdown products into a plant stem – the feeding of a plant by ants. Amer. J. Bot. 66:87–90.

Rickson, F. R. 1979b. Ultrastructural development of the beetle food tissue of calycanthus flowers. Amer. J. Bot. 66:80–86.

Rikitake, T. 1958. Oscillations of a system of disc dynamos. Proc. Camb. Phil. Soc. 54:89–106.

Rindos, D. 1984. The Origins of Agriculture: An Evolutionary Perspective. Academic Press, Inc., Orlando, Florida. 325 pp.

Risch, S. J. and F. R. Rickson. 1981. Mutualism in which ants must be present before plants produce food bodies. Nature 291:149–150.

Risch, S. and D. Boucher. 1976. What ecologists look for. Bull. Ecol. Soc. Amer. 52:8–9.

Robbins, W. J. 1952. Patterns formed by motile *Euglena gracilis* var. *bacillaris*. Bull. Torrey Bot. Club 79:107–109.

Roff, D. A. 1974. Spatial heterogeneity and the persistence of populations. Oecologia 15:245–258.

Room, P. M. 1972. The fauna of the mistletoe *Tapinanthus bangwensis* (Engl. & K. Krause) growing on cocoa in Ghana: relationships between fauna and mistletoe. J. Animal Ecol. 19:611–621.

Root, R. B. 1973. Organization of a plant-arthropod association in simple and diverse habitats: The fauna of collards *(Brassica olevaceae)*. Ecol. Monogr. 43:95–124.

Rosenblueth, A., N. Wiener, and J. Bigelow. 1943. Behavior, purpose, and teleology. Philos. Sci. 10:19.

Ross, D. M. 1971. Protection of hermit crabs (*Dardanus* spp.) from octopus by commensal anemones (*Calliactis* spp.). Nature 230:401–402.

Ross, G. N. 1966. Life history studies on Mexican butterflies: IV. The ecology and ethology of *Anatole rossi*, a Myrmecophilous metalmark (Lepidoptera: Riodinae). Ann. Ent. Soc. Amer. 59:985–1004.

274

Roughgarden, J. 1975. Evolution of marine symbiosis – a simple cost-benefit model. Ecology 56: 1202–1208.

Roy. 1962. Soil development and the availability of phosphorus and potassium to *Aspergillus niger*. Nature 195:475.

Rubinstein, M. F. 1975. Patterns of Problem Solving. Prentice-Hall, Inc., Englewood Cliffs, New Jersey. 544 pp.

Rushton, S. and A. J. Mautner. 1955. The deterministic model of a simple epidemic for more than one community. Biometrika 42:126–132.

Sagan, C. 1977. The Dragons of Eden. Random House, New York, New York. 263 pp.

Sahlins, M. D. 1976. The Use and Abuse of Biology: An Anthropological Critique of Sociobiology. The University of Michigan Press. Ann Arbor, Michigan.

Schadde, R. and J. H. Calaby. 1972. The biogeography of the Australo-Papuan bird and mammal faunas in relation to Torres Strait. pp. 257–300. In Bridge and Barrier: The Natural and Cultural History of Torres Strait (D. Walker, ed.) Canberra, Australian National University.

Schadde, R. and W. B. Hitchcock. 1972. Birds. pp. 67–86. IN Encyclopedia of Papua and New Guinea, 1 (P. A. Ryan, ed.) Melbourne University Press. Melbourne, Australia.

Schaefer, M. B. 1957. A study of the dynamics of the fishery for the yellowfin tuna population of the Eastern Pacific Ocean. Inter-Am. Trop. Tuna Comm., Bull. 12:87–137.

Schaffer, W. M. 1974. Selection for optimal life histories: The effects of age structure. Ecology 55:291–303.

Schaffer, W. M. 1977. A note concerning the invasion of a species into a community. Theor. Pop. Biol. 8:152–154.

Schaffer, W. M., and R. H. Tamarin. 1973. Changing reproductive rates and population cycles in lemmings and voles. Evolution 27:111–124.

Schaller, G. B. 1972. The Serengeti Lion. University of Chicago Press. Chicago, Illinois. 480 pp.

Schemske, D. W. 1980. The evolutionary significance of extrafloral nectar production by *Costus woodsonii* (Zingiberaceae): An experimental analysis of ant protection. J. Ecol. 68:959–967.

Schemske, D. W. 1982. Ecological correlates of a neotropical mutualism: ant assemblages at *Costus* extrafloral nectaries. Ecology 63:932–941.

Schmidt-Nielsen, K. 1975. Animal Physiology: Adaptation and Environment. Cambridge University Press, New York, New York. 699 pp.

Schorger, A. W. 1973. The Passenger Pigeon. University of Oklahoma Press. Norman, Oklahoma. 424 pp.

Schuster, P., and K. Sigmund. 1980. A mathematical model of the hypercycle, pp. 170–178. IN Dynamics of Synergetic Systems (H. Haken, ed.) Springer-Verlag, New York, New York.

Sebetich, M. J. 1975. Phosphorus kinetics of freshwater microcosms. Ecology 56:1262–1280.

Seddon, G. 1974. Xerophytex, xeromorphs and sclerophylls: the history of some concepts in ecology. Biol. J. Linn. Soc. 6:65–87.

Seigler, D. S. and P. H. Price. 1976. Secondary compounds in plants: Primary functions. Am. Nat. 110:101–105.

Seneta, E. 1973. Non-Negative Matrices. Allen and Unwin, London.

Sharp, G. D. 1978. Behavioral and physiological properties of tunas and their effects on vulnerability to fishing gear. IN The Physiological Ecology of Tunas (G. D. Sharp and A. E. Dizen, eds.) Academic Press. New York.

Shelton, W. L., W. D. Davies, T. A. King, and T. J. Simmons. 1979. Variations in the growth of the initial year class largemouth bass in West Point Reservoir, Alabama and Georgia. Trans. Amer. Fish. Soc. 108:142–149.

Shugart, H. H., D. E. Reichle, N. T. Edwards, and J. R. Kercher. 1976. A model of calcium cycling in an East Tennessee *Liriodendron* forest: Model structure, parameters, and frequency response analysis. Ecology 57:99–109.

Shugart, H. H., Jr. and D. C. West. 1977. Development of an Appalachian Deciduous Forest succession model and its application to assessment of the impact of the chestnut blight. J. Envir. Manag. 5:161–179.

Siccama, T. G. 1974. Vegetation, soil and climate on the Green Mountains of Vermont. Ecol. Monogr. 44:325–349.

Siljak, D. D. 1975. When is a complex system stable? Math. Biosc. 25:25–50.

Simberloff, D. S. 1976. Species turnover and equilibrium island biogeography. Science 194:572–578.

Simberloff, D. S. and E. O. Wilson. 1969. Experimental zoogeography of islands: The colonization of empty islands. Ecology 50:278–296.

Simberloff, D. S. and E. O. Wilson. 1970. Experimental zoogeography of islands: A two-year record of colonization. Ecology 51:934–937.

Simberloff, D. S. and L. G. Abele. 1976. Island biogeography theory and conservation practice. Science 191:285–286.

Simberloff, D. S., B. J. Brown and S. Lowrie. 1978. Isopod insect root borers may benefit Florida mangroves. Science 201:630–632.

Simon, H. A. 1952. A formal theory of interaction in social groups. Amer. Social. Rev. 17:202–211. Reprinted in H. A. Simon 1957, Models of Man, John Wiley and Sons, New York. pp. 99–114.

Singer, P. 1981. The Expanding Circle: Ethics and Sociobiology. Farrar, Straus and Giroux, New York, New York.. 190 pp.

Skellam, J. G. 1951. Random dispersal in theoretical populations. Biometrika 38:196–218.

Skellam, J. G. 1973. The formulation and interpretation of mathematical models of diffusionary processes in population biology. pp. 63–85. IN The Mathematical Theory of the Dynamics of Biological Populations (M. S. Bartlett and R. W. Hiorns, eds.) Academic Press. New York.

Slatkin, M. 1974. Competition and regional coexistence. Ecology 55:128–134.

Slobodkin, L. B. 1974. Prudent predation does not require group selection. Amer. Natur. 108:665–678.

Smith, A. 1776. The Wealth of Nations. (ed. E. Cannan). The Modern Library (1937), New York. 976pp.

Smith, N. 1968. The advantage of being parasitized. Nature 219:690–694.

Smouse, P. E. and K. M. Weiss. 1975. Discrete demographic models with density-dependent vital rates. Oecologia 21:205–18.

Southwood, T. R. E. 1977. Habitat, the templet for ecological strategies? J. Anim. Ecol. 46:337–365.

Springett, B. P. 1968. Aspects of the relationship between burying beetles *Necrophorus* spp. and the mite *Poecilochirus necrophori* Vitz. J. Anim. Ecol. 37:417–424.

Spurr, S. H. 1964. Forest Ecology. Ronald Press Co., New York. 325 pp.

Staddon, J. E. R. 1983. Adaptive Learning and Behavior. Cambridge University Press, Cambridge, England. 555 pp.

Stanley-Jones, D. 1970. The role of positive feedback, pp. 249–263. IN Progress of Cybernetics. Vol. 1 (J. Rose, ed.) Gordon and Breach Science Publishers, London.

Starmer, W. T. 1981. A comparison of *Drosophila* habitats according to the physiological attributes of the associated yeast communities. Evolution 35:38–52.

Stenseth, N. C. 1978. Do grazers maximize individual plant fitness. Oikos 31:299–306.

Stent, G. S. 1978. Paradoxes of Progress. W. H. Freeman and Company, San Francisco. 231 pp.

Stephenson, A. G. 1982. The role of the extrafloral nectaries of *Catalpa speciosa* in limiting herbivory and increasing fruit production. Ecology 63:663–669.

Stevens, P. S. 1974. Patterns in Nature. Little, Brown, and Company, Inc., Boston. 240 pp.

Street, P. 1975. Animal Partners and Parasites. Taplinger, New York. 209 pp.

Strobeck, C. 1973. N-species competition. Ecology 54:650–654.

Suthers, R. A. and R. A. Gallant. 1973. Biology: The Behavioral View. Xerox College Publishing. Lexington, Massachusetts. 556 pp.

Swift, M. J., O. W. Heal, and J. M. Anderson. 1979. Decomposition in Terrestrial Ecosystems. Studies in Ecology, Vol. 5. University of California Press, Berkeley, California. 372 pp.

Sykes, Z. M. 1969. On discrete stable population theory. Biometrics 25:285–293.

Takeuchi, Y., N. Adachi, and H. Tokumaru. 1978. Global stability of ecosystems of the generalized Volterra type. Math. Biosci. 42:119–136.

Takvanainen, J. O. and R. B. Root. 1972. The influence of vegetational diversity on the population ecology of a specialized herbivore, *Phyllotreta cruciferae* (Coleoptera: Crysomelidae). Oecologia 10:321–346.

Tamarin, R. H. (ed.) 1978. Population Regulation. Dowden, Hutchinson, and Ross. Stroudsburg, Pennsylvania.

Tansley, A. G. 1935. The use and abuse of vegetational concepts and terms. Ecology 16:284–307.

Temple, S. A. 1977. Plant-animal mutualism: Coevolution with Dodo leads to near extinction of plant. Science 197:885–886.

Terborgh, J. 1974a. Preservation of natural diversity: The problem of extinction prone species. BioScience 24:715–722.

276

Terborgh, J. 1974b. Faunal equilibria and the design of wildlife preserves. IN Tropical Ecological Systems: Trends in Terrestrial and Aquatic Research (F. Golley and E. Medina, eds.) Springer-Verlag. New York.

Terborgh, J. 1976. Island biogeography and conservation strategy and limitations. Science 193: 1029–1030.

Tester, C. F. 1977. Constituents of soybean cultivars differing in insect resistance. Phytochemistry 16:1899–1901.

Thompson, J. D. 1978. Effects of stand competition on insect visitation in two species mixtures of *Hieracium*. Amer. Midl. Natur. 100:431–440.

Tilman, D. 1978. Cherries, ants and tent caterpillars: timing of nectar production in relation to susceptibility of caterpillars to ant predation. Ecology 59:686–692.

Tinbergen, N. 1960. The evolution of behavior in gull. Scientific American 203:118–130.

Tinkle, D. W., H. M. Wilbur, and S. G. Tilley. 1970. Evolutionary strategies in lizard reproduction. Evolution 24:55–74.

Toates, F. M. 1980. Animal Behavior – A Systems Approach. John Wiley and Sons, Inc., New York. 299 pp.

Toates, F. M., and C. O'Rourke. 1978. Computer simulation of male rat sexual behavior. Medical and Biological Engineering and Computing 16:98–104.

Toffler, A. 1970. Future Shock. Random House, New York. 505 pp.

Transeau, E. N. 1935. The prairie peninsula. Ecology 16:423–437.

Travis, C. C., and W. M. Post. 1979. Dynamics and comparative statics of mutualistic communities. J. Theor. Biol. 78:553–571.

Travis, C. C., W. M. Post, D. L. DeAngelis, and J. Perkowski. 1980. Analysis of compensatory Leslie matrix for competing species. Theor. Popul. Biol. 18:16–30.

Trivers, R. L. 1971. The evolution of reciprocal altruism. Quart. Rev. Biol. 46:35–57.

Turing, A. M. 1952. The chemical basis of morphogenesis. Phil. Trans. Roy. Soc. London B 237:37–72.

Usher, M. B. 1979. Markovan approaches to ecological succession. J. Anim. Ecol. 48:413–426.

Valerio, C. E. 1974. A unique case of mutualism. Amer. Natur. 109:235–238.

Vance, R. R. 1978. A mutualistic interaction between a sessile marine clam and its epibionts. Ecology 59:679–685.

Vandermeer, J. H. and D. H. Boucher. 1978. Varieties of mutualistic interactions in population models. J. Theor. Biol. 74:549–558.

van der Pijl. 1969. Principles of Dispersal in Higher Plants. Springer-Verlag, New York. 154 pp.

VanderWall, S. B., and R. P. Balda. 1977. Coadaptations of the Clark's nutcracker and the Piñon pine for efficient seed harvest and dispersal. Ecol. Monogr. 47:89–111.

Varley, G. C., G. R. Gradwell, and M. P. Hassell. 1974. Insect Population Ecology. University of California Press, Berkeley.

Vaughan, D. S. and S. B. Saila. 1976. A method for determining mortality rates using the Leslie matrix. Trans. Am. Fish. Soc. 105:380–383.

Velarde, M. G. and C. Normand. 1980. Convection. Sci. Am. 243(1):92–108.

Vermeij, G. J. 1978. Biogeography and Adaptation: Patterns of Marine Life. Harvard University Press. Cambridge, Massachusetts. 332 pp.

Vermeij, G. J. 1983. Intimate associations and coevolution in the sea, pp. 311–327. IN Coevolution (D. J. Futuyma, and M. Slatkin, eds.) Sinauer Associates, Inc., Sunderland, Massachusetts.

Verwey, J. 1930. Coral reef studies. I. The symbiosis between damsel fishes and sea anemones in Batavia Bay. Treubia 12:305–366.

Vincent, T. L., and H. R. Pulliam. 1980. Evolution of life history strategies for an asexual annual plant model. Theor. Pop. Biol. 17:215–231.

Vine, I. 1973. Detection of prey flocks by predators. J. Theor. Biol. 40:207–210.

Vitousek, P. 1982. Nutrient cycling and nutrient use efficiency. Amer. Natur. 119:553–572.

Von Neumann, J. and O. Morgenstern. 1944. Theory of Games and Economic Behavior. (1967 edition) John Wiley and Sons, Inc., New York, New York. 641 pp.

Vuilleumier, F. 1970. Insular biogeography in continental regions. 1. The northern Andes of South America. Amer. Natur. 104:373–388.

Waggoner, P. E., and G. R. Stephens. 1970. Transition probabilities for a forest. Nature 255:1160–1161.

Walker, J. 1977. The Flying Circus of Physics: With Answers. John Wiley and Sons, Inc., New York. 295 pp.

Waser, N. M. and L. A. Real. 1979. Effective mutualism between sequentially flowering plant species. Nature 291:670–672.

Watson, R. K. 1972. On an epidemic in a stratified population. J. Appl. Prob. 9:659–666.

Watt, K. E. F. 1973. Principles of Environmental Science. McGraw-Hill Publishers. New York.

Way, M. J. 1963. Mutualism between ants and honeydew producing homoptera. Ann. Rev. Entom. 8:307–344.

Webley, D. M., M. E. K. Henderson, and I. F. Taylor. 1963. The microbiology of rock and weathered stones. J. Soil Science 14:102–112.

Weinberg, G. M., and D. Weinberg. 1979. On the Design of Stable Systems. John Wiley and Sons, Inc., New York. 353 pp.

Weis-Fogh, T. 1949. An aerodynamic sense organ stimulating and regulating flight in locusts. Nature 164–873.

Whelpton, P. K. 1936. An empirical method of calculating future population. J. Amer. Statist. Assoc. 31:457–73.

Whitcomb, R. F., J. F. Lynch, P. A. Opler, and C. S. Robbins. 1976. Island biogeography and conservation: Strategy and limitations. Science 193:1030–1032.

Whitehead, J. A., Jr. 1977. Cellular convection. Am. Sci. 44:248–263.

Whittaker, R. H. 1969. Evolution of diversity in plant communities. Symp. on Diversity and Stability in Ecological Systems. Upton, New York: Brookhaven National Laboratory.

Whittaker, R. H. 1970. Communities and Ecosystems. The MacMillan Company, Inc. New York. 158 pp.

Whittaker, R. H. 1975. Communities and Ecosystems. Macmillan, New York (second edition). 385 pp.

Whittaker, R. H. and S. A. Levin. 1977. The role of mosaic phenomena in natural communities. Theor. Pop. Biol. 12:117–139.

Wiebes, J. T. 1979. Co-evolution of figs. and their insect pollinators. Ann. Rev. Ecol. Syst. 10:1–12.

Wiens, J. A. 1976. Population responses to patchy environments. Ann. Rev. Ecol. Sys. 7:81–120.

Wiepkema, P. R. 1968. Positive feedbacks at work during feeding. Behavior 39:266–273.

Wilbur, H. M., D. W. Tinkle, and J. P. Collins. 1974. Environmental certainty, trophic level, and resource availability in life history evolution. Am. Nat. 108:805–17.

Wilcox, B. A. 1980. Insular Ecology and Conservation. IN Conservation Biology (M. E. Soule and B. A. Wilcox, eds.) Sinauer Associates, Inc. Sunderland, Massachusetts.

Williams, G. C. 1966. Adaptation and Natural Selection: A Critique of Some Current Evolutionary Thought. Princeton University Press. Princeton, New Jersey. 307 pp.

Williams, G. C. 1975. Sex and Evolution. Princeton University Press, Princeton, New Jersey. 200 pp.

Willis, E. O. 1974. Populations and local extinction of birds on Barro Colorado Island, Panama. Ecol. Monogr. 44:153–169.

Wilson, D. S. 1976. Evolution on the level of communities. Science 192:1358–1360.

Wilson, D. S. 1980. The Natural Selection of Populations and Communities. The Benjamin/Cummings Publishing Company, Inc. Menlo Park, California. 186 pp.

Wilson, E. O. 1975. Sociobiology: The New Synthesis. Harvard University Press, Cambridge, Massachusetts. 697 pp.

Wilson, E. O. 1978. On Human Nature. Harvard University Press, Cambridge, Massachusetts. 260 pp.

Winet, H. and T. L. Jahn. 1972. Effect of CO_2 and NH_3 on biovection: Patterns in *Tetrahymena* cultures. Exp. Cell. Res. 71:356–360.

Witzmann, R. F. 1981. Steroids: Key to Life. Van Nostrand Rheinhold Company, New York, New York. 256 pp.

Woodcock, A. and M. Davis. 1978. Catastrophe Theory. E. P. Dutton. New York. 152 pp.

Woodmansee, R. B. 1978. Additions and losses of nitrogen in grassland ecosystems. BioScience 28:448–453.

Worster, D. 1977. Nature's Economy: The Roots of Ecology. Sierra Club Books, San Francisco. 404 pp.

Wright, H. O. 1973. Effect of commensal hydroids on hermit crab competition in the littoral zone of Texas. Nature 241:139–140.

Wynne-Edwards, V. C. 1962. Animal Dispersion in Relation to Social Behavior. Hafner Press, New York. 653 pp.

Yamamura, N. 1976. A mathematical approach to spatial distribution and temporal succession in plant communities. Bull. Math. Biol. 38:517–526.

278

Zeeman, E. C. 1974. Primary and secondary waves in developmental biology, pp. 69–161. In: Lectures on Mathematics in the Life Sciences, American Mathematical Society.

Zeigler, B. P. 1977. Persistence and patchiness of predator-prey systems induced by discrete event population exchange mechanisms. J. Theor. Biol. 67:687–713.

Zimba, J. R., and A. Stachurski. 1976. Vegetation as a modifier of carbon and nitrogen transfer to soil in various types of forest ecosystems. Ekologia Polska 24:493–514.

Subject Index

A*cacia* tree   115
accelerated change   11
adaptive radiation   45, 46
ADP   53,54
adrenal gland   58
age-structured populations   127
aging of cells   52
Allee effect   12, 85
altruism   77–80, 96
Ampere's Law   33
anoxia in Chesapeake Bay   240
ant-aphid associations   101
ant-plant associations   102, 109, 115, 121
archipelagos   162
arms race, coevolutionary   10, 47, 48, 90
ATP   53, 54, 62
autocatalysis   28, 34, 39

bark beetles   202
behavioral isolation   117, 118
Bénard instability and cells   28–30, 189, 221
Bernoulli effect   31
blood clotting   55
— sugar   6
bouyancy, of fish   52
brush fires   210
bumblebees   63

cacao trees   204, 205
cancer   52
castes, insect   87–89
catastrophic failure   13
cells, metabolic feedback in   53
—, evolution of   42
cerebral cortex   56
chain reactions   35
chemical defense   68, 69
— evolution   34
cicadas, mimicry in   47
Clementsian model of succession   226–228
closed loop system   1, 2
cloud streets   30

cocktail party effect   12
coevolution   47–49, 114
colonial organisms   194–196
commensalism   99
communal breeding   78–80
community, ecological   225, 226, 242–244
— matrix   105
compensatory Leslie matrix   135–148
competition   20, 21, 121, 140, 145, 174–180,
    183–186, 196–199
— in patchy systems   174–186
—, limited   121–125
complexity of systems   10
compound interest law of plant growth   72, 223
congregation of colonial organisms   194–196
contagion, rate of   214–216
convective patterns   30
cooperation, evolution of   97, 98
cost-benefit models   114
courtship behavior   80–83
crystal formation   34
cultural evolution of humans   44

depensatory mortality   85
destructive positive feedback   52
diffusion in populations   189
diffusive instability   192
disease outbreaks   201–219
—, effects of spatial heterogeneity on   209–213
dissipative structures   30, 35, 226
DNA   42
dominance hierarchies   89, 90
drinking behavior   55, 56
droplet formation   34
dynamo theory   32–34

ecological community   225, 226, 242–244
economic collapse   13
economics, human   7, 8, 13, 61, 62, 77, 187, 188
ecosystem, definition   221
—, feedback networks in   233–235
edema   52
elastic toppling   13

eliasomes 109
emotions 58
endemic disease 215
endosymbiosis 118
energy allocation 62–69
environmental gradient 196–199
enzyme reactions 36, 53, 54
epidemics 201–219
epilepsy 58
ergodic theorem 131
erosion 236
estrogen 55
evolution of life 39–43
evolutionary individuals 118
— stable strategy (ESS) 150
exothermic reactions 27, 34
extinction 165
—, time to 170, 171

Faraday's Law 32
feedforward 5
feeding behavior 55, 56
female contribution to offspring 81–83
fire-dependent ecosystems 231–233
fish 70–72
fish schooling 93–95
fluctuation-dissipation cycle 39
force of infection 213
fragility of systems 12
fringe populations 46
fugitive species 167
fungal-insect association 204

game theory 95–98
genes 40, 41
genetic feedback loop 44
genetic fitness 62
graphical analysis 16–21
groups, dynamics of 87, 88
—, formation of 90–95
—, size of 90–95
growth depensation 70
—, rate of 70–72
— rate strategy 70–72
gull colonies 92

heat death, explosive 52
herbivore-autotroph interactions 236
heterosymbiotic 119
homeostasis 5–7, 59
homosymbiotic 120
hormonal feedbacks 55
host-symbiont models 109
host-vector models 217–219

hypercycles 41, 42
hypothalamus, posterior 55

immunization 213, 214
instabilities, Bénard 28–30, 189
—, density 189
—, diffusional 192
insular reserves 163–165, 180–185
infectious disease 201
intrinsic rate of increase 149
island biogeography 158–162
island clusters 162–163, 174–185

keystone mutualist 112
kinesis 57
kin selection 78–80, 97
K-strategy 75–77

landscape patterning 188
Law of Collapse 12
leaching of nutrients 236
learning behavior 55, 56
Le Chatelier's Principle 35
Leslie matrix model 121–131, 135
— — —, for interacting populations 137–145
locomotion of animals 56–58
locusts, flight of 56
—, swarming of 92, 93
logistic equation model 4, 127
Lotka's equation 149
lynx-hare cycles 6, 7

magnetic field of the Earth 32–34
malaria 218
male contribution to offspring 81–83
Markov chain models 228–230
mating behavior 80–83
meandering of rivers 187
measles 209
Metzler matrix 247
mimicry, Batesian 48
—, Müllerian 49
mind-body relationship 58
Morishima matrix 178
morphogenesis 51
moths, thermoreguation in 57
motor-sensory relationships 56, 57
movement of animals 56, 57
multicellular organisms 43
mutualism 9, 16–19, 40, 47, 174–180
— among pathogens 204
—, evolution of 114
—, facultative and obligate 100, 101

mutualism (cont.)
—, limits to 107
—, Lotka-Volterra models of 104
—, multispecies 111
—, stability condition 107

Na$^+$ permeability in nerve cells 58
nerve cells 57, 58
net reproductive rate, R 135, 143
newt courtship behavior 81
nitrogen deficiency in forest soil 241
nuclear evolution 25
— fusion 27–28
nutrient cycling 237–242

open loop cycle 1, 2
overcompensation 136, 139, 146

parasitic infection 218
parasitism 115
parthenogenesis 84
passenger pigeon 85
patch occupancy models 165–167
Pearl-Verhulst (logistic) equation 4, 127
persistence 114, 169, 171–173, 176, 177
phase transitions 34
phosphorus recycling 240, 241
phytoplankton patches 190, 191
positive linear systems 247–259
— — —, and Routh-Hurwitz criteria 259
— — —, bounds on roots 258, 259
— — —, comparative statics 252–254
— — —, discrete-time systems 251, 252
— — —, similarity transforms 254–258
— — —, stability 248–252
prey capture efficiency 63
principal minors of matrices 113
Prisoner's Dilemma 96, 97
proteins 42, 52

quasi-diagonal dominance 113

ramjet engine 53
reproductive investment 81–83, 150
— strategies 150–154
reserves, wildlife 163–165
reticular system 56
RNA 42, 54
Routh-Hurwitz conditions 110, 259
r-strategy 75–77

salmon migration and nutrient cycling 240, 241
sand dunes, succession on 222
saturation of mutualistic benefits 109
schooling of fish 93–95
sea urchin outbreaks 234–236
Second Law of Thermodynamics 10
secondary compounds 149
self-excited oscillations 31
self-replication 40
sexual reproduction in population models 83, 84
sexual selection 44, 45
similarity transforms 23, 254–258
S-I-S model 216
sleep 56
social insects 229, 87–89
social strategies 77–80, 95–98
sociobiology 75
spatial patterns, emergence of 187–200
— — along environmental gradient 196–198
species diversity 224–226
spruce budworm 202–204
stability 21–25, 170, 177–180
—, criteria for 24
stable age structure 131
stellar evolution 27, 28
stock market 63
stock-recruitment models 135
structures, stability of 13
succession, ecological 221–233
—, classical and contemporary hypotheses
swollen shoot virus 205
symbiosis 119

termites 194, 229
territorial defense 64–68
territories 198, 199
thermoregulation 57
threshold effects 11, 12
threshold population size 208–210
toxin production 68, 69, 234

vectors, disease 217–219
vortex formation 31

water waves 31
weathering 239
wildlife reserves 163–165
winter flounder 133
wolf territories 199

Author Index

Abele, L.G. 164
Abraham, R.H. 31
Abugov, R. 229
Adams, H. 11
Addicot, J.F. 99
Ahmadjian, V. 101
Aidley, D.J. 58
Albrecht, F. 107
Alexander, M. 204, 223, 239, 240
Allee, W.C. 12, 85
Allen, P.M. 114
Anderson, R.M. 204, 207
Archer, J. 90
Arman, P. 236
Armstrong, P. 167
Arndt, W.F. 204
Aron, J.L. 218
Ashby, W.R. 1
Atsatt, P.R. 103
Axelrod, R. 96–98

Bailey, N.T.J. 210, 217
Bajzer, Z.K. 52
Bakker, R.T. 45
Balda, R.P. 102
Barash, D. 44, 78, 81
Barbour, C.D. 164
Barlow, G.W. 198
Bartlett, M.S. 209, 217
Bateson, G. 8
Batra, L.R. 102
Batra, S.W.T. 102
Beattie, A.J. 102
Been, G.H. 236
Bender, E.A. 86
Bently, B.L. 102
Berg, R.Y. 102
Bernadelli, H. 127, 128, 131
Bernard, C. 5, 61
Berry, R.S. 35, 36
Berryman, R.A. vi, 24, 25
Beverton, R.J. 135
Biever, L.J. 242

Birch, L.C. 5, 8
Bishop, R.E.D. 31
Black, F.L. 210
Blackman, V.H. 72, 223
Blagden, C. 5
Bobisud, L.E. 103
Bock, K. 75
Bokhari, U.G. 236
Bonner, J.T. 43, 55, 91
Booth, D.A. 56
Bormann, F.H. 236
Bossert, W.H. 150
Botkin, D.B. 171
Boucher, D.H. 99
Bourgeois-Pichat, J. 132
Bowley, A.L. 131
Boyce, M.S. 149
Briand, F. 119
Brock, E.M. 239
Brock, T.D. 102
Brown, J.L. 164
Brush, S.G. 34
Buchsbaum, R. 119
Bullard, E.C. 32
Burke, E. 61
Bush, G.L. 45
Buss, L. 183
Butler, J.H. 188
Butzer, K.N. 236

Cairns-Smith, A.G. 42
Calaby, J.H. 164
Calow, P. 52
Calvin, M. 28, 35, 36
Cannan, E. 131
Cannon, W.A. 5
Caraco, T. 91
Careri, G. 34
Cates, R.G. 149
Cavanaugh, C.M. 101
Charlesworth, B. 148, 150
Chase, I.V. 81–82, 89–90
Cherret, J.M. 123

285

Chiang, H.C. 46
Clark, C.W. 93
Clark, L.R. 204–205
Clements, F.E. 222
Cohen, D. 154
Cohen, J.E. 131, 165
Cohn, R.M. 42
Colgan, P. 98
Colwell, R.K. 102
Comins, H.W. 137, 144–146
Connell, J.H. 183, 222, 224–225
Copasso, V. 216
Corning, P.R. 46
Coughenour, M.B. 244
Cowles, H.C. 222
Culver, D.C. 102, 110, 165

Darling, F.F. 91
Darlington, P.J., Jr. 14, 48, 90
Darwin, C. 10, 44, 58
Davidson, D.W. 103, 125, 126
Davies, N.B. 81
Davis, M. 93
Dawkins, R. 45, 47, 95
Dean, A.M. 108, 109
DeAngelis, D.L. 168, 174–175, 179
Debreu, G. 254
Deneubourg, J.L. 194
Desowitz, R.S. 103
DeVay, J.E. 204
Diamond, J.M. 158–161, 164–165
Dinsmore, J.J. 128
Dodson, S.I. 103
Donaldson, J.R. 241
Douce, G.K. 237
Doutt, R.L. 103
Drury, W.H. 223
Duncan, N. 92
Dworkin, M. 91
Dyer, M.I. 103, 236–237

Eastop, V.F. 217
Edlin, H.L. 210, 223
Ehrlich, A. 85
Ehrlich, P.R. 5, 8, 85
Eigen, M. 42
Eisenberg, J.F. 45
Eiskowitch, D. 101
Einstein, A. 187
Ek, A.R. 226
Ellul, J. 11, 14
Elsasser, W.M. 32
Emlen, J.M. 196
Esnouf, M.P. 55

Faegri, K. 99, 101, 102
Fagen, R.M. 150
Feder, H.M. 102
Fey, W.R. 222–224, 226, 239
Fisher, R.A. 44
Flannery, K.V. 8
Fraser, H.R. 55
Fritz, R.S. 101
Futuyma, D.J. 47–48

Gadgil, M. 150
Galil, J. 101
Gallant, R.A. 80
Galston, A.W. 61
Gehrz, R.D. 27
Geiselman, P.J. 55
Geist, V. 165
Ghiselin, M.J. 44, 61, 77
Gilbert, F.S. 161
Gilbert, L.E. 103, 111, 119, 165
Gilpin, M.E. 78, 133
Givnish, T.J. 13
Glynn, P.W. 102
Goh, B.S. 24, 104, 105
Goodman, D. 150
Goodwin, D. 85
Gordon, J.E. 13
Gosz, J.R. 241
Gould, J.L. 44
Gould, S.J. 44
Gourou, P. 231
Graham, K. 102, 204
Grant, K.A. 46
Grant, V. 46, 48
Grasse, P.-P. 194
Grime, J.P. 68–69
Guckenheimer, J. 138
Gutierrez, L.T. 22–224, 226, 239

Hairston, N.E. 5
Haken, H. 30, 42
Hall, C.A.S. 240
Hallam, T.G. 114
Halle, L.J. 11
Halliday, T.R. 81, 85
Hamilton, T.H. 162
Hamilton, W.D. 45, 91, 96–97, 118
Hamilton, W.J., III 81
Hanson, F.B. 170
Hardy, H.C. 12
Hargrave, B.T. 235
Harth, E. 55
Hartling, L.K. 102
Haselwandter, K. 101
Hassell, M.P. 137, 144, 145

Hastings, A. 165
Haugness, J.A. 99
Hay, M.E. 102
Heatwole, H. 100
Heinrich, B. 25, 57, 63, 77, 99
Heithaus, E.R. 99, 102, 109
Herstein, I.N. 254
Hess, K.W. 133
Hethcote, H.W. 216–217
Hirsch, M.W. 107
Hitchcocke, W.B. 164
Hocking, B. 101
Holt, S.J. 135
Homans, G.C. 86
Horn, H.S. 76–77, 157, 165–166, 228–229
Horst, T.J. 134
Howard, B.H. 101
Howe, H.F. 99, 102, 103
Hutchinson, G.E. 5, 239

Jackson, J.B.C. 183
Jacob, F. 10
Jacobs, J.A. 32
Jahn, J.A. 189
Janson, C.H. 102
Janzen, D.H. 49, 99, 101, 115, 118
Jeffers, J.N.R. 228
Jenny, H. 236
Johnson, M.P. 162
Jordan, C.F. 237, 244
Juday, C. 240
Judson, H.F. 1

Karakashian, M.W. 101
Karakashian, S.J. 101
Karlin, S. 183
Kassanis, B. 204
Katchalsky, A.K. 189
Kay, D. 205
Keast, A. 70
Keeler, K.H. 102, 114, 118
Keller, E.F. 184
Kierstad, H. 190
Kippenhahn, R. 27–28
Kleinfeldt, S.E. 101
Kline, M. 15
Knight, R.L. 244
Koch, A. 101
Koshland, D.E. 57
Krebs, C.J. 135, 136
Krebs, J.R. 47, 81
Krokhin, E.M. 241
Kropotkin, P.A. 78
Kushlan, J.A. 165

Lajmanovich, A. 216–217
Land, G.T.L. 59
Landau, H.G. 90
Lande, R. 45
Lange, R.T. 101, 102
Langley, L.L. 5
Larow, E.J. 237
Lefever, R. 51
Legget, W.C. 136
Leon, J.A. 150
Leontief, W. 13
Leshner, A.I. 90
Leslie, P.H. 127, 128, 131, 138
Levandowski, M. 189
Levin, B.R. 47
Levin, D.A. 129
Levin, S.A. 47, 157, 167, 175, 183
    188, 190–192, 243
Levins, R. 25
Lewis, E.G. 127, 128, 131
Likens, G.E. 236
Limbaugh, C. 102
Lindeman, R.L. 223
Loefer, J.B. 189
Loosli, C.G. 204
Lorenz, K. 81
Lovelock, J.E. 59
Ludlow, A.R. 52
Ludwig, D. 202–203
Luenberger, D.G. 247, 251, 258
Lugt, H.T. 31
Lumsden, C.J. 75
Lyttle, T.W. 45

MacArthur, R.H. 157–158, 165–166, 170
MacFarlane, R.G. 55
Mackie, R.A. 101
MacLean, W.R. 12
Majer, J.D. 199
Mangel, M. 93
Mann, K.H. 234–235
Margalef, R. 5, 222
Margulis, L. 101
Marler, P. 81
Martinez, H. 51
Maruyama, M. V, 7, 25
Mautner, A.J. 216–217
May, R.M. 25, 108, 109, 110, 158
    173, 213–214
Maynard Smith, J. 95–96, 150
Mayo, O. 45
Mayr, E. 45
McCleery, R.H. 81
McClendon, J.H. 28
McGregor, J. 183
McKey, D. 102, 119

McLaren, I.A.   5
McLaughlin, J.A.   101
McMahon, T.A.   13
McNaught, D.C.   237
McNaughton, S.J.   103, 237, 244
McNew, G.L.   202
Meade, P.S.   102
Mech, L.D.   199
Medawar, P.B.   148
Mefferd, R.B.   189
Meijer, W.   101, 111, 115, 165
Messina, F.J.   99, 123
Meyer, F.H.   101
Michod, R.E.   150, 152
Miles, J.   233
Milne, A.   5
Milsum, J.H.   vi
Mimura, M.   192
Minorsky, N.   15
Monod, J.   10
Monserud, R.A.   226
Mooney, H.A.   72–73
Morgenstern, O.   95
Murchie, G.   52
Murdoch, W.W.   5
Murton, K.K.   85
Muscatine, L.   99
Mutch, R.W.   231

Nakada, J.   103
Neill, W.E.   103
Nestler, E.J.   58
Neuhaus, R.J.   103
Nicholson, A.J.   5
Nicholis, J.S.   97
Nicolis, G.   39, 97, 194
Nisbet, I.C.T.   223
Nitecki, M.H.   47
Noble, J.V.   210
Normand, C.   29
Novin, D.   55
Noy-Meir, I.   223

O'Donald, P.   45
O'Dowd, D.J.   102
Odum, E.P.   5, 221–224, 236, 242, 244
Officer, C.B.   240
Okubo, A.   157, 188–189, 192–193
Olson, J.S.   222
Oosting, H.J.   196
Orgel, L.E.   52
Orias, E.   222, 224–225
Osman, R.W.   99
Oster, G.F.   87
Owen, D.F.   103

Parsegian, V.L.   56
Pasineti, L.L.   13
Pastor, J.   224
Patten, B.C.   242, 244
Peacocke, A.R.   39, 51
Pearson, K.   57
Pennycuick, C.J.   137
Perkowski, J.J.   148
Peterken, G.F.   165
Phillips, B.P.   204
Pianka, E.R.   104
Pickett, S.T.A.   165
Picton, H.D.   164
Pielou, E.C.   157, 231
Pierce, N.E.   102
Platt, J.R.   189
Plemmons, R.V.   249
Plowright, R.C.   102
Pomeroy, L.R.   236
Porter, J.W.   99
Porter, R.W.   58
Post, W.M.   24, 111, 116, 117, 173, 216,
    248, 251
Poundstone, W.   39
Powell, J.A.   101
Preston, F.W.   157
Price, G.R.   150
Price, P.H.   149
Price, P.W.   84
Prigogine, I.   30, 39–40, 51, 194
Pulliam, H.R.   68, 154

Quinlan, R.J.   123

Rai, B.   123
Rapoport, A.   44, 97
Rapp, P.E.   54
Ratcliffe, D.A.   85
Raven, P.H.   162
Read, D.J.   101
Real, L.A.   103
Rehr, S.S.   149
Rensch, B.   43
Ricker, W.E.   135
Ricklefs, R.E.   198
Rickson, F.R.   101, 102
Rikitake, T.   32
Rindos, D.   64
Risch, S.J.   99, 102
Ritts, R.E.   204
Robbins, W.J.   189
Roff, D.A.   157
Room, P.M.   102
Root, R.B.   183
Rosenblueth, A.   58

Ross, D.M. 102
Ross, G.M. 102
Roughgarden, J. 102, 114, 121
Roy, B.N. 239
Rubinstein, M.F. 15
Rushton, S. 216–217

Sagan, C. 58
Sahlins, M.D. 75
Saila, S.B. 133
Schadde, R. 164
Schaefer, M.B. 93
Schaeffer, W.M. 114, 150
Schaller, G.B. 91
Schemske, D.W. 102
Schmidt-Nielson, K. 51
Schuster, P. 42
Sebetich, M.J. 237
Seddon, G. 231
Segel, L.A. 192
Seigler, D.S. 149
Seneta, E. 131
Serio, G. 207, 217
Sharp, G.D. 93
Shaw, C.D. 31
Shelton, W.L. 70
Shugart, H.H. 171, 226, 237
Siccama, T.G. 196
Sigmund, K. 42
Siljak, D.D. 24
Simberloff, D.S. 103, 161, 164, 183
Simon, H.A. 86
Simpson, G.G. 44
Simpson, P.C. 56
Singer, P. 75
Skellam, J.G. 157, 188–189
Slatkin, M. 47, 165
Slobodkin, L.B. 47, 90
Smale, S. 107
Smith, A. 61, 77
Smith, N. 103, 111
Smouse, P.E. 138
Solbrig, O.T. 150
Southwood, T.R.E. 76
Springett, B.P. 123
Spurr, S.H. 231
Staddon, J.E.R. 57
Stanley-Jones, D. V, 53, 56, 58
Starmer, W.T. 102
Stengers, I. 194
Stenseth, N.C. 103
Stent, G.S. 11
Stephens, G.R. 228
Stephenson, A.G. 102
Stevens, P.S. 187
Street, P. 102
Strobeck, C. 173, 180, 250

Suthers, R.A. 80
Swaney, D.P. 244
Swift, M.J. 237
Sykes, Z.M. 133

Takeuchi, V. 255
Takvanainen, J.O. 103
Tamarin, R.H. 5, 150
Tansley, A.G. 221
Temple, S.A. 99, 102
Terborgh, J. 160–161, 165
Tester, C.F. 149
Thompson, J.D. 103
Thompson, J.N. 165
Tilman, D. 102
Tinbergen, N. 80–81
Tinkle, D.W. 150
Toates, F.M. 25, 52, 55–56
Toffler, A. 11
Transeau, E.N. 196
Travis, C.C. 24, 104, 105, 136, 141, 145
    148, 173, 248, 251
Trivers, R.L. 80
Tuckwell, H.C. 170
Turing, A.M. 39, 51, 192

Usher, M.B. 229

Valerio, C.E. 103
Vance, R.R. 102
VandeKerchove, G.A. 102
Vandermeer, J.H. 104, 114
van der Pijl, L. 102
VanderWall, S.B. 102
Varley, G.C. 202
Vaughan, D.S. 133
Velarde, M.G. 29
Vermeij, G.J. 10, 47–48
Verwey, J. 101
Vincent, T.L. 68
Vine, I. 91
Vitousek. P. 241
Von Neumann, J. 37, 95
Vuilleumier, F. 164

Waggoner, P.E. 226
Walker, J. 12, 31
Waser, N.M. 103
Watson, R.K. 216–217
Watt, K.E.F. 223
Way, M.J. 101
Webb, D.P. 237
Webley, D.M. 239

Weinberg, D.   6, 12
Weinberg, G.M.   6, 12
Weis-Fogh, T.   56
Weiss, K.M.   138
West, D.C.   171, 226
Westwood, N.J.   85
Whelpton, P.K.   131
Whitcomb, R.F.   165
Whitehead, J.A., Jr.   30
Whittaker, R.H.   99, 108, 109, 223, 226, 243
Wiebes, J.T.   101
Weins, J.A.   199, 243
Weipkema, P.R.   55
Wilbur, H.M.   149
Wilcox, B.A.   159–160, 163–164
Williams, G.C.   78, 84, 242
Willis, E.O.   165
Wilson, D.S.   102, 121, 233–234, 242–244
Wilson, E.O.   44, 75, 77, 79, 87, 89,
   158, 170, 183, 198

Winet, H.   189
Witzmann, R.F.   52
Wolf, L.L.   91
Woodcock, A.   93
Woodmansee, R.B.   236
Worster, D.   187, 233
Wright, H.O.   99, 123
Wynne-Edwards, V.C.   5, 77–78

Yamamura, N.   197
Yodzis, P.   119
Yorke, J.A.   216–217

Zahl, P.A.   101
Zeeman, E.C.   197
Zeigler, B.P.   165

290

Biomathematics

Managing Editor: S.A.Levin

Editorial Board: M.Arbib,
H.J.Bremermann, J.Cowan,
W.M.Hirsch, S.Karlin, J.Keller,
K.Krickeberg, R.C.Lewontin,
R.M.May, J.D.Murray, L.A.Segel

Volume 14: C.J.Mode

Stochastic Processes in Demography and Their Computer Implementation

1985. 49 figures, 80 tables. XVII, 389 pages. ISBN 3-540-13622-3

Contents: Fecundability. – Human Survivorship. – Theories of Competing Risks and Multiple Decrement Life Tables. – Models of Maternity Histories and Age-Specific Birth Rates. – A Computer Software Design Implementing Models of Maternity Histories. – Age-Dependent Models of Maternity Histories Based on Data Analyses. – Population Projection Methodology Based on Stochastic Population Processes. – Author Index. – Subject Index.

Volume 13: J.Impagliazzo

Deterministic Aspects of Mathematical Demography

An Investigation of the Stable Theory of Population including an Analysis of the Population Statistics of Denmark

1985. 52 figures. XI, 186 pages. ISBN 3-540-13616-9

Contents: The Development of Mathematical Demography. – An Overview of the Stable Theory of Population. – The Discrete Time Recurrence Model. – The Continuous Time Model. – The Discrete Time Matrix Model. – Comparative Aspects of Stable Population Models. – Extensions of Stable Population Theory. – The Kingdom of Denmark – A Demographic Example. – Appendix. – References. – Subject Index.

Volume 12: R. Gittins

Canonical Analysis

A Review with Applications in Ecology

1985. 16 figures. XVI, 351 pages. ISBN 3-540-13617-7

Contents: Introduction. – Theory: Canonical correlations and canonical variates. Extensions and generalizations. Canonical variate analysis. Dual scaling. – Applications: General introduction. Experiment 1: an investigation of spatial variation. Experiment 2: soil-species relationships in a limestone grassland community. Soil-vegetation relationships in a lowland tropical rain forest. Dynamic status of a lowland tropical rain forest. The structure of grassland vegetation in Anglesey, North Wales. The nitrogen nutrition of eight grass species. Herbivore-environment relationships in the Rwenzori National Park, Uganda. – Appraisal and Prospect: Applications: assessment and conclusions. Research issues and future developments. – Appendices: Multivariate regression. Data sets used in worked applications. Species composition of a limestone grassland community. – References. – Species' index. – Author index. – Subject index.

Springer-Verlag
Berlin
Heidelberg
New York
Tokyo

Volume 11: B.G.Mirkin, S.N.Rodin

Graphs and Genes

Translated from the Russian by H.L.Beus
1984. 46 figures. XIV, 197 pages. ISBN 3-540-12657-0

Contents: Graphs in the analysis of gene structure. – Graphs in the analysis of gene semantics. – Graphs in the analysis of gene evolution. – Epilogue: Cryptographic problems in genetics. – Appendix: Some notions about graphs. – References. – Index of genetics terms. – Index of mathematical terms.

Biomathematics

Managing Editor: S.A.Levin

Editorial Board: M.Arbib,
H.-J.Bremermann, J.Cowan,
W.M.Hirsch, S.Karlin, J.Keller,
K.Krickeberg, R.C.Lewontin,
R.M.May, J.D.Murray, L.A.Segel

Volume 10: A.Okubo

Diffusion and Ecological Problems: Mathematical Models

1980. 114 figures, 6 tables. XIII, 254 pages. ISBN 3-540-09620-5

Volume 9: W.J.Ewens

Mathematical Population Genetics

1979. 4 figures, 17 tables. XII, 325 pages. ISBN 3-540-09577-2

Volume 8: A.T.Winfree

The Geometry of Biological Time

1980. 290 figures. XIV, 530 pages. ISBN 3-540-09373-7

Volume 7: E.R.Lewis

Network Models in Population Biology

1977. 187 figures. XII, 402 pages. ISBN 3-540-08214-X

Volume 6: D.Smith, N.Keyfitz

Mathematical Demography

Selected Papers
1977. 31 figures. XI, 514 pages. ISBN 3-540-07899-1

Volume 5: A.Jacquard

The Genetic Structure of Populations

Translators: D.Charlesworth, B.Charlesworth
1974. 92 figures. XVIII, 569 pages. ISBN 3-540-06329-3

Volume 4: M.Iosifescu, P.Tăutu

Stochastic Processes and Applications in Biology and Medicine

Part 2: Models
1973. 337 pages. ISBN 3-540-06271-8

Volume 3: M.Iosifescu, P.Tăutu

Stochastic Processes and Applications in Biology and Medicine

Part 1: Theory
1973. 331 pages. ISBN 3-540-06270-X

Volume 2: E.Batschelet

Introduction to Mathematics for Life Scientists

3rd edition. 1979. 227 figures, 62 tables. XV, 643 pages
ISBN 3-540-09662-0

Springer-Verlag
Berlin
Heidelberg
New York
Tokyo